Birkhäuser

Trends in the History of Science

Trends in the History of Science is a series devoted to the publication of volumes arising from workshops and conferences in all areas of current research in the history of science, primarily with a focus on the history of mathematics, physics, and their applications. Its aim is to make current developments available to the community as rapidly as possible without compromising quality, and to archive those developments for reference purposes. Proposals for volumes can be submitted using the online book project submission form at our website www.birkhauser-science.com.

More information about this series at http://www.springer.com/series/11668

Frédéric Brechenmacher
Guillaume Jouve · Laurent Mazliak
Rossana Tazzioli
Editors

Images of Italian Mathematics in France

The Latin Sisters, from Risorgimento to Fascism

Birkhäuser

Editors
Frédéric Brechenmacher
Humanités et Sciences Sociales
École Polytechnique
Palaiseau
France

Guillaume Jouve
Université d'Artoise
Lens
France

Laurent Mazliak
Laboratoire de Probabilités et Modèles Aléatoires
Université Pierre et Marie Curie
Paris
France

Rossana Tazzioli
UFR Mathématiques
Université Lille 1: Sciences et Technologies
Villeneuve d'Ascq
France

ISSN 2297-2951 ISSN 2297-296X (electronic)
Trends in the History of Science
ISBN 978-3-319-40080-8 ISBN 978-3-319-40082-2 (eBook)
DOI 10.1007/978-3-319-40082-2

Library of Congress Control Number: 2016946011

Printed on acid-free paper

This book is published under the trade name Birkhäuser
The registered company is Springer International Publishing AG Switzerland
(www.birkhauser-science.com)

Contents

By way of Preface: Daladier's Stay in Italy in 1910. A Mirror of the French Look at Italy

Antonin Durand

The long 19th century in Italy was a century of nation building. This process was twofold, at the same time centripetal and centrifugal: it consisted of an integrating dimension, which meant a process designed to strengthen the new State and in an attempt to obtain international recognition from the rest of the world. A study on the way French mathematicians considered Italy should give elements to assess the reception of the second aspect. This small chapter does not deal with mathematicians, but it intends to bring a slightly different light on the same question: the actor I will focus on, Edouard Daladier, when he travelled through Italy, was a historian about to become a politician. As such he offers an interesting observation point to understand the link between the circulation of scholars across Italy and the representation they carried on the political evolutions of Italy. In that sense, Daladier's itinerary is not so different from that of other students who travelled in Italy, thus becoming vectors of knowledge on this country.

Before he became the head of the French radical party who was to become sadly notorious for being the French Prime Minister during the Munich Crisis in 1938, Edouard Daladier was a historian. He had passed the difficult competitive examination of *agrégation*, coming first with outstanding grades, and had started a specialization in the history of Italian unification, just before becoming in 1911 the mayor of Carpentras in the south of France, and starting a successful political career. In order to favour his research project in Italy, he had been offered a new scholarship created by the French patron of the sciences David David-Weill aimed at helping excellent students to travel abroad. This allowed him to spend one year in Italy, mainly in Rome even though he also travelled a lot during the last two months of his stay, in order to lead a research on Giuseppe Mazzini and the Italian republican movement during the *Risorgimento*. In 1944, while he was still a

A. Durand (✉)
École Normale Supérieure, 45, rue d'Ulm/29 rue d'Ulm, 75230 Paris Cedex 05, France
e-mail: Antonin.Durand@ens.fr

F. Brechenmacher et al. (eds.), *Images of Italian Mathematics in France*,
Trends in the History of Science, DOI 10.1007/978-3-319-40082-2_1

prisoner at Itter Castle in Germany, he remembered his stay after hearing that Rome had been taken by the Allied on June 4th 1944:

> After violent fighting in Rome suburbs, the 8th Army came into the city and every form of resistance stopped. What a great event, and what memories! I spent in Rome some years while I was preparing my thesis about the Italian cities[1] (Daladier 1991, pp. 289–290).

On several occasions, Daladier underlined the importance of this journey in his personal training and the great time he had spent in Italy. It was the first time he travelled abroad, which gave to this stay a special importance in his political training. It doubtlessly transformed his perception of international relations and of the role of Italy. His biographers underlined this aspect but did not focus on the first purpose of this stay which was to study Italian recent history, probably because he did not carry out his research (Du Réau 1993, p. 24). Though this stay was important in Daladier's career, it was never specifically studied, essentially due to the fact that the report he drafted after his stay is not conserved in his personal archives but in those of the Chancellery of the University of Paris.

This report was drafted in the particular context of 1911 that represents a turning point in Italian diplomatic history as well as in the relations between France and Italy. The foreign policy of Italy was deeply changing due to its first colonial war in Libya, with a clear ambition to be considered as a great power in Europe. The reactions in France about it were ambivalent: the emergence of a new competitor in North Africa was considered suspiciously but no one would risk being too severe about it out of fear to reinforcing the Italian implication in the Triple Alliance. What makes Daladiers's report particularly interesting is the link he tried to establish between his historical research on the Risorgimento and the new kind of nationalism he saw emerging in the beginning of the 1910s. While some people interpreted this claim for Italian grandeur as a continuation of the Risorgimento ideals, others saw a perversion of these ideals in this new form of nationalism. Daladier's point of view is interesting because this transition coincides with his own conversion from academic research to a political commitment. It reveals the accurateness of his perception of international relations as well as his understanding of the transformation of the question of nationalities. That is why this report offers an original French point of view, at the same time academic and political, on Italy. In order to give account of this originality, I will first insert it into the history of the David-Weill scholarship that transformed the way one should consider staying abroad in a French University. Then I will focus on Daladier's analysis of the Italian Risorgimento and on the birth of a myth of the Risorgimento in France. This will lead to the question of continuities and discontinuities between the Italian nation building in the 19th century and the emergence of a new kind of nationalism that underwent a first climax with the war in Lybia. Finally, I will focus on a third actor that was omnipresent in the Franco-Italian relation: Germany.

---

[1]«4 juin 1944. Après de violents combats dans les faubourgs de Rome, la huitième armée a pénétré dans la ville où toute résistance a cessé. Grandeur de cet événement et que de souvenirs! J'ai passé à Rome des années, quand je préparais ma thèse sur les cités italiennes.».

1 Opening French Perspectives: The David-Weill Scholarship

Edouard Daladiers's stay in Italy in 1910 was made possible thanks to an important donation to the University of Paris from David David-Weill, a rich banker and art collector. This donation was part of a more general trend in France that consisted in sponsoring student travels throughout the world in order to create new contacts between the French élite and foreign countries (Durand forthcoming). Some French students had admittedly already established strong relations with Greece and Italy through the rising network of the *Écoles françaises* in Athens and Rome. The first had been established in 1846 in a philhellene context (Basch 1995), and the second was created a few months after the French defeat of 1870, first as an extension of the *École française d'Athènes* and then an autonomous institution (Poncet 1872). The foundation of the *École française de Rome* was the first manifestation of a will to establish a French scientific presence in Italy, and it was not by chance that it closely followed the settlement of a German Archeological Institute in the new capital of the Italian Kingdom. As such, they are part of what Guillaume Tronchet has called the 'French universitary imperialism' (Tronchet 1920).

However, the creation of travel scholarships by private sponsorship from the very end of the 19th century, following the initiative of Salomon de Rothschild in 1896, followed a different logic: their purpose was not only to bolster French researchers who needed to work in a particular field, but it aimed at establishing new contacts between the French élite and the rest of the world. French geographer Vidal de la Blache put the stress on this aspect in an article he devoted to the most famous scholarship of this kind, the «bourse autour du monde», founded by Albert Kahn in 1898:

> Everyone can see how the principles of this foundation are different from those of the permanent missions improperly called «écoles» that exist in Athens Rome, Cairo and more recently Hanoi: these excellent institutions also request long trips; both those of Athens and Rome served very efficiently science and education. But they investigate the past, a past that has admittedly much to do with the present; but the present is only at its margin. These scholarships offer, on the contrary, a quick excursion throughout men and things of present time.[2]

[2]«On voit combien les idées dont s'inspire cette fondation diffèrent de celles qui ont présidé à l'institution de ces missions permanentes assez improprement appelées écoles, qui existent à Athènes, à Rome, au Caire et depuis quelques années à Hanoi. Ces institutions excellentes supposent aussi de longs voyages; les deux plus anciennes ont rendu, non seulement à la science, mais à l'enseignement même secondaire, de très réels services. Mais leurs investigations portent sur le passé; un passé qui certainement n'est pas sans rapport avec le présent: le présent, toutefois, n'est qu'en marge. [...] Ces voyages autour du monde sont, au contraire, des excursions rapides à travers les hommes et les choses du temps présent.» «Les bourses de voyage autour du monde», Rapport de Paul Vidal de la Blache présenté au Congrès international d'expansion économique mondial, Sect. 1, enseignement, Archives nationales de France, AJ/17/7020, pp. 2–3.

The David-Weill scholarship that was offered to Edouard Daladier and that was first attributed in 1909 had a similar purpose; it did not only aim at representing France in Italy, or at broadcasting French culture in the world, but the recipient of the scholarship was also supposed to experience the atmosphere of another country in order to open his or her mind. David David-Weill, who founded his scholarship by donating every year several hundred of thousand francs to the University of Paris, intended to encourage the French élite to travel through Europe. His intentions were specified in a note presented to the Council of the University of Paris:

> No research project will be requested from [the candidates] because, in the donor's mind, the purpose is to help distinguished professors, not to gather elements for scientific research, but merely to live in a different atmosphere in order to become able to share their experience with their future students.[3]

In that sense, David-Weill's ambition was very similar to Albert Kahn's when he created a few years earlier a series of scholarships aimed at helping outstanding students to travel across the world. He developed his project in a note conserved by the University of Paris:

> Every thoughtful mind, preoccupied with our country's future, is confronted with serious challenges. We can feel that these problems cannot be solved only with purely theoretical studies. We need to give account of the exact role that every nation plays in the world, try to understand their aspirations, and see where they are leading us, whether they drive to some violent blasts or if some kind of conciliation is possible. While abstract discussions can only assess possibilities and probabilities, a direct contact with things and men should give, maybe not certain knowledge, but at least firm, lively and transferable impressions. To institute this experience, to establish contact with ideas, feelings and life of different people, who would be better than young people chosen among our nation's intellectual and moral élite, not old enough to be already prejudiced, but mature enough to be able to watch and understand.[4]

---

[3]«On ne leur demandera pas les travaux ou les recherches qu'ils comptent entreprendre, car, dans la pensée du donateur, il s'agit de permettre à des professeurs distingués, non de réunir les éléments d'un travail scientifique, mais simplement de vivre dans une autre atmosphère et de faire plus tard profiter leurs élèves de l'expérience dont ils n'auront pas manqué de s'enrichir.» Proceedings of the Paris University Council, December, 27th 1909.

[4]«De graves problèmes se posent aujourd'hui à tous les esprits réfléchis, soucieux de l'avenir de notre pays. Ces problèmes, nous sentons bien que des études purement théoriques ne suffiront pas à les résoudre. Nous devons chercher à nous rendre un compte exact du rôle que les diverses nations jouent à la surface du globe, déterminer leurs aspirations diverses, voir où ces aspirations les mènent, si elles doivent les conduire à des chocs violents, ou si elles pourraient se concilier les unes avec les autres. Tandis que des discussions abstraites ne pourraient dégager ici que des possibilités ou des probabilités, un contact immédiat avec les choses et les hommes devra donner, sinon des certitudes, au moins des impressions fermes, vivantes, communicables. Or, pour instituer cette expérience, pour entrer en communication sympathique avec les idées, les sentiments, la vie enfin des différents peuples, à qui s'adresser sinon à des jeunes gens choisis dans l'élite intellectuelle et morale de la nation, pas assez âgés pour avoir déjà des idées préconçues, assez murs d'esprits cependant pour savoir regarder et comprendre.» Archives nationales de France, AJ/16/7020.

As a young *agrégé*, already interested in politics, Daladier fitted particularly well to this ambition to draw a form of continuity between investigation on the past and involvement in present. In his report, he underlined this aspect:

> Thanks to my historical studies, I prided myself on being able to understand immediately the long dispatches on the walls of the banks that sum up for us the everyday life of foreign nations. As a matter of fact, I had to acknowledge that I was completely ignorant of foreign countries. I may have known them but I had no feeling of them.[5]

So far, he had been informed about the political evolutions in Italy through the press that he described as 'the thick wall, probably higher than the Great Wall of China, that limits the horizon of most cultured French people'.[6]

2 France and the Italian Risorgimento

As a specialist in recent history, Daladier was particularly interested in the echoes that the Risorgimento could have in the Italy of his time. He developed his ambition in his report, explaining that he wanted at the same time to 'study the main documents in the very country and to approach some of the men who were involved in political struggles'.[7] His sympathies for the radical movement in France may explain why he became interested in the republican movement in Italy; but he was rapidly struck by the difference between the reality he discovered in Italy and the self-centered approach of Italian Risorgimento that predominated in France. Of course, he was aware that the Second Republic's shilly-shallying between support to the liberal process in Piedmont and struggle against the Roman Republic had created a deep gap between the Italian republican movement and France. At the same time, he underlined the impact of the struggles for the Italian independence in French society around 1870. In 2011, a great exhibition took place at the Army Museum of the Invalides in Paris on France and the birth of independent Italy. On this occasion, Gilles Pécout stressed that 'liberal France became fascinated about Italian independence and unity that asserted itself against the young king of the Two Sicilies, the King-Pope and the adamant Austrian Emperor'[8] (Pécout 2011, p. 29).

[5]«Je me flattais, grâce à mes études historiques, de comprendre dès la première lecture ces longues dépêches de l'étranger qu'affichent les grands établissements de crédit et qui résument pour nous la vie quotidienne des nations. En fait, je dus reconnaître que j'ignorais à peu près tout de l'étranger, puisqu'en ayant peut-être la connaissance, je n'en avais certainement pas le sentiment.» Archives nationales de France, AJ/16/7004.

[6]«l'épaisse muraille, plus haute certes que celle de la Chine, qui limite souvent l'horizon de la majorité des Français cultivés?» *ibid.*

[7]«Je voulais me consacrer à l'histoire du Risorgimento: il était donc nécessaire d'étudier dans le pays lui-même les principaux documents et d'approcher quelques uns des hommes qui prirent part aux luttes politiques».

[8]«La France libérale et progressiste se passionne alors (en 1860) pour l'indépendance et l'unité italienne qui s'affirment contre le jeune roi des Deux Siciles, contre le Pape-roi et contre l'inflexible empereur d'Autriche.» Archives nationales de France, AJ/16/7004.

Daladier, who was not born yet in 1860, belonged to a different generation, but he was struck by the fact that this fascination for the Italian Risorgimento drove many French toconsider the political situation in Italy through the prism of what they knew about 1860:

> People often believe in France that current Italy is only a mature version of 1860 Italy. In their mind, the political ideas in the name of which the Risorgimento men fought also inspired their followers.[9]

The consequence of this perspective effect was twofold. First, it led to exaggerating French influence in the rebirth of Italy: Daladier mentioned ironically what he called the 'romantic policy' led by Napoleon III, and tended to play down the importance of French diplomacy in the Italian wars of independence, stressing instead the role of the ideas of 1789. 'Obviously, he concluded, French ideas did not play a less important role than French diplomats and armies.'[10] Thus, Daladier was reluctant to simply consider France as a model for Italy.

On the second hand, the omnipresence of references to the Risorgimento when dealing with Italian questions drove the French politicians to miss the deep transformation that Italian politics had undergone since Unity. Daladier made very clear that for him the Risorgimento was over:

> In current Italy, slight traces of this intellectual movement that was so noble and so human hardly remain.[11]

Indeed, the political generations had been completely renewed, and, most of all, the new kind of nationalism that was emerging was very different from the nation building that had been the core of Italian politics around Unity. So when Daladier tried to identify in the Italian political life of the early 1910s possible inheritors of the Risorgimento movement, he quickly discarded the socialist party that he considered 'divided in several groups, with no doctrine nor project, without a proper head'.[12]

---

[9]«On croit volontiers en France que l'Italie contemporaine n'est que l'Italie de 1860 parvenue à sa maturité. Les idées politiques au nom desquelles les hommes du Risorgimento combattirent inspireraient encore leurs successeurs.» *ibid.*

[10]«Il est certain que la part des idées françaises dans la formation, de l'unité italienne est au moins aussi grande que celle de nos diplomates et de nos armées.» *ibid.*, p. 4.

[11]«Dans l'Italie contemporaine, c'est à peine si quelques traces légères de ce mouvement intellectuel, si noble et si humain, subsistent.» *ibid.*

[12]«[…] divisé entre plusieurs groupes ennemis, sans doctrine ni programme, sans chefs qui soient bien à lui», *ibid.*, p. 7.

3 The Birth of a New Italian Imperialism

On the contrary, he had to acknowledge that the party 'whose birth is the most interesting event of the few last years'[13] was the nationalist party. He felt very early the growing influence of this movement:

> It regroups most of the young men who will soon run Italy. Even his adversaries struggle against its language excesses without questioning their doctrines.[14]

Daladier perceived very accurately the continuities and discontinuities between the Risorgimento liberal patriotism and the new form of aggressive nationalism that was emerging on the eve of WWI. On the one hand, this party shared with the founders of united Italy the idea that their country needed to become a respected actor of international relations. On the other hand, its conception of nationality was very different. Daladier summed up the novelty of this movement by writing:

> Its originality consists in a resolute denegation of any problem of internal politics. They have been solved by the men of the Risorgimento; there are no such things as religious problems, or constitutional issues. There is admittedly a social question. But neither the deputies of postmen – as the nationalists call the socialist members of Parliament – nor the trade unions, nor the employers' or State's paternalism can solve it: employees have nothing to win from these false supporters but a derisory increase of their wages that might worsen the cost of life. The only solution is in foreign politics. 'Proletarian' nations like Italy can only cope with misery and suffering through conquests and lucrative wars.[15]

Thus, Daladier described very well the transition between an integrating form of nationalism, aimed at unifying a divided country and based upon political liberalism, and an aggressive form of nationalism that needed a direct confrontation with other countries to assert its strength. While war was a reluctantly accepted necessity for liberal nationalists, it was the very basis of aggressive nationalism.

The war between Italy and the Ottoman Empire had caused deep surprise in France. Daladier explained this surprise by the misinterpretation by some French politicians of the nature of this new nationalism. As he wrote:

---

[13]«dont l'apparition est l'événement politique le plus intéressant de ces dernières années» *ibid.*

[14]«Il groupe la plupart des jeunes hommes qui dirigeront demain la politique italienne. Ses adversaires enfin ne combattent que les excès de langage de quelques uns de ses membres sans critiquer d'ailleurs leurs doctrines», *ibid.*, p. 8.

[15]«Son originalité consiste à nier résolument qu'il existe des problèmes de politique intérieure. Ils ont été résolus par les hommes du Risorgimento; il n'y a pas de problème religieux, il n'y a pas de question constitutionnelle. Il existe, il est vrai, une question sociale. Mais ce ne sont point les «députés des facteurs»—c'est ainsi que les nationalistes appellent les députés socialistes—, ni les syndicats ouvriers, ni le «paternalisme» des patrons ou de l'État qui la résoudront: les salariés ne peuvent gagner à l'intervention de tous ces faux défenseurs qu'un accroissement dérisoire de leur salaire ce qui, d'ailleurs, aggrave encore la cherté de la vie. Il n'y a qu'une solution et c'est dans la politique extérieure qu'il faut la chercher. Les «nations «prolétaires» comme l'Italie ne peuvent trouver de remèdes aux misères et aux souffrances que dans une politique de conquêtes, dans les guerres lucratives.» *ibid.*, p. 9.

> The expedition to Tripoli is a satisfaction given by the government to this party that had been requesting it for two years through its newspapers and numerous conferences that I attended.[16]

In the context of a war that aimed at reasserting Italian ambitions in Africa, Daladier tried to explain to his French fellows the reasons of the war and to make the Italian point of view understandable for a French audience. And the best way he found to clear up the Italian project was to compare it to the German one, a few weeks after the Agadir crisis that had jeopardized the balance of European power in Africa. German expansionism had been a French preoccupation since the 1870, and this obsession had been reactivated by the attempted «coup de force» of Wilhelm II in July 1911. It was then very natural for Daladier to make a parallel between the Italian and German nationalism and expansionism in Africa:

> The origins of this party […] can be found in the economical development of current Italy, in the growing German influence, and in the Pan-German ideas that directly inspired the Italian nationalist program.[17]

4 Considering Germany Through Italy

As his main reason for choosing Italy as a destination, Daladier explained that it was 'a country where one can read the secular struggle between occidental and central Europe'.[18] This was significant of a general tendency in France to consider Germany through Italy, and to see Italy as a territory of the competition between France and Germany. On that aspect, like many intellectuals in France, Daladier was aware of the diminishing influence of France in Italy while the attractiveness of the German model was growing:

> Under the influence of the German model, scientific methods are spreading very fast in Italy. They are greeted by more and more students.[19]

[16]«L'expédition de Tripoli est une satisfaction accordée par le gouvernement à ce parti qui depuis deux ans ne cessait de l'exiger dans ses journaux, dans de nombreuses conférences auxquelles nous avons-nous-même assisté.» *ibid.*, p. 8.

[17]«On trouverait les origines de ce parti—dont l'apparition est l'événement politique le plus intéressant de ces dernières années—dans le développement économique de l'Italie contemporaine, dans le progrès de l'influence allemande et la diffusion des idées pangermanistes dont le programme nationaliste italien n'est qu'une traduction à l'usage de l'Italie.» *ibid.*

[18]«[…] un pays où se lire encore la lutte séculaire entre les influences rivales de l'Europe occidentale et de l'Europe centrale». *ibid.*

[19]«Sous l'influence de l'imitation allemande, les méthodes scientifiques se propagent en Italie avec une extrême rapidité. Elles ont la faveur des étudiants dont le nombre d'ailleurs croît sans cesse.» *ibid.*

This remark is very significant of a more general preoccupation in France that has been studied by Claude Digeon who named it the 'German crisis of the French thought' (Digeon 1959): the idea that the defeat of 1870 was not only a military rout but also the symptom of a French cultural decline was an obsession to every French intellectual since then. During their stay abroad, because of their numerous contacts with young people of various origins, French students were in a particularly privileged position to gauge the continental strategical situation on the eve of the Great War. A good instance is given by the young mathematician Maurice Janet's travel to Germany. He stayed in Göttingen for some months in 1912 and his diary (Mazliak 2014) shows how he was impressed by the warming evolution of some parts of German society he perceived.

Italy was an interesting ground to assess the French influence in a disputed zone of influence. From that point of view, Daladier could only confirm a diagnostic made by many of his predecessors: France was losing its influence in Italy, while the prestige of the German Universities was higher than ever.

At the same time, Daladier's main originality consisted in trying to go beyond this competitive aspect and looking at Italy in a new way by taking seriously its ambition to assert itself a great power. As such, it could not be merely considered as a potential ally against Germany and France had to take into consideration its own interests.

> Since 1870, we think that people in the world are to be divided depending on their affinity with France or Germany and that these two countries' rivalry inspires foreign policies of all other countries – similarly to the Ancien Régime duelists who saw every people they met as potential witnesses.[20]

France, he thought, was naïve in its way of considering Italy in a double sense. On the one hand, it underestimated Italy's real weight in international relations. On the other hand, it missed the real reasons that pushed Italy to enter the Triple Alliance to everyone's surprise:

> The French tend to believe that the Triple Alliance is some kind of «*combinazione*» in which Italy has only involved its immediate and temporary interests. 'Latin brothers' only speak to the 'Barbarians of Germany' in order to avoid a war against Austria. But when they use their reason and their deep feelings, they would never invite Gretchen to the ball – but our Marianne.[21]

[20]«En outre, depuis 1870, nous pensons que les peuples se groupent suivant leurs affinités avec la France ou l'Allemagne, que la rivalité de ces deux pays inspire la politique extérieure des autres puissances—semblables en cela aux duellistes de l'Ancien régime qui au moment de vider leur différend voyaient dans tous les possibles passants autant de seconds.» *ibid.*

[21]Et les Français croient volontiers que la Triple-Alliance est une sorte de «*combinazione*» où se sont engagés les intérêts immédiats essentiellement temporaires de l'Italie. Les «frères latins» ne causent avec les «Barbares de la Germanie» que pour éviter une guerre autrichienne; mais quand ils consultent leur seule raison, et leurs sentiments profonds, ce n'est point Gretchen—c'est notre Marianne qu'ils viennent inviter au bal.» *ibid.*

By opposing the dream of a union between the 'Latin brothers' to the pragmatic rapprochement of the Triple alliance, Daladier mocked the illusions of some French politicians who did not understand the reason for the Italian choice of an alliance with Germany. Pierre Milza has shown that this choice, made in 1882, was the result of a coalition between a nationalistic left that wanted to assert Italian positions in North Africa and a conservative right that worried about a possible contamination of the French republican movement in Italy (Milza 1981). The situation had admittedly evolved at the turn of the century, in particular because Italy needed to access the French credit market. But the building of the category of a Latin brotherhood, or sisterhood, that was recently studied by Christophe Poupault, was largely exploited by both sides (Poupault 2009, 2013; Fraixe et al. 2014): the French hoped to put Italy out of its German alliance, and Italy wanted to reassure France without giving up its relations with Austria and Germany. Thus, the reference to Latinity in French and Italian discourse must not be essentialized: there is an obvious dimension of opportunity in this reference that was invoked as a support to different diplomatic strategies. And Daladier pointed out very early the blindness of some French who thought that this allusion to Antiquity guaranteed a community of interests between France and Italy. He could later experiment by himself the complete reversibility of this reference when he saw as French Prime minister the degradation of the relations between France and Italy at the end of the 1930s: at this time, Mussolini knew perfectly how to justify his gallophobia and the rupture of the Latin sisterhood by sending back the French to their Gallic origins:

> In the antic world, Gaul tended to suffocate the rising forces of Rome, invaded Italy several times, and formed an alliance with Hannibal. Only with Julius Caesar's conquest they had to adapt to the Roman superiority that gave them a guarantee over the Rhine.[22]

Daladier was thus particularly skeptical about the idea of a Latin sisterhood that prevented France from understanding the Italian ambitions over Europe and the Mediterranean. In the French conception of this sisterhood, Italy was always the little sister, which was unacceptable for Italy.

Daladier's stay in Italy and the report he drafted afterwards are thus significant of many aspects of the way French historians and politicians tended to consider Italy on the eve of WW1. They reveal a growing internationalization of the French University thanks to the support of important patrons such as David David-Weill. They also show how the Italian Risorgimento shaped the French representations of their neighbors, assimilating Italian nationalism to an integrating and liberal form of patriotism and largely ignoring its evolutions. On the contrary, Daladier's observations led him to consider the emergence of a new nationalist party as a true rupture from the nation building process and to focus on the international competition that Italy was entering by going to war in Libya. For that reason, he invited the French to reconsider their attitude towards Italy and to stop considering it only

[22]Nel mondo antico, i Galli tendono a soffocare le forze nascenti di Roma, invadono a più riprese l'Italia, si alleano con Annibale e solo dopo la conquista di Cesare si adattano alla superiore civiltà di Roma, che li «garantiva sul Reno» Benito Mussolini, «Segnalazione», *Il Popolo d'Italia*, 161, June, 10th 1939, reproduced in *Opera omnia di Benito Mussolini*, vol. 29, p. 296.

as grounds for their competition with Germany. This conclusion is all the more interesting as it came from a young historian about to become a major actor of French international relations: it underlined at the same time the obvious rise of Italy as a great power and the ambivalence of the French reaction, at the same time tempted to treat Italy as a little sister in a mythical Latin sisterhood and worried about the emergence of a potential rival in Europe and in the Mediterranean.

References

Basch Sophie, *Le Mirage grec. La Grèce moderne dans la littérature française depuis la création de l'École française d'Athènes jusqu'à la guerre civile grecque (1846-1946)*, Paris-Athènes, Hatier, 1995.

Daladier Edouard, *Journal de captivité (1940-1945)*, Paris, Calmann-Lévy, 1991.

Digeon Claude, *La Crise allemande de la pensée française(1870-1914)*, Paris, Presses universitaires de France, 1959.

Durand Antonin, « L'odeur de l'argent. Dons et legs dans le financement de l'Université de Paris (1885-années 1930) », *Revue d'histoire moderne et contemporaine*, forthcoming.

Fraixe Catherine, Piccioni Lucia and Poupault Christophe (ed.), *Vers une Europe latine. Acteurs et enjeux des échanges culturels entre la France et l'Italie fasciste*, Bruxelles/Paris, Peter Lang, 2014.

Mazliak Laurent, *Le voyage de Maurice Janet à Göttingen*, Paris, Editions matériologiques, 2014.

Milza Pierre, *Français et Italiens à la fin du XIXe siècle. Aux origines du rapprochement franco-italien de 1900-1902*, Rome, École française de Rome, 1981.

Pécout Gilles, « L'Unité des italiens de 1849 à 1870 : une histoire française ? » *in Napoléon III et l'Italie. Naissance d'une nation (1848-1870). Catalogue de l'exposition du musée de l'Armée*, Paris, Nicolas Chaudun, 2011, pp. 29–46.

Poncet Olivier, « La querelle des origines. Émile Burnouf, Albert Dumont et la création de l'École française de Rome (1872-1875) » *in* Michel Gras, Olivier Poncet (ed.), *Construire l'institution. L'École française de Rome, 1873-1895*, Rome, Publications de l'École française de Rome, 2014, p.

Poupault Christophe, « Amitié « latine » et pragmatisme diplomatique. Les relations franco-italiennes de 1936 à 1938 », *Relations internationales*, 2013/2, 154.

Poupault Christophe, « Les voyages d'hommes de lettres en Italie fasciste. Espoir du rapprochement franco-italien et culture de la latinité », *Vingtième Siècle. Revue d'histoire*, 2009/4, 104.

Réau Élisabeth du, *Edouard Daladier*, Paris, Fayard, 1993.

Tronchet Guillaume, « L'État en miettes. La fabrique de l'impérialisme universitaire français (années 1920-1930) », *Hypothèses 2010. Travaux de l'École dorcotale d'histoire de l'Université Paris 1 Panthéon-Sorbonne*, Paris, Publications de la Sorbonne, 2011, pp. 281–292.

Introduction: The Latin Sisters and Mathematics

Frédéric Brenchenmacher, Guillaume Jouve, Laurent Mazliak and Rossana Tazzioli

F. Brenchenmacher
Ecole Polytechnique, Palaiseau, France
e-mail: frederic.brechenmacher@polytechnique.edu

G. Jouve
Université d'Artois, Arras, France
e-mail: guillaume.jouve@espe-lnf.fr

L. Mazliak (✉)
Université Pierre et Marie Curie, Paris, France
e-mail: laurent.mazliak@upmc.fr

R. Tazzioli
Université Lille 1, Villeneuve d'Ascq, France
e-mail: rossana.tazzioli@math.univ-lille1.fr

F. Brechenmacher et al. (eds.), *Images of Italian Mathematics in France*,
Trends in the History of Science, DOI 10.1007/978-3-319-40082-2_2

O sisters ! your heart is the eternal Host
Whose fecundity in itself humanity feels!
As this Sordello named by Dante
Give again, always, your beloved ideal
To nourish the charmed world
I see in your sky this original star
The symbol of every light
The one who saluted Greece at its beginnings
And who promised the world to the Latin ideal[1]

The somewhat grandiloquent words of the previous poem, written by the French writer Jean Aicard (1848–1921), are a tribute to the visit of the Italian sovereigns to Paris, an important diplomatic event organized in 1904 to mark the rapprochement between Paris and Rome. These words express a major rhetorical argument used at the beginning of the 20th century to justify the alleged proximity between Italy and France: the Latin ideal, heir of Greek classical wisdom, considered as a counterweight to other cultural influences in Europe, primarily the German influence. Since the end of the 18th century, tales about barbarian invasions led to new reflections about the distinction between the Roman civilization and barbarism. Expressions such as "Latin language" and then "Latin race" appeared during the 19th century, and publications related to various research areas followed where the word "Latin" is opposed to Barbarian or Germanic.[2] In the first years of the new century, Italy was accepted by her transalpine neighbour as a partner, unavoidable to be, if not already equal, and this may be seen for instance through the increasing interest of French academics or students in Italian intellectual production, as was illustrated by Antonin Durand's text introducing this book.

That was also true for mathematicians. The mathematician Vito Volterra (1860–1940) was much involved in the effort of developing cultural relations between Italy and France from the beginning of the 20th century– Italian-French associations and journals, exchanges of students, and invitations of scholars.[3] Volterra used his relevant political and institutional responsibilities for increasing this kind of activity. Italian and French mathematicians regularly evoked Latin ideals with reference to Italy and France. Such rhetoric attained a peak during the Great War when it was recurrently used as a motto for propaganda, especially during the nine month period when Italy was neutral. In a letter to his French colleague Gaston Darboux (1842–1917), Volterra wrote on 7 September 1914:

[1]Vous, ô soeurs! Votre coeur est l'hostie éternelle / Que l'humanité sent féconde en elle ! / Comme ce Sordello, que le Dante a nommé, / Donnez encor, toujours, votre idéal aimé / En pâture au monde charmé./ Je vois dans votre ciel cette étoile première, / Symbole de toute lumière / Celle qui salua la Grèce à son matin / Et qui promet le monde à l'idéal latin.

[2]An essay on latinity in Europe that focalises on cultural relations between France and Italy during the Fascist regime is (Fraixe et al. 2014).

[3](Mazliak and Tazzioli 2009), especially pp. 181–184.

> I think Italy must take its place next to her Latin sister, France, and its allies against Austria and Germany. This is its role and mission.[4]

An echo came from France a few months later, when the Alsacian mathematician Paul Appell (1855–1930) wrote to Volterra on 13 January 1915:

> We see with great pleasure, in France, that we have the sympathy and moral support, pending the loving cooperation, from the noble Italian nation that has known and still knows the Austrian oppression, as we have suffered the German oppression in Alsace and Lorraine. Italy, like France, [...] a civilization of freedom, justice and humanity...[5]

Both quotations paint the picture of France and Italy as two Latin sisters struggling for freedom and justice against the Barbarian invaders.

That was a remarkable evolution after many years of a difficult relationship between Italy and France in the last part of the 19th century. Italy, as an independent country, was a newcomer on the European stage. The kingdom was born in the 1860s and completed its unification after the suppression of the Pope's states with the conquest of Rome in 1870. Due to a mixture of political pragmatism and the sloppiness of French diplomacy, Italy in the 1880s chose to turn her gaze towards the other newly arrived political entity of the European concert, Germany. The latter had just severely defeated France and had undergone formidable development since the 1820s. During the fifteen years between 1881 and 1896, and especially when Francesco Crispi was Prime Minister, Italian "Gallophobic" politics was openly close to Germany and hostile to France.[6] A Triple Alliance between Germany, Austria-Hungary and Italy was signed in 1882. Italy sought support against France in order to be able to start her colonial policy shortly after having given up her North-African ambitions in Tunisia to the French.[7] The political and cultural model the Italians tried to follow was that of Germany.

The Italian attitude towards Germany was often depicted by the French press during this period. Italy's politics were officially displayed in France as treason, in light of the help the French offered in the fights of the Risorgimento. Moreover, social backwardness generated an important emigration of poor Italian workers, especially to the south of France and social tensions were important. Some dramatic events, such as the murder of several Italian workers in the "salines of Aigues-Mortes" in 1894 on a background of severe competition for employment

---

[4]The letter is kept in Volterra's Archive, Biblioteca dell'Accademia dei Lincei, Rome. The original French is:« A mon avis l'Italie doit prendre sa place à côté de sa sœur latine : la France, et de ses alliés contre l'Autriche et l'Allemagne. C'est son rôle, et sa mission ».

[5]Letter by Appell to Volterra (in Volterra's Archive, Biblioteca dell'Accademia dei Lincei, Rome). The original French is: « Nous voyons avec grand plaisir, en France, que nous avons les sympathies et l'appui moral, en attendant la coopération affectueuse, de la noble nation Italienne qui a connu et qui connait encore l'oppression autrichienne, comme nous avons subi l'oppression allemande en Alsace et Lorraine. L'Italie, comme la France, [...]une civilisation de liberté, de justice et d'humanité... ».

[6]Concerning Italian policy towards France in the late part of the 19th century see for instance (Pécout 1997).

[7]On Italian colonial policy see for instance (del Boca 1976).

between local and foreign workforces and a violent thrust of xenophobia, contributed to make the situation all the more difficult.[8] Crispi's fall in March 1896 after the Italian defeat of Adua in Abyssinia and the subsequent rapprochement with France inaugurated a new era in the relationship between the two countries, celebrated in 1904 by the ceremonies mentioned above.

The construction of the French perspective on new Italy after the Risorgimento has become an important theme in cultural history after Pierre Milza's thesis in 1982.[9] Milza describes how the collective mentality in France about Italy evolved between rejection, sympathy and attraction over fifty years. Italy's own cultural construction was caught in the crossfire of German and French influences, both being rather strong in the country though, as said before, the former was certainly more efficient. As Milza remarks, at the turn of the century, German philosophy, especially Neo-Hegelianism and Nietzschean ideas, were strongly represented in Italy, as well as German studies in history and philology. However, Italian cultural presence in France was rather scattered. Even in 1914, when he had to write a report for the David-Weill foundation about his travel grant, the young mathematician René Gateaux (1889–1914) rightly regretted that Italy and the Italian language were so poorly known in France while, on the other hand, France and French were widely present in Italian society.[10]

Yet, Milza observed an increasing number of scientific missions by French academics to Italy from the 1880s, a proof that the new country was gradually being considered with greater attention.[11] However, it is worth observing that the academics (*savants*) considered by Milza in fact only include archaeologists, geographers or physicians. Mathematicians (and other scientific specialists such as physicists of biologists) are totally absent from his considerations. Christophe Poupault's remarkable study on the travels of the French to Italy during the Fascist period has the same limitation.[12] In general, it was above all the literary specialists who benefited from an investigation by the historians of international politics. In his thesis, Angelo Morabito provides a fresh image of post-Risorgimento Italy in France by focusing on Italian heroes, such as Mazzini and Garibaldi, and the most beloved writers and poets in France.[13] Much remains thus to be done to explore how other actors of the French intelligentsia met the appearance of this new partner beyond the Alps.

---

[8](Noiriel 2010). The author writes that this massacre was the most important «pogrom» in contemporary French history (p. 9).

[9](Milza 1981). Milza studies the image of Italy and Italians in the collective mentality of the French through three main vectors: travel diaries in Italy, daily press, school textbooks (see in particular, vol. 1, pp. 354–473).

[10]See (Mazliak 2015).

[11]Travel diaries concern all Italy, but especially Sardinia, Pompei, and Sicily; see (Milza 1981) note 8, pp. 343–345.

[12](Poupault 2011).

[13](Morabito 2012).

On the French mathematical scene, Italy was also considered as being distant at the beginning. Some clear-sighted personalities seemed however to realize as soon as in 1870 that in Europe, the powerful German mathematical community and the quick Italian mathematical development represented a major challenge for French mathematicians who, since the legendary times of Napoleon, thought themselves placed at an unreachable height. Gaston Darboux, for instance, wrote in 1870 to Jules Houël (1823–1886) that the Germans clearly surpassed the French on the mathematical scene and that if things would continue in that way, the Italians would soon do the same.[14]

Many documents prove the self-consciousness of Italians about their new importance in European mathematics at the turn of the century. In his analysis of the evolution of mathematics in Italy since 1839, during the first congress of the SIPS in 1907 Valentino Cerruti (1850–1909), a professor of rational mechanics at the University of Rome, provided an enthusiastic picture of the situation:

> It has often been lamented that in Italy it was never possible to create a mathematical tradition, or that continuity was missing even when some great mind succeeded in generating love for studies of geometry. This fact, true to some extent, has many causes that have been said and repeated many times. But among them, there is a very simple one that has exerted and can still exert an even greater influence than one usually mentions. In a poor country, where motion and transformation of wealth are reduced to minimal proportions, where the main problems on which prosperity and power of the state depend are not dealt with, the practical usefulness of mathematics does not go beyond the most essential elements of everyday life and the interest for higher and secret questions can only be cultivated by a few privileged minds. This was the situation in Italy in the first half of the nineteenth century. But things are different today. The fantastic industrial renewal, the colossal works for the arrangement and the establishment of air, land and sea communications, the rebirth of building construction, the exploitation of natural energy in all its forms, the organization of the great services of the State, the creation of new economic institutions to cope with the continuous social changes required and still require solving complex problems, for which the use of mathematics in their most subtle dimensions is so necessary and valuable that many people must get a scientific training in order to use science as a means and not as an end. These technical imperatives have in turn a positive impact on defenders of science for itself, both because of the direction they give to their research and because of the improvement of the working methods. Indeed, abstract mathematics in the new life of Italy no longer represents only a ceremonial or decorative doctrine, but has become a valuable economic factor. One may hope that the climb that has lasted for more than half a century with great success would not slow down but would go on and increase with new vigour. Our Company will reach this aim by gathering and organizing studies and researchers. At the beginning, it aimed to form a national political awareness. It pursues now an ideal no less noble and better suiting its character: creating a national scientific awareness.[15]

[14]The undated letter is contained in Dossier Hoüel, Archives de l'Académie des Sciences in Paris, and published in (Gispert et al. 1987). In particular, the following passage of the letter is significant: « Je pense que vous êtes du même avis, les Allemands nous enfoncent par le nombre, là comme ailleurs. Je crois que si cela continue les Italiens nous dépasseront avant peu. » (p. 160).

[15](Cerruti 1907) (p. 106–107 for the quotation).

It was however difficult for the French audience to look with new eyes at Italy and without a priori considerations dictated by the traditional role devoted to the country: a museum, keeping the treasures of Antique or Renaissance culture, and also a conservatory for the Roman Catholic church, though the complicated relationship between new Italy and the Papacy induced contradictory feelings on that matter.

To what extent latinity, whatever this rather vague concept meant for those who used it, had played a positive or a negative role is an interesting point. Clearly used for propagandistic purposes during the Great War, as already mentioned, the term was also used before to mark the distance between the two countries. In the national legend forged in the 19th century, if France could obviously not refuse to Rome the privilege of having brought its civilization to the Gauls the subsequent evolution of French culture was mostly due to other factors. In this vision, in contrast to Italy, latinity in France had been adapted by people from other cultures and mixed with a creative efficiency. This produced what the French thought to be the *golden proportions* of French classicism. Moreover, a cult for centralization had been developed among the French elite at least since the 17th century and was considerably enforced during the Revolution; in contrast the Peninsula contained many urban centres, Florence, Turin, Venice, Naples, Palermo and of course Rome—though the presence of the Papacy created there a complex and delicate situation- and this seemed to be counterproductive for the diffusion of science. It was on the whole not a very favourable situation for the development of a new view on the recent Italian evolution.

It took generally quite a long time for the French to regard their transalpine neighbour as a modern country, with modern economy and interests in modern research, in particular science, including mathematics. It seems quite possible to adapt, *mutatis mutandis*, what Christophe Roux writes about the French view on Italian political sciences after World War 2 to the situation in mathematics during the 19th century considered in the present book.

> Only by a seeming paradox, that feeling of closeness has nevertheless not induced an attempt for an exhaustive knowledge of her transalpine neighbour in France. The universal admiration for Italian culture has indeed produced different effects in different countries. The Anglo-Saxons—a central reference due to their international leadership in most branches of humanities and social sciences, including political science—have shown their fascination for the country by strongly promoting scientific knowledge, as exemplified by the numerous groups of historians, anthropologists and researchers in political science that had been sent to the Peninsula, even though some of them may have come there with political matters in mind, particularly during the Cold War. In France, there was generally no fascination for some figure of radical alterity, but on the contrary a feeling of proximity fed by intuition, which did not prompt to explore her neighbour further. From that situation partly stems the gap between the impression of knowing and the relative lack of knowledge.[16]

The present book deals with the question of the new view from the French side on Italian mathematics during and after the Italian unification in the 19th Century.

[16](Roux 2003).

Mathematics had played a specific role in France for years, and the astounding prestige of the French mathematicians of the beginning of the 19th century (with names such as Laplace, Cauchy, Poisson and so many others), as mentioned earlier, reassured their successors of the dominant position of the French mathematical school on the European stage. The shock of the defeat in the Franco-Prussian war in 1870 forced many of them to admit the formidable expansion of German science. As it is clear from Darboux's letter we quoted above, at that time, the question was raised to foresee to which height the Italian mathematical community was able to climb. The chapters of the book provide various examples illustrating how French mathematicians gradually became aware of what was happening beyond the Alps.

The present publication is based upon the contributions to a conference held in Lille and Lens (France) in November 2013, which was devoted to images of Italian mathematics viewed from France, during the period starting just before the Italian Risorgimento and lasting until the stabilization of the fascist regime (roughly from 1815 to 1928, a little more than a century). Though several works on mathematical relations between France and Italy have already been published, they generally concentrate on how French mathematics were a model for Italian mathematicians—the French model concerning not only mathematical research, but also sometimes French institutions. Examining in detail how French mathematicians looked at Italian mathematical research, as well as Italian universities, institutions, scientific journals and conferences, helps giving a more faithful picture of some episodes in the history of mathematics. French mathematicians were inspired by the research developed in the peninsula on several occasions, and from this emerged a subtle network of relations between the two countries, where mathematical exchanges fit into the changing and evolving framework of Italian political and academic structures. Far from being a one-way relationship, this shows a multiplicity of actors and places, which became crucial in the dynamics of some mathematical specialities—by bringing out specific methods to solve problems, or becoming reference points for such methods. The book tries to take into account not only the disciplinary aspects, by considering mathematical problems faced simultaneously by mathematicians of the two countries, but also the institutions to which these mathematicians belonged, the journals in which they published, the conferences they attended.

To conclude this introduction, let us briefly describe the contents of the eight chapters. Their order roughly follows the chronology.

The first chapter, written by **Pierre Crépel**, is devoted to describing how Italian mathematicians were presented in the French biographical dictionaries of the 19th century. Until the 1870s, a huge number of biographical dictionaries were published in all Europe, such as those by Michaud and Hoefer. Pierre Crépel provides a general picture of the presence of Italians in these publications, and who were the main authors of these biographical notices. He also discusses if these biographical notices are representative of the image French mathematicians had of their Italian colleagues. The study focuses on Italian mathematicians who died between the late 18th century and 1860s and follows the evolution of their biographies both in different dictionaries and in different editions of the same works.

The second chapter, by **Frédéric Brechenmacher**, depicts some Italian mathematicians through the eyes of the scientific correspondence of Camille Jordan. This correspondence constitutes a very rich—and mostly unexploited-material on the evolution of mathematics from 1860 to 1900. Italians are so well represented in this correspondence that one may mistake Jordan himself for an Italian mathematician. As a matter of fact, Jordan took on several "Italian" roles during his career. For instance he contributed to the *Annali di Matematica pura ed applicata* early on in the 1860s, and was quickly considered by Brioschi and Cremona as a "scientific brother" as well as the "future" of mathematics. Jordan was not only elected corresponding member of several Italian institutions, such as the Lombard institute for sciences, but was also appointed as a delegate of (and anonymous reporter for) the Circolo mathematico de Palermo in France. In describing the various Italian roles Jordan took on during his career, this chapter aims at questioning the relevance for analyzing the evolution of mathematics of national categories such as France and Italy.

The third chapter, due to **Aldo Brigaglia**, concentrates on the question of the scientific relationships between Emile Picard and the Italian algebraic geometers, especially Federigo Enriques, Guido Castelnuovo and Francesco Severi. Even though their research fields sometimes intersected, they developed different mathematical languages, which often made mutual understanding difficult.

The chapter provides moreover an overview of the relationship between Picard and Bianchi in number theory, and also investigates the influence of Corrado Segre and Guido Fano on Élie Cartan, who also had a profound influence on Italian algebraic geometers, especially on Beniamino Segre.

The fourth chapter, written at six hands by **Angelo Guerraggio**, **Frédéric Jaëck** and **Laurent Mazliak** deals with the fascinating personality of Vito Volterra. In 1887, the latter published several papers introducing the notions of "function depending on another function" and of "function of lines" and opened thus a new chapter for mathematical analysis. Fifteen years later, Jacques Hadamard discovered in these works relevant tools to tackle some problems about partial differential equations and integral equations with a new approach. He subsequently published several papers in which he introduced Volterra's notion of function of line under the name of "functional" and used Volterra's formalism to deal with variational problems with a moving boundary. After 1904, Volterra and Hadamard began a regular correspondence and their scientific convergence can be observed in the session they jointly organized at Heidelberg's international congress that year. At the same time, Maurice Fréchet, who had just completed his studies at the École Normale Supérieure in Paris, was looking for a suitable theme for his PhD. Following Borel's advice, he wrote to Volterra to get his opinion about the relevance of extending some aspects of his theory of calculus for functions of lines. When he began to work on Volterra's results and when he discovered some related works, such as Arzelà's papers, Fréchet understood the interest of a more abstract approach, which would lead him to his study of abstract spaces and to his 1906 thesis.

The two following chapters are both concerned with Peano's activity and its reception in France. They consider, however, two rather different aspects of this activity. The first chapter, due to **Erika Luciano**, studies the reaction of the French community to the Italian treatises of calculus. From the 1880s indeed, the teaching of calculus in Italian universities showed significant elements of modernity and originality, especially concerning questions linked to the foundations of mathematics. Some works were at the origin of the renewal of university textbooks, including for instance the *Fondamenti per la teorica delle funzioni delle variabili reali* of Ulisse Dini and three treatises by Giuseppe Peano collecting his university lectures (1884, 1887, 1893). These treatises had a strong influence both on the university teaching practice and on the methodological choices of most Italian university books until the 1920s. The chapter illustrates how this educational trend had a significant impact on France. The French mathematical community appreciated the "rigorous" Italian treatises and seized this educational model. In particular, the circulation of these in France is scrutinized, in order to examine the phenomenology of the use of Peano's treatises by his French colleagues, in the *Grandes Écoles* and in *Classes préparatoires*. The other chapter, written by **Paola Cantù**, describes the influence of the interdisciplinary approach developed by the Peano School on the interest acquired by French philosophers for logic and mathematics. In Italy, during the second part of the 19th century, the tension between a specialized and an interdisciplinary approach to knowledge emerged both in the discussions about the didactic reform of universities (and secondary schools) and in scientific research. In Turin, the group of mathematicians who studied or worked under the supervision of Giuseppe Peano between the end of the 19th century and the beginning of the 20th century developed an original interdisciplinary approach: on the one hand, they were highly specialized in logic and the foundations of mathematics, but on the other hand, they had multidisciplinary interests in linguistics, psychology, history of sciences and philosophy, philology, and politics. The chapter considers the influence of Peano's school on an inversion of the tendency that emerged at the beginning of the 20th century in France, when philosophers started to be attracted by the writings of mathematicians who algebraized or formalized logic. In particular, it investigates the relation between the interdisciplinary and "pragmatic" conception of Peano's school and the philosophical developments due to Louis Rougier and Jean Nicod. From the perspective of the history of ideas, it is particularly relevant to compare Rougier's conventionalist interpretation of Poincaré and the probabilistic approach developed by Nicod with the ideas on the same topics expressed by Giovanni Vailati, a brilliant member of Peano's school.

The seventh chapter by **Rossana Tazzioli** studies how French mathematicians judged Tullio Levi-Civita (1873–1941) and his scientific contribution to hydrodynamics. Before the First World War, Levi-Civita was already a well-known mathematician in Italy and abroad, especially in France. Professor at the University of Padua since 1898, he had published important works on tensor calculus, the theory of relativity, hydrodynamics, and the three-body problem. In 1911 he was elected a corresponding member of the Academy of Sciences of Paris. Nevertheless, from the 1920s, the relationship between Levi-Civita and his French colleagues became stronger. In particular, Levi-Civita was a privileged partner of the director of the École Normale of Paris, Ernest Vessiot, and the study periods of French students in Rome were becoming more numerous and considered fundamental for their future research. In this chapter, Rossana Tazzioli analyzes the influence of Levi-Civita's hydrodynamics in France, especially on Henri Villat's works, and highlights the reasons, both institutional and scientific, which led to this special relationship between Levi-Civita and his French colleagues.

And finally, the last chapter, written by **Annalisa Capristo**, examines the situation of the most important international mathematical event of the 1920s, the international congress of Bologna in 1928. This conference took great significance both in the scientific and political spheres. Since this was the first congress after the First World War that was open to mathematicians from all countries, its organization was complicated by the conflicts within the International Mathematical Union and the International Research Council with regard to the admission of mathematicians from the former Central Powers. Eventually, this led some renowned French scientists to stay away. For the Italian authorities the Bologna Congress was seen as an exceptional opportunity to showcase the scientific strength of Fascist Italy, so it has also to be considered for its propagandistic significance, and implications. The chapter aims to analyze the organizational and political aspects of the 1928 Bologna Congress, with particular attention to the presence and absence of French mathematicians.

The aforementioned contributors come from different backgrounds: they are historians of mathematics, philosophers of science, and specialist historians of Italy or of intellectual relations. Some of them gathered in an interdisciplinary research group, and have already collaborated in the organization of conferences in France and Italy, and by publishing articles and books on related themes (see for instance (Mazliak and Tazzioli 2009; Brechenmacher 2011; Durand et al. 2013; Mazliak and Tazzioli 2015)). The present publication is hence in consonance with this new perspective of the historiography of science of the 19th and 20th centuries, aiming at reconsidering the issue of nationalities under the different pertinent points of view and in its complexity, often omitted in the research on history of mathematics.

Frédéric Brechenmacher,
Guillaume Jouve,
Laurent Mazliak,
Rossana Tazzioli.
Palaiseau-Lens-Paris-Lille, February 2016

References

A. del Boca, *Gli Italiani in Africa orientale*. Vol. 1 : Dall'Unità alla marcia su Roma, Laterza, 1976.

F. Brechenmacher (ed.), Numéro spécial Galois, *Revue d'histoire des mathématiques*, vol. 17, 2011.

V. Cerruti, Le matematiche pure e miste nei primi dodici congressi della Società Italiana per il Progresso delle scienze, Attidella SIPS, 1907, p. 94–107.

A. Durand, L. Mazliak and R. Tazzioli, *Des mathématiciens et des guerres*, CNRS Editions, 2013.

C. Fraixe, L. Picciano, C. Poupault (ed. by), *Vers une Europe latine. Acteurs et enjeux des échanges culturels entre la France et l'Italie fasciste*, Peter Lang, 2014.
H. Gispert, La correspondance de G. Darboux avec J. Hoüel. Chronique d'un rédacteur (déc. 1869-nov. 1871), *Cahier du séminaire d'histoire des mathématiques*, vol. 8, 1987, p. 67–202.
L. Mazliak, The ghosts of the Ecole Normale. Life, death and legacy of René Gateaux, *Statistical Science, Institute of Mathematical Statistics*, (3) 30, 2015, p. 391–412.
L. Mazliak, R. Tazzioli, *Mathematicians at war. Volterra and his French colleagues in World War I*, Springer, 2009.
L. Mazliak and R. Tazzioli (editors), *The calculated victory. Italian mathematicians and the Great War*. Lettera Matematica International, Volume 3, issue 1. June 2015 (http://link.springer.com/journal/40329/3/1/page/1).
P. Milza, *Français et italiens à la fin du XIXème siècle*, Rome, école française de Rome, Paris, De Boccard, 1981, 2 vol.
A. Morabito. *La construction nationale italienne dans le miroir français. Représentations croisées des « Pères de la Patrie italienne » en France du Printemps des Peuples à la Grande Guerre (1848–1914).* Thèse. Université de Paris-Est, Università di Pisa, 2012.
G. Noiriel, *Le massacre des Italiens, Aigues-Mortes, 17 août 1893*, Fayard, Paris, 2010.
G. Pécout, *Naissance de l'Italie contemporaine 1770–1922. Des origines du Risorgimento à l'Unité : comment l'Italie est devenue une nation*, Nathan, Paris, 1997.
C. Poupault, *A l'ombre des Faisceaux. Les voyages français dans l'Italie des Chemises noires (1922–1943).* Thèse. Université de Paris 10, Università La Sapienza Roma, 2011.
C. Roux. *Franchir les Alpes. Regards vers la science politique italienne dans les années de la transition. Pôle Sud*, 19, 2003, p. 3–26.

Italian Mathematicians as Seen by French Biographical Dictionaries in the 19th Century

Pierre Crépel

1 Why This Subject?

In the aftermath of the Enlightenment and the French Revolution, between 1800 and 1840, France was habitually considered—and believed itself to be—the country at the centre of scientific endeavour. She hardly saw coming the rise to power of a Germany in tatters and sometimes still considered "barbaric". Italy, for its part, was in the first half of the 19th century a country without unity and judged to be distant from the lustre of the Renaissance.

The understanding of this moment when Paris was preparing, unbeknownst to itself, the loss of its scientific hegemony, requires diverse studies drawing not only from French and German, but also from Belgian, Swiss, Italian, and other sources. We do not presume to deal with as a whole this work in progress which is being explored by some of the most remarkable historians of science. We will be content to shine a light on a specific yet significant point: how did French scientists, polygraphs, and popularisers perceive "Italian" mathematics in the first half of the 19th century? We have already tackled an analogous question concerning the reception of Italian probability and statistics in France in the literature during the period 1820–1840 (Crépel 2012). We will consider this time how biographical dictionaries present (and judge) Italian mathematicians of the 18th and the beginning of the 19th centuries. We will concentrate on the epoch of great and famous monumental works such as the *Biographie universelle ancienne et moderne* of the brothers Michaud, an endeavour which extends at least from 1811 to 1865.

This chapter was translated by James Johnston a few days before his tragic death in January 2016.

P. Crépel (✉)
Institut Camille Jordan UMR 5208, Université Lyon 1 - Bâtiment Braconnier,
21 Avenue Claude Bernard, 69622 Villeurbanne Cedex, Lyon, France
e-mail: crepel@math.univ-lyon1.fr

F. Brechenmacher et al. (eds.), *Images of Italian Mathematics in France*,
Trends in the History of Science, DOI 10.1007/978-3-319-40082-2_3

2 Biographical Dictionaries of the 19th Century

The beginning of the 19th century is a "golden age" of biographical dictionaries. In addition to the numerous augmented reprints of those of the 18th century, such as Chaudon or Feller, there were biographies written by contemporaries, abridged dictionaries said to be "portable", etc. All of these works have their specific characteristics and ideologies: "philosophical", pro-Jesuit or pro-Jansenist, eclectic… An edition could dedicate a good part of its preface to criticising the orientations of its predecessor. But this is not the place to navigate through these issues, which are interesting in themselves (Chappey 2013). We will examine our subject from the perspective of the most famous and impressive of these serial works, Michaud, even if it means introducing some points of comparison with others.

The *Biographie universelle ancienne et moderne* of the brothers Michaud, above all of Louis-Gabriel Michaud (1773–1858), called the "younger", has no equal. The preface of vol. 52 affirms, without modesty but with a certain justice: "never a literary work, even that of the Encyclopedia, was executed with the same means and such great effort". The first edition in 52 volumes (1811–1828), and then the supplements extending to vol. 85 in 1862, included more than 350 contributors, more than 50,000 pages in-8° of more than 3300 characters… At the point when the supplements had reached only the letter L (vol. 72), L.-G. Michaud and the family Desplaces issued a new edition, with various additions, which included 45 volumes in large format (1843–1865). These dictionaries were always plagiarised, more or less shamelessly. Michaud's, which certainly had lifted materials liberally, attracted the envy of the publishing house Firmin-Didot, which issued a competing edition under the direction of Dr. Hoefer. A land-mark court case resulted from 1852 to 1855. Besides this, the large encyclopedias (such as Pierre Larousse's a bit later) developed important biographical articles.

There certainly exist particular studies on the dictionaries of the 19th century, but we do not have available analogous work instruments (nor even approaching) to those relative to the *Encyclopedia* of Diderot and d'Alembert, such as the table of the pastor Mouchon, the article inventories "Schwab-Rex", the biographical repertory of authors "Kafker". Whilst we wait for a new generation to start a collective work of scope, we must limit ourselves to a couple of preliminary probes, of which this article, amongst others, constitutes a modest attempt.

The articles of Michaud 1 and Michaud 2 are in general signed by a set of initials, which correspond to the names of authors which appear at the beginning or the end of the volume. One can make of list of them and search for the manuscripts in public libraries. To attack these collections would certainly go beyond our objective, but one can remark certain interesting examples: Beuchot at the National Library of France, Weiss at the Municipal Library of Besançon, Maury at the Library of the Institute of France, Bréghot de Lut at the Academy of Science, Letters, and Arts of Lyons, etc.

According to various notices, preliminary discourses, or even within the articles regarding its deceased contributors, the *Biographie universelle* Michaud 1 and 2 presents, incidentally or in more systematic fashion, the work of its editors. One also often finds a mention of these authors in competing biographical dictionaries, be it to possibly engage in veiled criticism, or because these authors had also written articles in these dictionaries. One must also consult the various biographical dictionaries written by contemporaries in the first two thirds of the 19th century. There exist various texts which are explicit to this end, but let us take up the Notice which opens the last volume (52), published in 1828, of the first edition of the *Biographie universelle* of Michaud.

3 Italian Mathematicians of the 18th and the Beginning of the 19th Century

If numerous French historians have directed their efforts towards Galileo and his disciples, they have not done the same for Italian mathematicians of the 18th and the beginning of the 19th century. The Italy of that period was regarded as a mere shadow of her former self. The French hardly deigned to read Italian, and it was rare for scientists to devote themselves to Italian science. Without a doubt, Italian exiles or French journalists who were interested in Italy could try to write a couple of panoramas in the journals. We have cited two assessments of this type of letters and sciences in Italy, published in the important *Revue encyclopédique* by Francesco Salfi and Adriano Balbi (Crépel 2012). In the 20th and 21st centuries, there are numerous and sufficiently systematic works on this era by Italian historians of mathematics, yet one must deplore the fact that French, English, or German historians only are interested in these in passing.

For the purposes of this article, we have considered a couple of dozen Italian mathematicians. Every choice has an arbitrary aspect, and ours are no exception. We have limited ourselves to the associates and correspondents of the Academy of Sciences of Paris, to a list of about thirty scientists whose names came back the most often, and finally to a list, different enough, of about forty mathematicians who figure in an abridged "portable" biographical dictionary (that is in a single volume) of the 19th century.

Here is the list of the associates and correspondents of the Royal Academy of Sciences (before 1793):

Vincenzo Viviani (1622-1703), foreign associate, 23 November 1699, first occupant
Eustachio Manfredi (1674-1739), foreign associate, 24 August 1726, replacing Hartsoeker
Giovanni Poleni (1683-1761), foreign associate, 24 August 1739, replacing Manfredi
-
Ruggiero Boscovich (1711-1787), correspondent of Mairan, 4 May 1748
Paolo Frisi (1728-1784), correspondent of Fouchy, 8 August 1753
Alfonso Malvezzi (1730-1804), correspondent of Condorcet, 5 May 1773
Antonio Maria Lorgna (1735-1796), correspondent of Lalande, 28 June 1771

And the list of the associates and correspondents of the Institute of France (after 1795):

Giuseppe Piazzi (1746-1826), correspondent, 9 Pluviôse, year XII, foreign associate, 18 August 1817
Pietro Paoli (1759-1839), correspondent, 9 December 1811
Guivanni Plana (1781-1864), correspondent, 2 October 1826, foreign associate 20 January 1864
Guglielmo Libri (1803-1869), correspondent, 31 December 1832, member, 18 March 1833

In the printed minutes of the Academy of Sciences, during the period 1795–1835, the name which comes back the most often is that of Plana, not for his scientific importance, but for his contributions in the domain of astronomy.

For the other lists, one will refer to the tables figuring in the annexes.

4 Articles on Italian Mathematicians in Michaud

It is first necessary to note that the 52 volumes of Michaud 1 appeared in alphabetical order, at an irregular rhythm of three per year, on average, between 1811 and 1828. An analogous phenomenon took place for the 30 volumes (56–85) of the *Suppléments* which appeared between 1834 and 1862. As his dictionary only covered deceased persons, one must cross-reference the dates of volumes for such or such letter of the alphabet with the dates of their death. Thus, for example, since in 1813, they were already in the middle of the letter C, a scientist such as Teodoro Bonati, deceased in 1820, could not have his article in Michaud 1, but he had it in the *Supplément*, whose first volume only came out in 1834. In order to analyse the contents of the articles, one must equally take account of the life, the death, and the kind of participation (short-lived or not) of the contributors all through this half-century.

Among the 29 scientists with whom we concern ourselves here, four died too late to have an article in Michaud, three simply did not have one. Among the twenty-two remaining, one sees that only Galileo (mentioned here for the sake of memory) and two of his disciples had a biography done by star-contributors such as Biot and Lacroix. Weiss wrote half of the articles; the others are scattered amongst several authors about whom we will say a word below. To better understand this division and the content of the articles, it is necessary to enter into the "biographical manufacturing", that is, to see from the inside how the articles are attributed to the various authors of Michaud according to the competences, their availability, and their ability to stand the authoritarianism of the management and the low pay. Since only a collective and systematic study of great scope would be able to clarify this intricate ensemble, we will content ourselves with a couple of paths linked to our subject. In order to be able to write an article on an Italian mathematician, it is best to be able to read Italian and know something about mathematics. Let us see then whom Michaud had at his disposal.

The Authors of Articles on Mathematicians in Michaud

Described somewhat summarily, the team assembled by Michaud included stars, drudges, and diverse others. The editor boasted often of his "catches": Benjamin Constant, Chateaubriand, Cuvier, Guizot, Mme de Staël, Suard, etc. For mathematics and astronomy, he put forward the contributions of four scientists: Biot, Delambre, Lacroix, le baron Maurice (a friend of Lagrange's), and a little later, de Prony.

However, one must not let oneself be deceived: these stars only wrote a couple of dozen biographies, and not necessarily those of the most eminent scientists; the greater part of the articles were owing to librarians, archivists, polygraphs, and compilers, whom one would be wrong to disdain, for they accomplished a considerable amount of work. Even if these articles contain errors, one must first of all appreciate the amount of research that they represent before criticising them. Among the authors of biographies of mathematicians, one finds also authors little known about whom we will say a few words.

Under the Restoration, tertiary education, already little appreciated, and judged somewhat useless and too excitable by Napoleon, was frankly repressed: the faculties of sciences and letters were closed in 1816; l'École Nomale Supérieure was suppressed in 1822. The energies of scientists were liberated for journals and dictionaries, since many generations educated in mathematics or other subjects found themselves somewhat at a loose-end, not being able to use their competences, as was the case for Cournot, Deflers, Ferry, Nicollet… (see Bru and Martin, in Cournot 2011).

One might expect that the articles on Joseph Fourier (deceased in 1830), Laplace (deceased in 1827), Legendre (deceased in 1833), Poisson (deceased in 1840), Prony (deceased in 1839), be signed by prestigious mathematicians of the middle of the century. But this was not at all the case. They were in fact all written by the somewhat obscure Valentin Parisot (1800–1861) … and they are not all that mediocre. He had been admitted at the École Normale Supérieure to study letters in 1816, defended a thesis in literature in 1820 on Aeschylus, Shakespeare and Schiller, yet he had a rather chaotic career, namely in Grenoble and Rennes; marginal and caustic, he was interested in everything. But we digress. It was a matter of penetrating the diversity of the authors of article in Michaud, even those about mathematicians.

Who Speaks Italian?

Among the contributors cited, for example, in vol. 52, one notes: "For the Literature and History of Italy, Messieurs Ginguené, de Sismondi et de Angelis" (p. viii). Pierre-Louis Ginguené (1748–1816) was namely famous for his *History of Italian Literature*, 9 vol. in-8° (1811–1819), which was being edited during the publication of the first volumes of Michaud 1, the last volume being completed, after the death of Ginguené, by Francesco Salfi (1759–1832). But Ginguené did not know much about mathematics!

As is known, in the first half of the 19th century Italy was a country divided into numerous small states, subject to all kinds of turbulence. In particular, at the fall of the Napoleonic Empire, Metternich and the Austrians submitted the country to repression, the *carbonari* were chased away and many Italian authors, liberals or revolutionaries, took refuge in France, where they sought to employ their talents. It was the case for Francesco Salfi, who had to flee from Naples to Genoa, and then to Paris during the Revolution, then back to Lombardy where he found a professorship, then to Naples, and had to exile himself again under the Restoration. Pietro de Angelis (1784–1859), who was implicated in the government of Murat, had to go into exile in Paris from 1819 to 1826, and then ended his life in Argentina, where he is now better known than in Europe. Adriano Balbi, originally from Venice where he was born in 1784, had actually studied mathematics. Gaspard de Grégory, born in the Piedmont in 1768, also had a chaotic carrier. All these persons were in search of more or less stable employment, even more or less precarious: dictionaries and journals, which flowered under the Empire, and above all under the Restoration, were an enticing prospect.

Charles Weiss

We have particularly turned our attention to the case of Charles Weiss (Besançon 1779–1866), since he wrote half the articles on mathematicians. As a librarian from 1812 until his death, he manifested a phenomenal activity, and noted everything. His papers are conserved; his manuscripts at the Municipal Library of Besançon include his correspondence, his diary, and his notes. He has been studied relatively little, but a researcher from Québec, Eldon Kaye, has drawn up an inventory with numerous transcriptions of his correspondence (Kaye 1987), and an historian from Besançon, Suzanne Lepin, has published the beginning of his Diary up to 1842 (Lepin 1972–1997). His writing is not easy to decipher, but that of his correspondents is in general much easier to read.

Charles Weiss was the principal contributor to Michaud, haven written around 5400 articles, which amounted to about to one per day for twenty years. From 1813 to 1826, Claude-Marie Pillet (1771–1826), one of Michaud's assistants for the first edition, harassed him every month in order to extract about thirty articles from him, and he raised his voice when it did not go fast enough, as we will see below (Ms 1890). One can observe how Weiss was solicited and how he replied. The correspondence gives us an idea about his network of librarians, booksellers, and friends from whom he asked for information of all kinds. We will be able to see, at least partially, how Weiss made inquiries about Italian mathematicians whose articles he was responsible for, and how he tried to have the articles written by those more competent in this domain than him.

The Michaud Factory

Moreover, the publication of a giant dictionary required a rigorous and often authoritarian organisation. Louis-Gabriel Michaud employed that we could call "foremen" to do this work. Volume 52 says a word about this, which we will reproduce here:

> MM C.-M. Pillet, Gence, Beuchot, de Fortia, Lefebvre, Miger and Philbert have successively reviewed and completed the bibliography, read all the proofs, verified the references, citations, and filled the lacunæ; These men have aided and assisted the editor in all of his efforts in the most efficacious manner; and Mr. Weiss, whose knowledge in literary history is so extended, has not contented himself to write a large number of articles of all kinds, but has reviewed and corrected with the greatest care all the proofs, and verified all the texts.

The Compagnie Michaud worked constantly in haste and under tension. The rhythm was frantic; putting things into alphabetical order was no joke. If a contributor threatened to delay the machine, he was replaced. Death, sickness, quarrels, and indolence did not have a place: one went forward, one managed, one forced one's way through, if need be.

One can imagine, as an example, the kind of regime that Weiss was subject to by the beginning of a letter from Pillet on 20 October 1818:

> "Dear Friend,
> Your letter of the 15th of this month brought me only 8 articles (le Laelius to Lafaye); all of them were overdue, and you are surprised that the Biography is going slowly!! The reason is that all the articles of all the contributors were overdue: these articles have been asked of you since the August 16th, that is, 60 days ago. Here is a new request of 45, that is 22 (from Lainez to Larue) for vol. 23, and 23 (from Legendre to Lemuet) for vol. 24, which will be printed at the same time as the other. Thus, they are needed just as urgently as the others.

Pillet was moreover in the same boat as Weiss under Michaud; he wrote numerous articles and passed the same instructions and pressures of his superior onto his inferiors.

Return to the Italian Mathematicians

Michaud was therefore able to easily find contributors who read and spoke Italian and others who had a knowledge of mathematics, but in general they were not the same people.

In order to understand the choices and omissions of the persons to be considered, the contents of the articles and their insufficiencies, it is indispensable to grasp the concrete conditions under which the contributors worked. Let us consider what comes out in the articles on the Italian mathematicians. To be succinct: they have a certain charm, not a little bit of the unexpected, and often contain information which one finds elsewhere with difficulty; but as a whole, they do not constitute a lucid and relevant panorama of a century and a half of Italian mathematics. But there is nothing surprising in this: as most of the authors are not mathematicians, how could they truly have the intuition of what is important or promising in the work of a scientist? They were dependent on available articles or ideas that were "lying about". However, as they were sufficiently erudite they had available information that was not widespread, and thus little known. They give us, concerning the life or the bibliography of their subjects, facts which sometimes escaped more recognised scientists.

Without entering into the details, a couple of citations from the correspondence of Weiss will let us understand what we might expect from such articles better than long discourses. One will not doubt the lucidity of Weiss, who, for a scientist such as Bonati, recognises the importance of the subject and the necessity of having it treated by a true specialist; in which he succeeds, as the article is signed by de Prony:

> The art. *Dolci* which Mr. Michaud asked of you, I don't know why, is not very important: I suggested that he ask it of Mr. Guillon in Milan, who would have found the materials at his fingertips. (letter from Cl.-M. Pillet to Weiss, 13 December 1813, Ms 635)

Ginguené was sick. Pillet took advantage of it to ask of Weiss a certain number of articles on Italian figures. When Ginguené had recovered, Pillet retracted the request (without having asked him whether he had already written them). (letter from Pillet to Weiss 9 February 1816, Ms 635)

> Grandi (Guido) m. 1742. Regarding this former person, if you're lacking the materials, let me know first and I will ask the article of Mr. Millin [...] mr. G-é [Ginguené] took him off his list. (letter from Pillet to Weiss, 8 June 1816, Ms 635)
>
> You asked for the art. *Bonati* the famous fluid mechanic, from Mr. de Prony? I have an article on him in Italian. And I will do the art. If necessary, but it would be better to ask it of someone specialised in the subject. It's a very important art. Bonati died 2 January 1820 (letter from Weiss to Michaud, 27 November 1834, Ms 634)

By studying the articles by Nicollet, Aimé Guillon, De Angelis, Ginguené, Parisot or Walckenaer in the same spirit, one can distinguish the strengths and the weaknesses of the facts and interpretations that they contain.

5 Concerning the Articles in Other Dictionaries

Simply for the sake of future paths of research, we will consider other dictionaries. We will evoke (briefly) the *Biographie portative universelle* [Universal Portable Biography] (BPU) of (1844), Hoefer, and the Grand Larousse. By using methods similar to those above, we may also gradually find information about the various contributors to these works: in their dictionaries themselves, in those of others, and in dictionaries of their contemporaries.

Among all the possible collections of biographies, we have chosen the BPU (which may seem to be anecdotal and of little importance) for two reasons. First of all, it is a work in one volume of close to 2000 very dense pages, with numerous short articles (the editor boasts that Michaud only included 23,500, whereas they have 28,400); secondly, it includes more Italian mathematicians, of which some are otherwise little known. Moreover, this dictionary includes an index of individuals by discipline and by century. We will reproduce this list in Annex 3, where we have joined a table of correspondence with Feller and Michaud. Finally, as the articles themselves are quite short, we will reproduce them integrally in Annex 4. This *Bibliothèque universelle portative* does not present itself as eclectic or neutral; rather it is "partisan". For example, it is overtly *robespierriste*, very hostile to

Marie-Antoinette, lukewarm on Lafayette or Napoleon. In the sciences, it defends Descartes, except for his "vortices", praises Newton, but castigates his egoism and his penchant for mystery in his publications and his works: for the editor, it is a way to show the moral superiority of the French spirit over that of across the Channel! As the reader will be able to see in Annex 4, these short articles have the inconvenience that they present both important and unimportant characters in the same light. But one must take things are they are and "positivise"; why not take advantage of it to learn of the existence of diverse scientists who did not lack merit? Most of them have a corresponding article in Michaud which is much more developed and instructive.

The *Nouvelle Biographie Générale* of Hoefer is of another kind altogether. It is overtly in competition with, and essentially after Michaud. The editor and the authors have at their disposal the articles of their forerunners, from which they let themselves be inspired and which they improve upon. The texts are frequently not signed, or are simply by initials which are not otherwise made explicit, but there exist a couple of signatures by quite obscure authors (E. Merlieux, Guyot de Fère, Apollin Briquet…). A study of the competences of the editors and the conditions of the composition would be necessary, in addition to the comparison of the scientific and bibliographical contents of the articles. This remains to be done. Annex 2 and gives the references in the two dictionaries to the articles of the 29 mathematicians considered.

We will not present the *Grand Dictionnaire Universel du XIXe siècle* (The Great Universal Dictionary of the 19th century) of Pierre Larousse, in 15 volumes (1866–76), without bringing attention to two supplements. The preface, very lengthy, gives an overview of previous dictionaries, including "biographical works" (p. XLVII–LII), and gives assessments which are quite personal, and a bit rough, on social, philosophical, and political questions. More complimentary towards Michaud (even if it is critical) than towards Hoefer, Larousse hardly speaks of biographies of scientists. But this does not keep it from containing various articles on them, and for Italian mathematicians, almost as many as Michaud and Hoefer. We indicate the references in Annex 1 but we have only cast out a few probes; the exercise would be interesting, knowing that the contributors to Larousse, such as Maximilien Marie for mathematics, were dissident authors in their time.

6 Conclusions and Perspectives

With the exception of Lagrange, originally from Turin, but who quickly left Italy after 1766 and did not come back, one considers today the Italian mathematicians of the 18th century to be Giovanni Poleni, Paolo Frisi, Mascheroni, Malfatti, Fagnano, Manfredi, the Riccati, etc. The *Dictionary of Scientific Biography,* which is an authority in these matters, also includes Lorgna, Caldani, and Frullani. The judgement in this regard has not much changed over the last two centuries, but one will note that there was little interest for Malfatti and Malvezzi in the 19th century, and that for Jacopo Riccati came quite late.

We took as our point of departure the research from the French perspective of the first half of the 19th century on Italian mathematicians. We have considered perspectives of French mathematicians of the period and after, but also of diffusers of science and history, popularisers, compilers, historians of science or men of general culture. Our wanderings have led us to a couple of national or political issues, to the scientific or linguistic competences of the contributors, to the attentive examination of the letters of the alphabet and the dates of death, and to the circumstances of the concrete organisation of a commercial enterprise with ill-payed collaborators.

This somewhat broken landscape does not prevent us from accentuating a couple of major points. Italian mathematicians were not held in very high esteem by the French. Beyond a couple of personal relationships which should not be overlooked, they remained in general blinded by a cultural and linguistic hegemony. However, aspiring to universality, they sought to take account of what was happening the world over, and for the essential points, they delegated this work to various people whose knowledge of languages and of science allowed them to grasp a preliminary panorama and gave them a couple of paths to pursue their enquiries.

As a good part of this article, "meagre as well as chimerical", took place at the time of Victor Hugo in "Besançon, old Spanish city", let us wish that it is not "abandoned by all, [including his father]" and that

> Cast like a seed at the whims of the wind,

it will find a group of young and old researchers, professionals or dynamic amateurs, supported by digital innovation, possibly ready to sacrifice their careers a little bit to create efficacious work instruments for large biographical dictionaries: lists of authors (with their articles, their personal biographies, the location of manuscripts), instruments of comparisons (and plagiarisms) among the dictionaries, between the successive editions of a single dictionary, etc. Only the spirit of these work instruments and of projects ready to be extended, such as ENCCRE for the *Encyclopedia*

http://enccre.academie-sciences.fr/

will allow us to pass the stage of patch-work for a transversal subject of the type studied here.

Annexes

(1) A list of 29 Italian mathematicians, with their presence or absence in 9 dictionaries
(2) The same list of 29 mathematicians in Michaud and Hoefer, with the authors of the articles and references
(3) The list of 42 Italian mathematicians having an article in the *Bibliothèque universelle portative*—table of correspondence with Feller and Michaud
(4) The short articles on these 43 mathematicians in the *Bibliothèque universelle portative*

See Annexes 1, 2 and 3.

Annex 1 29 Italian mathematicians in 9 biographical dictionaries

M\D->	Chaudon-Elandine	Chaudon-Goigoux	Michaud 2	Hoefer	Lalanne BPU	Feller 2	Larousse 19e	Larousse 20e	DSB
Galileo	x	x	x	x	x	x	x	x	x
Cavalieri (1598–1647)	x	x	x	x	x	x	x	x	x
Torricelli (1608–1647)	x	x	x	x	x	x	x	x	x
Viviani (1622–1703)	x	x	x	x	x	x	x	x	x
Manfredi (E.) (1664–1739)	x	x	x	x	x	x	x	–	x
Grandi x (1671–1742)	x	x	x	x	x	x	x	x	x
J. Riccati (1676–1754)	–	–	x	x	x	–	x	x	x
Fagnano (1682–1766)	x	x	x	x	x	–	x	x	x
Poleni (1683–1761)	x	x	x	x	x	x	x	–	x
Zanotti (1692–1777)	x	x	x	?	x	x	x	x	–
V. Riccati (1707–1775)	x	x	x	x	x	x	x	–	x
Boscovich (1711–1787)	x	x	x	x	x	x	x	x	x
Bonati (1724–1820)	/	–	x	x	x	–	x	–	–
Frisi (1728–1784)	x	x	x	x	x	x	x	x	x
Malvezzi (1730–1804)	/	–	–	–	–	–	–	–	–
Malfatti (1731–1807)	/	–	–	–	–	–	–	–	x
Lorgna (1735–1796)	–	x	x	x	x	–	x	–	–
G. Fontana (1735–1803)	/	x	x	x	x	x	x	–	–
Piazzi (1746–1826)	/	/	x	x	x	x	x	x	x
Mascheroni (1750–1800)	–	x	x	x	x	x	x	x	x
Paoli (1759–1839)	/	/	–	–	–	–	–	–	–
Ruffini (1765–1822)	/	/	x	x	x	x	x	–	x
Brunacci (1768–1818)	/	/	x	x	x	x	x	–	–
Plana (1781–1864)	/	/	/	/	/	/	x	–	x

(continued)

Annex 1 (continued)

M\D->	Chaudon-Elandine	Chaudon-Goigoux	Michaud 2	Hoefer	Lalanne BPU	Feller 2	Larousse 19e	Larousse 20e	DSB
Bordoni (1789–1860)	/	/	/	/	/	/	–	–	–
Mossotti (1791–1863)	/	/	/	/	/	/	–	–	x
Piola (1794–1850)	/	/	–	–	/	/	–	–	–
Frullani (1795–1834)	/	/	x	–	–	–	x	–	–
Libri (1802–1869)	/	/	/	/	/	/		x	–

Legend x = yes; − = no, / = not applicable at the time

Annex 2 29 Italian mathematicians in Michaud and Hoefer

M\D->	Michaud 2	Author	Commentary	Hoefer	Author
Galileo	15, 411a–418a	Biot		19, 251–268	H. and J.
Cavalieri (1598–1647)	7, 264b–265b	Lacroix		9, 281–283	E. Merlieux
Torricelli (1608–1647)	41, 703b–705b	Biot		45, 506–508	S.R.
Viviani (1622–1703)	43, 689a–690b	Weiss		46, 323–325	Unsigned
Manfredi (E.) (1664–1739)	26, 336a–337b	Weiss		33, 197–198	E.M.
Grandi x (1671–1742)	17, 331a–333a	Weiss		21, 641–643	Z.
J. Riccati (1676–1754)	~			42, 131–132	P.
Fagnano (1682–1766)	13, 328b	Nicollet		17, col. 11–12	Unsigned
Poleni (1683–1761)	33, 610a–612a	Weiss		40, 598–599	P.
Zanotti (1692–1777)	45, 391a–b	Weiss		?	
V. Riccati (1707–1775)	35, 553a–b	Weiss		42, 132–133	E.M.
Boscovich (1711–1787)	5, 115a–166a	Walckenaer		6, 768–771	Apollin Briquet
Bonati (1724–1820)	4, 685b–687a	Prony		6, col. 549	Unsigned
Frisi (1728–1784)	15, 205b–208b	Aimé Guillon		18, 8901–891	Unsigned
Malvezzi (1730–1804)	–			–	
Malfatti (1731–1807)	–			–	
Lorgna (1735–1796)	25, 108a–b	Weiss		31, 656–657	P.
G. Fontana (1735–1803)	14, 336a–337b	Aimé Guillon		18, 105–106	Guyot de Fère
Piazzi (1746–1826)	37, 168b–171b	De Angelis and Audiffret		40, 38–40	P.
Mascheroni (1750–1800)	27, 179a–180a	Weiss		34, 127–129	E. Merlieux
Paoli (1759–1839)	~ 32, 88b	Weiss		–	
Ruffini (1765–1822)	37, 49b–51a	De Angelis		42, 866–868	Unsigned
Brunacci (1768–1818)	6, 10b–12b	Prony		7, 598–599	Unsigned

(continued)

Annex 2 (continued)

M\D->	Michaud 2	Author	Commentary	Hoefer	Author
Plana (1781–1864)	/			/	
Bordoni (1789–1860)	/			/	
Mossotti (1791–1863)	/			/	
Piola (1794–1850)	–			–	
Frullani (1795–1834)	15, 251a–b	De Grégory and Weiss		–	
Libri (1802–1869)	/			/	

Annex 3 *Biographie portative et universelle*, 1844 (+ Correspondence with Feller and Michaud), by Ludovic Lalanne, Louis Renier, Thomas Bernard, Charles Laumier, S. Choler, Jean Mongin, Ernest Janin, Auguste Deloye, Charles Friess, *Extract from «§ XXXVII. Geometers and Mathematicians»* , col. (~) *1877–1879, with correspondance with the articles of Feller 2—Weiss–Busson (1848–1850) et Michaud 2—Desplaces (1843-[1865])*

18th cent.	Feller 2 (8 vol. + sup.)	Michaud 2 (45 vol.)
Agnesi	*1, 74a	1, 233a–b (Beuchot)
Araldi	–	2, 137b–138a (Weiss)
Asclépi	–	–
Belgrado	*1, 528b–529a	3, 530b–531b (Ginguené)
Benvenuti (C.)	*1, 559b	3, 677b–678a (Ginguené)
Boscovich	2, 127a–b	5, 115a–116a (Walckenaer)
Briga	*2, 224b	5, 529a (C.M. Pillet)
Brunacci	*2, 257b	6, 10b–12b (Prony)
Burgundio	*2, 290b–291a	6, 188a–b (C.M. Pillet)
Cagnoli (Ant.)	*2, 324b	6, 341b–342a (Valentin Parisot)
Caldani (M.)	–	6, 382b–383a (Weiss)
Canovai	*2, 379a	6, 576b–577b (Ginguené)
Chelucci	–	–
Cocoli	2, 680b	8, 517b518a (Aimé Guillon)
Cossali	*3, 55a	9, 293a–294b (Weiss)
Crivelli (the eld.)	–	9, 507b–508a (Weiss)
Fergola	–	13, 571b (Z. = anonymous)
[Fontana (Grég.)]	*3, 575a–b	14, 336a–337b (Aimé Guillon)
Fontana (Mariano)	*3, 575b–576a	14 337b–338b (Aimé Guillon)
Frisi	*3, 658a–b	15, 205b–208b (Aimé Guillon)
Galian[i] (Dom. Cel.)	–	15, 399b–400a (Gence)
[Galiani (Ferdinando)]	4, 17a–b	15, 400a–405b (Ginguené)
Ghersi [Gherli]	Sup., 71a–b	16, 388b–389a (Weiss)
Gismondi	–	16, 5881–589a (de Grégory)
Grandi (Guido)	*4, 186b	17, 331–333a (Weiss)
Intieri	–	20, 355b–356a (Weiss)
Lecchi	5, 174a	23, 514a–b (Aimé Guillon)
Lorgna	–	25, 108a–b (Weiss)
Luino	–	25, 460b (Aimé Guillon)
Manfredi (the)	5, 453b–454a	26, 336a–337b (Weiss) (the 2 art.)
Maraldi	5, 464a–b	26, 410b–411b (Weiss)
Marinoni	5, 515b	–
Morozzo	– (brother only)	29, 359a–b (de Grégory)
Niccolai (J.–B.)	–	30, 435a–b (Weiss)
Piazzi	*6, 503b–504b	33, 146b–171b (de Angelis et Audiffret)
[Poleni]	6, 617a	33, 610a–612a (Weiss)
Riccati [Vinc.]	7, 228a	35, 553a–b (Weiss

(continued)

Annex 3 (continued)

18th cent.	Feller 2 (8 vol. + sup.)	Michaud 2 (45 vol.)
Ruffini	*7, 364b–365a	37, 49b–51a (de Angelis)
Torelli	*8, 174b–175a	41, 680a–b (de Angelis)
Valperga	–	42, 530b–532b (Ugoni)
Vassali Eandi	Sup, 146a–b	42, 677a–b (de Grégory)
Venini	*8, 288a	–
19th cent.		
Collalto	*2, 693b	8, 584a–b (Weiss)

Annex 4 Biographie Portative et Universelle, 1844. Articles

Articles taken from the list «Geometers and Mathematicians», found col. (~) 1877–1879 of the "Analytical, Chronological and Alphabetical Table", at the end of the volume, found at § XXXVII

18th s.

Agnesi (Ma.-Gaétane), famous mathematician, Milan 1718–1799. At the age of 19, she defended 191 theses printed under the title of *Propositiones philosophicae*. - Her mathematical works have been translated into French under the title: *Traités élément. du calcul différentiel et du calcul intégral*, 1775, in-8°.

Araldi (M.), physiologist, mathematician, Modena 1740–1813.

Asclepi (Jo.-Ma.), Italian jesuit, mathematician, physicist, Macerata 1706–1776.

Belgrado (Ja.), jesuit, mathematician, physicist, and philosopher; Udine 1704–1789.

Benvenuti (C.), jesuit, physicist, and mathematician; Livorno 1716, Warsaw 1789. - *Synopsis physicae generalis*, in-4°; *De lumine dissertatio*, in-4°.

Boscovich (Rog.-Jo.), jesuit scientist, physicist, mathematician, astronomer, negociator; Republic of Ragusa 1741–1787. One owes to him many discoveries in astronomy and optics. *Philosophiae naturalis theoria*, 1758, in-4°.

Briga (Melchior della), jesuit, mathematician; Cesena (Papal states) 1686–1749. - *Scientia eclipsium ex commercio Sinarum illustrata*, 3 vol., in-f°.

Brunacci (Vicenzo), Italian geometer; Florence 1768–1818.–*Corso di matematica sublime*, 4 vol.

Burgundio (Horace), jesuit, mathematician, poet; Brescia 1679–1741. - *Telescopium geodeticum.*

Cagnoli (Ant.), mathematician and astronomer, director of the observatory of Milan; Verona 1743–1816. - *Traité de trigonométrie; Des sections coniques; Notions astronomiques; Mémoire sur la figure de la terre.*

Caldani (…). = His brother Marie, mathematician, 1735–1808. He was, according to d'Alembert, the foremost geometer and algebrist of Italy. - *Della proposizione Bernoulliona fra il diametro e la circonferenza del circolo.*

Canovai (Stanislas), Italian mathematician and erudite; Florence 1740–1811.

Chelucci (P.-Paulin), mathematician; Lucca 1682–1754. - *Institutiones analyticae; Institutiones arithmeticae.*

Cocoli (Domin.), mathematician; Brescia 1747–1812. - *Geometria et trigonometria; Elementi di statica.*

Cossali (le P. P.), mathematician; Verona 1748–1815. - *Storia dell'algebra.*

Crivelli (Father J.), geometer, physicist; Venice 1691–1743. - *Eléments de physique*, 2 vol. in-4°.

Fergola (N.), Geometer; Naples 1753–1824.

[absent from the list, but having an article:] Fontana (Fél) […]. = His brother, the Father Gregory, famous mathematician, professor of mathematics at the chair of Pavia; 1735–1803. - *Analyseos sublimioris opuscula; la Dottrina degli azzardi applicata ai problemi.*

Fontana (le P. Mariano), mathematician; Casalmaggiore (Milanese) 1746–1808. - *Corso di dinamica*, 3 vol. in-4°.

Frisi (Pa.), famous mathematician and physicist; Milan 1728–1784. - *Disquisitio mathematica in causam physicam figurae et magnitudinis telluris nostrae*, 1751; *Cosmographiae physicae et mathematicae*, 2 vol. in-4°.

Galiani (Dom. Cel.), theologian, philosopher, linguist, mathematician, and antiquary; Foggia (Puglia) 1681–1753. He was successively named Archbishop of Taranto, and then Archbishop of Thessalonika. Benedict XIII and the Emperor Charles VI, the King of Naples appealed to him numerous times at a mediator. = L'abbé Ferdinand, nephew of the former […].

Ghersi (Odoardo), geometer; Guastalla (Duchy of Parma) 1730–1780.

Gismondi (C.-Jo.), Piedmontese mineralogist and mathematician; Mentone (near Nice) 1762, Rome 1824.

Grandi (Guido), Camaldolese brother and distinguished geometer; Cremona 1671–1742. - *Sistema del mondo*, 1716, 2 vol. in-4°; *Elementi geometrici piani e soledi*, Venice, 1759, in-8°.

Intieri (Bart.), economist, mathematician, and mechanic; born Pistoia (Tuscany) 1672, died Naples 1757. - One owes to him the invention of silos for the preservation of harvests, and the discovery of a process to prevent the counterfeit of banknotes. He established a chair of political economy at Naples.

Lecchi (J.-Ant.), engineer, physicist, and mathematician; Milan 1702–1776.

Lorgna (Ant.-Ma.), geometer; born Verona 1730, died 1796.

Luino (F.), mathematician; Milan 1740–1792.

Manfredi (Eust.), famous geometer, astronomer, writer; Bologna 1664–1739. - *Ephemerides motuum caelestium*, 1715–25, 4 vol. in-4°. = His brother Gabriel, mathematician; Bologna 1681–1761. - *De constructione aequationuum*, in-4°; *Considerazioni sopra alcuni dubii*, in-4°.

Maraldi (Ja.-Ph.), mathematician, astronomer, member of the Academy of Science, nephew of Cassini, Nice 1665-1729. = His nephew, J.-Dominique […]

Marinoni (J.-J.), mathematician, architect, astronomer; Udine 1676–1755.

Morozzo (the count C.-L.), physicist, mathematician; Turin 1743–1804. = His brother Joseph, cardinal; Turin 1758–1842.

Niccolai (J.-B.), priest, mathematician; Venice 1726–1793. - *Nova analysis elementa*, 1791, 2 vol. in-4°.

Piazzi (Jo.), Theatine brother, mathematician, and astronomer; born Ponte (Valteline) 1746, died in Naples 1826. He discovered the planet Ceres, and made a catalogue of 7646 stars. The government employed him to established a uniform metric system for Sicily, where he had constructed an observatory (1791). - *Leçons d'astronomie; Catalogue des étoiles; Mémoire sur la planète Cérès*, 1802.

[classed at a physicist] Poleni (J.), famous physicist, antiquary; Venice 1683–1761. - *De motu aquae mixto libri duo*, 1717, in-4°; *De castellis per quae derivantur aquae fluviorum*, etc., 1718, in-4°.

Riccati (Vinc. de), Italian jesuit, mathematician, engineer; Castel-Franco 1707–1775. - *Institutiones analyticae*, 3 vol. in-4°. = His brother Jourdain, musician, mathematician; 1709–1790.

Ruffini (Pa.), doctor, mathematician; Valentano (Duchy of Castro) 1765–1822. He refused during the French invasion a position in the administration. - *Teoria generale dell'equazioni.*

Torelli (Jo.), writer, geometer, Verona 1721–1781. - *Edition* of Archimedes, 1792. Valperga di.

Caluso (Th. des comtes Masino), astronomer, mathematician, orientalist, writer; Turin 1737–1815. - *Litteraturae copticae rudimenta*, 1783.

Vassali-Eandi (Ant.-Ma.), mathematician; Turin 1761–1825. - *Physicae elementa; Lettres sur le galvanisme.*

Venini (l'abbé F.), mathematician, poet writer; Milan 1737–1820.

19e s.

Collalto (Ant.), physicist, mathematician; born Venice, died 1820. - *Identità del calcolo differenziale*, etc.

Bibliography

Biographical Dictionaries and Abbreviations

We will indicate, in a separate bibliography, a couple of dictionaries with the following abbreviation.

Chaudon-Delandine (1805), *Nouveau dictionnaire historique, ou histoire abrégée de tous les hommes (...)*, Bruyset et Buynand, Lyon, 8e éd., 9 + 4 vol.

Chaudon-Goigoux (1821–1823), *Dictionnaire universel, historique, critique et bibliographique*, Ménard et Desenne, Paris, 10e éd., 30 vol.

Michaud 1 (1811-1828), puis (1834-1862), *Biographie universelle ancienne et moderne*, Michaud frères, Paris, 85 vol.

Michaud 2 (1842-[1865]), *Biographie universelle ancienne et moderne*, Desplaces, Paris, 45 vol. (Many variations of the title page).

Hoefer (1852-1866), *Nouvelle biographie générale*, Firmin-Didot, Paris, 46 vol. Feller 2 (1847-1850) ; *Biographie universelle ou Dictionnaire historique* … Leroux et Jouby, etc. Paris, 8 vol.
BPU-Lalanne (1844), *Biographie portative universelle*, Dubochet, Paris, 1 vol.
Larousse 19e (1866-1876, and 1878 et s.d.), *Grand Dictionnaire Universel du XIXe siècle*, Paris, 15 vol. + 2 vol. of "Supplément".
Larousse 20e (1960-1964, puis 1969), *Grand Larousse encyclopédique*, Paris, 10 vol. + 1 vol. of "Supplément".
DSB (1971-…), *Dictionary of Scientific Biography* (ed. C.C. Gillispie).

Other References Manuscripts

Principal manuscripts used: Municipal Library of Besançon, Charles Weiss collection, Ms 634 (letters from Weiss to Michaud), 635 (letters from Claude-Marie Pillet to Weiss), 1755, 1758, 1771–1772, 1890.

Printed Works

Chappey JL (2013), *Ordres et désordres biographiques,* Champ Vallon, Seyssel.
Cournot A (2011), *Œuvres complètes. Ecrits de jeunesse et pièces diverses*, t. XI (par B. Bru et T. Martin), Vrin, Paris, 2 vol.
Crépel P (2012) in Pepe 2012: 3–14.
Pepe L (2007), *Rinascita di una scienza. Matematica e matematici in Italia (1715-1814)*, CLUEB, Bologna.
Pepe L (a cura di) (2012), *Europa matematica e Risorgimento italiano*, CLUEB, Bologna.
Kaye E (1987), *Les Correspondants de Charles Weiss,* Ed. du Préambule, Longueuil (Québec).
Lepin S (1972–1997), *Charles Weiss: Journal de 1815 à 1842,* Annales universitaires de Franche-Comté 132, 257, 420, 645.

The 27 Italies of Camille Jordan

Frédéric Brechenmacher

1 Introduction: Images of a French Mathematician with Italy in the Background

Following the path of his uncle, the symbolist painter Pierre Puvis de Chavanne, Camille Jordan (1838–1922) undertook the Grand Tour, a traditional trip across Europe for upper-class young men of means. The primary value of the Grand Tour, it was believed, laid in the exposure of these young men to the cultural legacy of both classical antiquity and the Renaissance. Italy was the essential destination of such journeys (Hornsby 2000), which were often sources of inspiration for French painters. Puvis de Chavannes, for instance, commented that it was in Italy that he had discovered decorative painting. In a painting of 1888, 'Inspiration chrétienne,'[1] he contrasted the idle serenity of pagan antiquity with the creative activity of an artist in an Italian convent at the beginning of Renaissance. But what about 'Italian mathematics' as a source of inspiration for mathematicians? From the perspective of a French mathematician such as Jordan who began his scientific activity after the Risorgimento, was the geographical and national unity of Italy connected to a specific type of mathematics?

To be sure, Jordan saw mathematicians such as Francesco Brioschi, Luigi Cremona, and Giovanni Guccia as 'Italian mathematicians.' Jordan himself undertook several trips to Italy during his career, often with his whole family.

[1]National Gallery of Art, Washington.

Ce travail a bénéficié d'une aide de l'Agence Nationale de la Recherche: project CIRMATH (ANR-14-CE31-0010).

F. Brechenmacher (✉)
LinX École Polytechnique Université Paris-Saclay, 91128 Palaiseau Cedex, France
e-mail: frederic.brechenmacher@polytechnique.edu

F. Brechenmacher et al. (eds.), *Images of Italian Mathematics in France*,
Trends in the History of Science, DOI 10.1007/978-3-319-40082-2_4

Whether he went to Turin, Milan, Rome, or Palermo, Jordan always requested his Italian correspondents—especially Cremona and Guccia—to introduce him to 'the other mathematicians of Italy.'[2] Yet national and disciplinary categories were not so easily mixed by Jordan. Not only did he never actually use the expression 'Italian mathematics' but he was even appointed himself a representative of 'Italian mathematicians' by the Mathematical circle of Palermo, which had fashioned itself as Italy's first national mathematical society.

Moreover, Jordan can hardly be considered a typical 'French mathematician.' His contemporary Felix Klein saw him as 'almost German' because of his algebraic approach to mathematics (Klein 1921, p. 51). Several of his successors—especially Jean Dieudonné, who took responsibility of the edition of his collected work—described Jordan as isolated on the French mathematical scene because his research on group theory was believed to have mainly influenced foreign mathematicians such as Cremona, Alfred Clebsch, Klein, and Sophus Lie.[3] Yet this retrospective presentation of Jordan as almost foreign to French mathematics is in stark contrast to the fact that he followed the typical career trajectory of an elite French scientist trained in the 1850s. Originating from an upper class family in Lyon, Jordan followed the path of his father and grand-father by 'integrating' École polytechnique in 1855. A few years later, he chose the 'École des mines' as his school of application—a typical choice for the best students of polytechnique. In the 1860s, Jordan started his mathematical research alongside his career as an engineer, from which he slowly walked away when he took up academic positions during the 1870s. Again, the parallel development of academic and engineering careers was a typical situation in France before the rise of the image of mathematicians as university professors in the 1880s. After having been appointed 'examinateur' at polytechnique in 1873, Jordan became 'Professeur d'analyse' in 1876—a prestigious chair that had been occupied by mathematicians such as Joseph Louis Lagrange, Augustin Louis Cauchy, Charles Sturm, Joseph Bertrand, and Charles Hermite. He was elected to the Académie des sciences in 1881, then appointed to the Collège de France in 1883 and, in 1885, took over the direction of one the most

[2]In addition to the Grand Tour, Jordan's correspondence documents three trips to Italy: to Milan in May 1874 (where he met with Cremona in person for the first time), to Rome in April 1889, and to Sicily in June 1889. In a letter dated 18th April 1874, Jordan wrote to Cremona: "Non seulement je serais bien heureux de faire votre connaissance de visu, mais je me proposerais de recourir à votre obligeance pour me mettre en relation avec les autres mathématiciens de l'Italie. Quoique plusieurs d'entre eux me soient fort connus de réputation, je ne sais pas bien où les trouver." This request to Cremona was repeated in 1889 at the occasion of a long familial visit to Rome, which was followed by a familial trip to Palermo where Jordan was introduced by Guccia to Sicilian mathematicians.

[3]See Julia's introduction to (Galois 1962), and Dieudonné's introduction to the first volume of (Jordan 1964). Some recent works have nevertheless shown that this presentation of Jordan as isolated on the French scene is biased by retrospective perspectives on group theory (Brechenmacher 2011). Moreover, the key role played by Jordan's *Cours d'analyse de l'École polytechnique* in the evolution of analysis highlights that the latter can hardly be reduced to the 'great algebraist' whose masterpiece, the *Traité des substitutions et des equations algébriques*, unfolded the group-theoretical content of Galois's work.

important mathematical journal in Europe, the *Journal de mathématiques pures et appliquées*. By the mid-1880s, Jordan thus occupied a key position in the Parisian mathematical scene.

The relationships between Jordan and Italy provide us the opportunity to investigate not only various images of the mathematical activities in Italy viewed from France, but also various images of Camille Jordan as a French mathematician. From 1867 to 1895, such issues are especially documented in Jordan's scientific correspondence.[4] Italian scientists such as Brioschi and Cremona were among the first mathematicians to establish a scientific correspondence with Jordan in 1867, and the latter maintained strong epistolary contacts with Italy throughout his career. This correspondence allows us to investigate the relevance of the juxtaposition of disciplinary and national categories underlying the notions of 'Italian mathematics' and 'French mathematics.' To be sure, Camille Jordan's career took place during a time of nation building in Europe, as exemplified by the Italian Risorgimento, in which several mathematicians actively participated, and by the important consequences of the Franco-Prussian war of 1870 on the institutional organization of sciences in France. On the other hand, however, our main source of investigation points to the last vestiges of the epistolary networks of the European République des lettres.

In the present paper, we shall follow the chronological organization of Jordan's correspondence. A structure in four sections reflects four time-periods in the evolution of Jordan's contacts with mathematicians in Italy. In the first section, we shall focus on Jordan's first scientific contacts following his publication of two papers in the *Annali di Matematica Pura ed Applicata* in 1867–1868. These two publications gave rise to a scientific correspondence between Jordan, Brioschi and Cremona from 1867 to 1870. As we shall see in the second section of this paper, this correspondence revolved around the issue of the link between geometrical equations and abelian functions, which played a key role in the early reception of Jordan's 1870 *Traité des substitutions et des équations algébriques*. Jordan's contacts with Italy shifted from a scientific correspondence to more institutional issues after 1870. The first half of the present paper therefore deals with more mathematical issues than the second half in which we successively discuss the connections between Jordan's *Cours d'analyse* and Italy in the 1880s, and the interplay between Jordan's key position in the Parisian mathematical scene and the development of the international dimension of Guccia's Circolo mathematico de Palermo at the turn of the 1890s.

We will see that both Jordan's images of Italy and the Italians' images of Jordan as a French mathematician point not to a unified picture of Italy but a diversity of images. Jordan's correspondence with the 'mathematicians of Italy' highlights several types of institutions, such as academies, mathematical journals, and international scientific societies. These institutions present various interplays between local, national, and international activities. These interplays will lead us to question

[4]Jordan's correspondence is archived at École polytechnique under the reference VI2a2X1855. For a general presentation of this correspondence, see (Billoux 1985).

the relevance of national categories at the smaller scale of interpersonal epistolary debates on precise mathematical issues.

The title of the present paper thereby aims at paying tribute to the main mathematical issue debated by Jordan and his Italian correspondants: the 27 lines on a cubic surface. It also suggests the diversity of images of Italy that can be perceived through the eyes of Camille Jordan. As we will argue, Jordan's scientific correspondence actually documents a plurality of Italies.

2 The Académie des Sciences and the *Annali*

In this first section, we shall investigate Jordan's first contacts with Italian mathematicians from 1867 to 1874. These contacts were initiated by two of his contributions to the *Annali di Matematica Pura ed Applicata* shortly after Cremona and Brioschi took over the direction of this journal.[5] As we will see, the Académie des sciences de Paris (to which we will henceforth refer as the Académie, for short) played a key role in these first contacts between Jordan and Italian mathematicians.

2.1 Mathematical Journal and National Mathematics

Before investigating the context which led Jordan to contribute two papers to the *Annali*, let us first discuss the national dimensions of the two main institutions we shall consider in this section: academies and mathematical journals.

On the one hand, the Académie had played a national role in the organisation of scientific activities in France since its creation in the 17th century. On the other hand, however, this institution was also embedded in a network of other academies that had been created all over Europe over the course of the 18th and 19th centuries, including those of Lombardy and Turin, and the mathematical society of Palermo, to which Jordan would be elected as a corresponding member in 1870, 1896, and 1889 respectively. By the 1860s, the Académie had lost its monopoly over both the evaluation and the diffusion of scientific research in France. Several specialized mathematical journals had indeed been created in the 19th century. Yet we will see that the Académie maintained a very close relationship with major research journals, such as the French *Journal de mathématiques pures et appliquées* (the *Journal* for short), and the Italian *Annali*. This relationship revolved in large part, though in a quite subtle way, around the system of prizes which had provided the Académie with a tool for promoting and evaluating research on specific subjects since the 18th century.

Most of the research mathematical journals that were created over the course of the 19th century contributed strongly to the structuring of national mathematical scenes. To better understand what this means, we can consider the examples of the

[5]On the history of the *Annali* see (Brigaglia 2014).

Annales de mathématiques pures et appliquées founded by Gergonne in 1810, and the *Journal de mathématiques pures et appliquées* created by Liouville in 1836. While the former was still explicitly entrenched in a network of European learned societies quite typical of the 18th century République des lettres (Gergonne 1810), the latter opened with the following *Avertissement* (Liouville 1836, p. 5):

> It is now up to geometers, and especially to French geometers, to guarantee the prosperity of this enterprise. [...] We shall dare to claim that their reputations is at stake: the fall of a useful Journal that they would have refused to support would be a disgrace for them as well as for France.

To be sure, the 'French' nature of Liouville's journal did not imply that its contributions were limited to French authors.[6] Even though most of the contributors were French, the contribution of prominent foreign mathematicians was also crucial for the *Journal* to fashion itself as a vector of mathematical knowledge in France. The claim that the *Journal* was a publication of national interest was even strengthened when Henri Résal succeeded Liouville as editor in chief in 1875 (Resal 1875). Yet after the creation of the French mathematical society—which started issuing its own *Bulletin* in 1872—the *Journal de mathématiques pures et appliquées* could hardly define itself any longer as the flagship of French mathematical research. As a matter of fact, when Jordan took over the direction of this publication in 1885, he reset the identity of the *Journal* as an elite publication supported by both 'French and foreign eminent geometers.' As we will later see in greater detail, however, the editorial board named by Jordan was nevertheless limited to a small group of French mathematicians from the Académie. A strong connection with the Académie had actually been at the core of the national identity of the *Journal* since its creation.[7] Before contributing to Liouville's journal, a French mathematician typically had to win the recognition of the members of the Académie, often by winning one of the academic prizes the Académie regularly sponsored, such as the emblematic Grand prix des sciences mathématiques. Moreover, most publications in Liouville's journal provided lengthier descriptions of results that had been previously stated in the short notes published by the *Comptes rendus de l'Académie des sciences de Paris* (to which we will now refer as the *Comptes rendus* for short).

In Italy, the creation of the *Annali di scienze matematiche e fisiche* by Barnaba Tortolini in Rome in 1850 was strongly connected to the Risorgimento. In 1858, this journal had evolved into a specialized mathematical journal and was renamed as the *Annali di Matematica pura ed applicata*. On the one hand, the 'Italian' nature of this journal was quite analogous to the 'French' nature of Liouville's journal, or to the 'German' nature of Crelle's journal: even though most of its contributors were Italians, the editors of the *Annali*, Tortolini, Enrico Betti, Cremona, Brioschi and Angelo Genocchi, had managed to attract contributions from major foreign

[6]On the history of Liouville's journal see (Verdier 2009).

[7]All the successive directors of the *Journal* and all the members of its editorial board were members of the Académie.

mathematicians such as Bertrand, Arthur Cayley, or Bernard Riemann. But, on the other hand, we have seen that the national nature of Liouville's journal was strongly connected to the *Académie* whose national and centralized dimension did not have any equivalent in Italy. In contrast with Liouville's journal, the *Annali* had actually aimed ever since its creation at developing connections with all European academies.[8] We shall now see that the Academy of Paris actually played a key role in the contributions to the *Annali* of young French mathematicians such as Jordan.

2.2 A Specific Practice of Publication of Mathematics Between the *Annali* and the Académie

The *Annali* is the only Italian journal to which Jordan contributed before the late 1880s,[9] with two papers published successively in 1867 and 1868. This raises the following questions: why did Jordan publish in the *Annali*? Why did he contribute with two, and only two, papers? Why did he publish these papers in a row, and how did he choose the issues he tackled in his papers?

The first contribution, "Mémoire sur la stabilité de l'équilibre des corps flottants" (Jordan 1867), deals with the stability of equilibrium of floating bodies. This is the only paper Jordan ever devoted to fluid mechanics. The second paper, "Mémoire sur les groupes de mouvements" (Jordan 1868a), is more coherent with Jordan's other mathematical works, especially the series of papers he had published on polyhedrons since 1863. In contrast with the algebraic nature of the series of notes published in the *Comptes rendus* and with the three papers published in Crelle's journal,[10] the memoir sent to the *Annali* focuses on the mechanical issue of the classification of the groups of motions of polyhedra.

Both contributions to the *Annali* were therefore connected to mechanics. To be sure, mechanics was one the main branches of the mathematical sciences in the 19th century in France, and one that was quite coherent with the training at École polytechnique. As a matter of fact, Jordan's two contributions to the *Annali* were both strongly connected to the works of two influent professors at polytechnique, i.e. Bertrand, his professor of analysis, and Auguste Bravais, his professor of physics.[11] Yet mechanics was hardly Jordan's predilection; throughout his career, he actually published no more than four memoirs on issues of stabilities of solid

[8]Each issue of the *Annali* especially reported on European academic activities (prizes, publications, etc.).

[9]As shall be seen later, Jordan contributed only to one other Italian journal, with a paper published in the *Rendiconti del Circolo Matematico di Palermo* in 1889.

[10]See (Jordan 1866c, b, 1868b).

[11]Jordan was especially influenced by Bravais's works. The latter had devoted a thesis to the stability of floating bodies in 1840. Jordan's approach to the groups of motions of polyhedra followed up Bravais's crystallographic investigations devoted to classifying the various ways in which a system of molecules can be superimposed upon itself.

bodies. Why, then, did Jordan choose to devote his two contributions to the *Annali* to mechanics?

The first letter sent by Jordan to Cremona on the 20th of February 1867 sheds some light on this issue. In this letter, Jordan expressed his desire to contribute to the *Annali*. Yet he could not propose a collaboration as deep as he would have liked because, one the one hand, of his duties as an engineer, and on the other hand, of the very nature of his mathematical activities. As he explained, "For many years I have devoted all of my leisure time to research on algebraic equations, the considerable scope of which is hardly suited for a publication in a periodical journal." He felt that this body of research deserved a monograph, and would actually lead to the publication of Jordan's famous *Traité des substitutions et des équations algébriques* in 1870. Yet from 1864 to 1870, Jordan's works on substitutions groups and algebraic equations would also be published in several periodicals such as the *Comptes rendus*, Liouville's journal, and Crelle's journal. Jordan was apparently more reluctant to publish his works on substitutions in the *Annali* than in other periodicals.

Jordan may have believed that substitution groups would reach a limited audience in Italy. More importantly, Jordan likely feared that a contribution in the *Annali* would have a more limited audience in France than one in Liouville's journal or in Crelle's journal, and that such a memoir would therefore remain isolated from his other papers on substitutions groups. As a consequence, a publication in the *Annali* would have emerged as an effort to develop the theory of substitutions as an autonomous field of investigation, while Jordan was currently struggling to develop as many applications of his group theoretical approach as possible. Recall that group theory would not be considered a coherent mathematical theory before the turn of the 20th century. In the 1870s, several prominent mathematicians such as Hermite and Leopold Kronecker would actually blame Jordan for the excessive generality, and even formal nature, of his research on substitution groups (Brechenmacher 2007). The 1867 letter to Cremona shows that Jordan was already very much aware that his research on substitutions groups would benefit from other fields of applications than the Galois theory of algebraic equations. As we shall see in greater detail in Sect. 3, from 1867 to 1870 Jordan's series of notes in the *Comptes rendus*, as well as his longer papers in both Liouville's and Crelle's journals, precisely aimed at proving the relevance of his investigations on substitution groups by applying them to cubic surfaces and abelian functions. We will also see that the terrible experience of Giuseppe Janni with substitutions in the 1870s would eventually demonstrate that Jordan may have made a good choice when he decided not to publish his research on substitutions in Italy.

The memoir Jordan eventually opted to send to Cremona sheds light on a specific form of circulation of mathematical papers between the *Annali* and the Académie. Three years before Jordan shipped to Italy his first contribution on the stability of equilibrium of floating bodies, this very topic had been selected for the "Grand prix des sciences mathématiques" of the Académie for 1864. Competing for the prizes of the Académie was crucial for a young polytechnician who aimed at making a career in mathematics. Already when he had worked on his thesis, Jordan had focused on the problem that had been set for the Grand prix of 1860, i.e. the problem of the number of values of rational functions under the action of substitutions.[12] Yet in 1860 the Académie members felt that none of the submitted memoirs was worthy of the prize, and the competition was therefore cancelled. This situation was not unusual: during the 1860s the Grand prix was actually awarded only once, in 1868, to a posthumous memoir of Edmond Bour.

Even though Jordan did not win an award in 1860, the research he had developed for the Grand prix would nevertheless have major consequences for his approach to mathematics: it triggered his interest in the theory of substitutions and led him to a new understanding of the works of Évariste Galois (Brechenmacher 2011). In contrast, while the 1864 Grand prix on floating bodies would have little influence on Jordan's subsequent works, his memoir was one of the two submissions that were officially 'encouraged' by the Académie, and Jordan was eventually awarded half of the amount of the prize. Among the members of the committee of academicians elected to evaluate the submissions for the prize, Hermite and Ossian Bonnet were regular contributors to the *Annali*, and Bertrand had been invited to publish in the very first issue of Tortolini's journal in 1858. In fact, Bertrand actually played a key role in the publication in the *Annali* of Jordan's submission for the 1864 Grand prix.[13]

The second memoir published by Jordan in the *Annali* has a similar history. The geometric theory of polyhedra had been set as the problem for the 1861 Grand prix, though it was eventually postponed to 1863. Jordan competed for this prize with a memoir entitled 'Recherches sur les polyèdres.' Once again, the Académie members decided not to award a prize for this competition. Once again, Bertrand was a member of the evaluation committee. Between 1865 and 1868, he actively supported the publication of the main results of Jordan's memoir in the form of a series of notes in the *Comptes rendus* of the Académie. In 1868, Bertrand even wrote an official report to the Académie supporting the publication of the entirety of Jordan's memoir in the *Recueil des savants étrangers*. Yet by that time, Jordan had already published most of his results on polyhedra in three contributions to Crelle's journal. It is in this context that Jordan settled for sending his remaining original results on polyhedra to the *Annali*.

The context of the publication of Jordan's memoirs in the *Annali* highlights a specific publication practice in connection with the system of mathematical prizes. The Grand prix, in particular, generated more memoirs than the Académie could publish. In order to be published by the Académie, a memoir had to actually win the Grand prix, though as we have seen, this prize was very seldomly awarded. As for the many memoirs that did not win, they were nevertheless evaluated by the Académie, often quite positively, and a few of them were even officially 'encouraged.' These memoirs were hardly welcome in Liouville's journal however, because of its strong connection to the Académie. For these papers, the possibilities

[12]This problem was one of the main lines of development of the theory of substitutions in the 19th century. It is tantamount to finding the possible orders for subgroups of the symmetric group.

[13]On the central role played by Bertrand in the Parisian mathematical scene, see (Zerner 1991).

of publication in France were therefore limited to the *Journal de l'École polytechnique* and a few other journals of engineers.[14] Foreign mathematical journals such as the *Annali* were therefore ideal places to publish these 'byproducts' of academic prizes, as was made explicit by Jordan in his second letter to Cremona, dated 8th May 1867:

> I am honored to send you the Mémoire sur la stabilité de l'équilibre des corps flottants that I presented 2 years ago to the Académie de Paris for its grand prix de mathématiques. The Academy granted this memoir its approval with an encouragement of 1500 francs. I have been invited to publish my work in the *Annales des Mines*, but this collection is exclusively devoted to applied sciences and is therefore unknown to mathematicians. I would therefore prefer the publicity of a Mathematical Journal such as yours, and I beg you to look into this memoir if you would be willing to accept it.
>
> I am establishing the conditions for the stability of equilibrium through a complete integration of the differential equations of the infinitely small motions of the floating body. This leads me to a condition that was already known, but the accuracy of which has been much contested recently, which is the reason why the Académie chose this question for the competition.

The role played by the evaluation of the Académie in Cremona's decision to publish Jordan's contribution is also made explicit in a letter dated 25th May 1867:

> M. Brioschi and I are very grateful to you for sending us your excellent memoir on the stability of the equilibrium of floating bodies. We are pleased that you have thought of sending to the Annali such a beautiful work; we know its high value thanks to the judgment of the Académie. Your manuscript is therefore welcome [...].[15]

The specific publication practice described above illuminates the connections between the newly founded mathematical journals, such as the *Annali*, and the much older academic organization of mathematics. This practice indeed required the active collaboration of French academicians who were in charge of setting up and evaluating the Grand prix, such as Bertrand. As a matter of fact, Bertrand appears to have played a key role in promoting to the editors of the *Annali* the works of a few promising young French mathematicians. Already in the beginning of the 1860s, Émile Mathieu, another young polytechnician, had sent to the *Annali* the memoir he had submitted for the 1860 Grand prix.[16] Both Jordan and Mathieu had followed Bertrand's lectures at École polytechnique, and their mathematical

[14]The *Journal de l'École polytechnique* seems to have played an important role in the publication of the memoirs generated by academic prizes. It was there that Jordan had published his paper for the 1860 prize on the number of values of functions, and that Bour's submission to the prize of 1865 on partial differential equations was published after his death in 1866 (the prize was eventually attributed to Bour posthumously in 1868). Yet it seems that this journal could not be used repeatedly; publishing there was no longer an option for Jordan in 1867.

[15]M. Brioschi et moi nous vous prions d'agréer nos sincères remerciements pour l'envoi de votre excellent mémoire sur la stabilité de l'équilibre des corps flottants. Nous sommes heureux que vous ayez pensé à destiner à nos Annali ce beau travail dont nous connaissons la haute portée d'après le jugement de l'Académie. Votre manuscrit est donc le bienvenu [...].

[16]As indicated above, Jordan had also competed for this prize but had settle for publishing his memoir in the *Journal de l'École polytechnique*.

careers were supported by their former professor in the early 1860s, even though Bertrand would eventually favor Jordan over Mathieu. Bertrand had been a correspondant of the editors of the *Annali* since its creation as a mathematical journal in 1858. When Cremona and Brioschi took over the direction of the journal in 1867, they solicited their network of correspondents for potential contributors, to whom they would send a call for papers and a circular on the new editorial policy of their journal. Jordan was one of the first to receive this circular at the beginning of 1867, probably through Bertrand's recommendation.

The connections between the *Annali* and the Académie also involved subtle national issues. On one hand, from the perspective of Bertrand, it was because the *Annali* was an 'Italian' journal that it provided a solution for publishing papers that could not be published in France. On the other hand, the young authors supported by Bertrand were explicitly considered promising 'French' mathematicians by the editors of the *Annali.* This national framework is explicit in Cremona's reference to the report the French government had commissioned in 1867. Bertrand was in charge of the section of this report devoted to the progress of mathematics in France. To the presumed despair of Mathieu, Jordan was the only mathematician of his generation whose works were praised for their novelty and fecundity. As Cremona phrased it:

> We have seen with a great pleasure that M. Bertrand speaks very highly of your research on the theory of equations in his Report, and especially of some results of the foremost importance contained in a book currently in press.[17]

The connections between the Académie and the *Annali* also explain why Jordan published no more than two memoirs in this journal. By the mid 1860s, Bertrand had enabled Jordan to publish his results in the notes of the *Comptes rendus de l'Académie des sciences de Paris* on a regular basis. In turn, these academic communications afforded the possibility for Jordan to further develop his results by publishing lengthier memoirs in Liouville's journal, as well as in the German *Journal für die reine und angewandte Mathematik.*

2.3 Le Journal de MM. Cremona et Brioschi and il Giornale de Jordan

The changes in the editorial policy of the *Annali* after 1867 is another important context in which Jordan's two contributions took place. As mentioned above, in 1867 Cremona and Brioschi took over the direction of *Annali,* which had until then been edited by Tortolini. In addition to moving the journal from Rome to Milan and to naming a new editorial board, Cremona and Brioschi also developed the

[17]Nous avons vu avec le plus grand plaisir que M. Bertrand parle dans son Rapport très avantageusement de vos recherches sur la théorie des équations, et surtout de certains résultats de la plus grande importance contenus un ouvrage sous presse.

international dimension of the journal by issuing a call for papers that circulated through their networks of foreign correspondents.

While the *Annali* had published no more than 29 papers from foreign contributors between 1858 and 1866, 50 papers from foreign mathematicians were published from 1867 to 1869 (while Italian authors made only 19 contributions during that same period). Most foreign contributors published no more than once or twice in the *Annali*, most certainly in response to the sollicitation of Cremona and Brioschi.[18] The most numerous foreign contributors were the French (16 different authors from 1858 to 1870), followed by the Germans (13 authors), the group of which quickly increased after 1867. From 1858 to 1870, the group of French contributors was largely dominated by polytechnicians (Bertrand, Mannheim, Bonnet, Hermite, Mathieu, Jordan, Méray) with the presence of a smaller, though quite active, group of Jesuit abbots (Jullien, Le Cointe, Aoust, Pepin).

It is likely that the successful internationalization of the 'journal de MM. Brioschi et Cremona,' as Jordan called it, played the role of a model when Jordan took over the direction of the *Journal de mathématiques pures et appliquées* in 1885. As a matter of fact, Jordan's first decision as an editor in chief was to establish an editorial board similar to that of the *Annali*. The institution of such an editorial board played a crucial role in the new identity Jordan forged for his *Journal*. As noted above, following the creation of the *Bulletin* of the French mathematical society, the *Journal* had lost its national identity as the leading journal for the publication of mathematical research.[19] With the constitution of an editorial board, Jordan explicitly aimed at defining his *Journal* as the most elite French mathematical publication venue, which was to occupy a prominent position in the international space of mathematics (Jordan 1885). This elite identity was constructed by institutionalizing the links the *Journal* had always maintained with the Académie des sciences and by developing its international dimension. The editorial board initially consisted of Georges Halphen, Edmond Laguerre, Maurice Lévy, Amédée Mannheim, Résal, and Émile Picard, all of whom were members of the Académie des sciences,[20] with the exception of Mannheim, who had nevertheless won the prestigious Poncelet price of the Académie in 1872. Moreover, while only two new members would be named to the editorial board before 1921, their nomination would quickly follow their election to the Académie (Henri Poincaré was elected to the Académie in 1887, and named to the board in 1888 when

[18]The foreign mathematicians who published more than 3 papers in the *Annali* from 1858 to 1870 formed a small group of contributors from Paris (i.e., Hermite (6 contributions), Jonquières (4 contributions), Le Cointe (4 contributions), Bonnet (3 contributions)), Zürich (Schläfli 6 contributions), Dublin (Roberts (5 contributions)), Cambridge (Cayley (3 contributions)), London (T. Archer Hirst (3 contributions)), and Bonn (Woepcke (3 contributions)).

[19]Before 1870, the other major journals published in France and devoted to mathematics were either linked to a specific institution, as with the case of the *Comptes rendus* and the *Journal de l'École polytechnique*, or were devoted to a specific public, such as the *Nouvelles annales de mathématiques*, which targeted the students training for the competitive exams of the École polytechnique and École normale.

[20]Picard would actually be elected to the Académie a few years later, in 1889.

Halphen died; Georges Humbert was elected to the Académie in 1901, and named to the board in 1905 after the death of Mannheim). Furthermore, all members of the editorial board were polytechnicians, and most were involved in the teaching of mathematics at École polytechnique as well, with the exception of Picard, who had studied at École normale supérieure and was named to the board at the request of Hermite, a former polytechnician.

Jordan's aim to develop the international dimension of his *Journal de mathématiques* is reminiscent of the actions of the directors of the *Annali*. As Brioschi and Cremona did before him, Jordan attracted foreign contributions through his network of correspondants. From 1885 to 1894, the proportion of foreign contributions increased to 25 % of the papers published in the *Journal de mathématiques*, while this proportion decreased continuously after Jordan stopped actively soliciting papers from his correspondants. As with the case of the *Annali*, most foreign contributors published no more than one or two memoirs in the *Journal*.[21]

But even though Jordan seems to have looked up at Brioschi and Cremona as models of successful editors in chief, his eyes were turned toward Germany rather than Italy. The material edition of the *Journal* was changed to follow the norm established by Crelle's journal upon its creation in the 1820s, and which had slowly diffused across Europe ever since: the publication of one installment every trimester, with four per year on this strict schedule, in contrast to the quite erratic rhythm of publication of the *Journal* when it was directed by Liouville. Moreover, Jordan's efforts to make his journal more 'international' were mainly directed toward Paul Gordan, Rudolf Lipschitz, and a few other mathematicians in Germany. In contrast, he never solicited any Italian mathematician and, as a result, the *Journal* did not publish a single Italian contribution between 1885 to 1902.

Jordan very likely aimed at placing his journal on the level of the journals he had himself favored when publishing his research abroad. The nature of the competition between France and Germany however, had nevertheless changed since the time when Jordan had begun to publish in Crelle's journal and Clebsch's *Mathematische Annalen* in the late 1860s. The outcome of the 1870 Franco-Prussian war had been widely viewed in France as the demonstration of a loss in scientific leadership to Germany. In this context, Jordan was presented on several occasions as one of the most promising hopes for French Science, because of his ability to challenge German mathematicians in the very fields, such as algebra, that they were considered to have mastered. In 1873 for instance, Maximilien Marie alluded to the loss of Alsace-Lorraine when commenting on some of Jordan's results on equations solvable by elliptic functions: "this, at least, has been taken back from the Germans" (Marie 1873, p. 943), and in 1874 Jordan engaged in a fierce controversy with Kronecker, one of Berlin's most influential mathematician (Brechenmacher 2007). Even in Italy, as we will discuss in Sect. 4, the general and diplomat Luigi Menabrea considered Jordan's *Traité des substitutions et des équations algébriques*

[21]Most of the contributors who published more than 4 papers were French mathematicians who had prominent positions in the Académie des sciences.

as the proof that the 'cruel hardship' France had suffered had not "extinguished the sacred fire of science," and that "France still has a bright future." When Jordan took up the direction of the *Journal*, Jules Tannery commented that "everyone welcomes you as the director of Liouville's Journal. That the journal recovers the éclat it had on the past is undoubtedly a matter of national interest for France."[22] In this context, it was undoubtedly in Jordan's best interest to attract German contributions to his *Journal*. Engaging in a competition with Italy, on the other hand, may have given credence to the fears expressed in the aftermath of the defeat of 1870 that 'even Italy' was catching up with France in the mathematical sciences.[23]

An other important context was that of the emergence of an 'international space of mathematics' in the 1880s. This concept, as promoted by elite journals such as *Acta mathematica* (Turner 2011), was very different from the ideals of the 18th century République des lettres. This international space was conceived as a concert of nations, whose powers were reflected by their respective scientific 'contributions,' a model that would structure most elite international organizations, from international congresses of mathematicians to olympic games. Of the 105 memoirs that were published in Jordan's *Journal* from 1885 to 1894, 85 were authored by French contributors and 10 by German mathematicians; no other country contributed with more than 3 papers.[24] The comparison with a non elite French mathematical journal such as the *Nouvelles annales de mathématiques* highlights that the absence of Italian contributions in Jordan's journal was not due to a lack of interest in publishing in France. Between 1885 and 1894, 26 % of the contributions to the *Nouvelles annales* were foreign, almost exactly the same proportion as that of Jordan's *Journal*, but Italian mathematicians represented the largest group of foreign contributors with 27 papers (25 % of all foreign contributions).

To be sure, the *Nouvelles annales* was not a research journal such as Jordan's, even though it did publish original contributions. For Italian mathematicians wishing to publish original research in France, the *Comptes rendus* of the Académies des sciences still maintained the European ideal of the République des lettres. It is there, and not in his *Journal*, that Jordan presented the papers sent to him by his Italian correspondants such as Guccia and Giovanni Bordiga. Let us consider the example of Bordiga, a Professore di Matematiche superior at Institute Paolo Sarpi in Venetia. Bordiga wrote to Jordan on 15th March 1886 a letter in which he stated a theorem on a surface in a space of n dimensions, which he connected to a paper published by Jordan in 1875 (Jordan 1875). Bordiga's paper was promptly presented by Jordan at the Académie, and therefore published in the *Compte rendus*

[22]Tannery to Jordan, 27/1/1885, "Je crois que tout le monde se félicite de ce que vous avez bien voulu vous charger de la direction du journal de Liouville il y a assurément un intérêt français à ce que ce journal retrouve l'éclat qu'il a eu."

[23]In 1870, Darboux wrote to Houël: "Je crois que si cela continue les Italiens nous dépasseront avant peu. Aussi tâchons avec notre *Bulletin* de réveiller ce feu sacré et faire comprendre aux Français qu'il y a un tas de choses dans le monde dont ils ne se doutent pas, et que si nous sommes toujours la Grrrande nation, on ne s'en doute guère à l'étranger."

[24]This situation changed completely after 1885; see (Brechenmacher 2009).

(Bordiga 1882). About 3 weeks later, Bordiga sent a second paper to Jordan, which, once again was swiftly presented to the Académie (Bordiga 1886).

3 Italian Reactions to Jordan's *Traité* Des Substitutions et Des équations Algébriques

Let us now focus on the scientific correspondence between Jordan, Brioschi, and Cremona. This correspondence was initiated at the occasion of Jordan's contributions to the *Annali*, but it was not directly concerned with the content of these contributions. Instead, it related to the 'geometrical equations,' especially the equation of the 27 lines on a cubic surface and its connection with the equation of the trisection of hyperelliptic (or abelian) functions.

The 'geometrical equation' associated with a configuration of n objects is an algebraic equation of degree n; its roots correspond to the objects of the configuration, and the relations existing between the roots correspond to the incidence relations existing between the objects. For instance, the equation associated with the nine inflection points of a cubic curve is the algebraic equation of degree 9, the roots of which are the abscissas of the inflection points. The equation of the 27 lines on a cubic surface had been discovered by Cayley and George Salmon, and had been investigated by Jakob Steiner. It is the equation obtained by eliminating three of the four parameters in the system of equations which describes a line lying entirely upon a given cubic surface.[25] Hyperelliptic functions, on the other hand, are functions of two complex variables (u, v), usually defined by pairs $\lambda_0(u, v)$, $\lambda_1(u, v)$, and which have four periods per variable. The problem of the trisection of the periods is to determine $\lambda_0(u/3, v/3)$ and $\lambda_1(u/3, v/3)$ with the help of $\lambda_0(u, v)$ and $\lambda_1(u, v)$, when u and v are given by integral linear combinations of the periods of λ_0 and λ_1. The problem depends on two equations in two unknowns (one for $\lambda_0(u/3, v/3)$ and the other for $\lambda_1(u/3, v/3)$); the elimination of these two unknowns generates an algebraic equation: the equation of the trisection.[26] Jordan's correspondence with Brioschi and Cremona revolved around the connection that the former had discovered between the equation of the 27 lines and the equation of the trisection.

Brioschi first comments on Jordan's works in April 1867 show that, by that time, the latter had already written the paper on geometrical equations that he would publish in 1869.[27] This chronology highlights, once again, the context of Jordan's quest for academic recognition. Jordan had competed several times for the Grand prix. Yet from 1860 to 1865, he had won no more than an official 'encouragement' from the Académie for his memoir on floating bodies. His research on geometrical equations targeted the recognition Jordan desperately needed for pursuing an academic career. In 1867, the Académie had indeed set for its Grand prix the problem

[25]For a discussion of several types of geometrical equations see (Lê 2015a).

[26]For more details on this issue, see (Lê 2015b, p. 147–155).

[27]See Jordan (1869a,b, 1870a).

of "obtaining some noteworthy progress in the theory of algebraic surfaces." By 1868 only a single memoir had been submitted to the Académie and the price was consequently cancelled. A new problem was set up for the next Grand prix of 1871: "to investigate the equations related to the determination of the singular modules, for which the formula of transformation of elliptic functions leads to a complex multiplication." By 1871 not a single memoir had been submitted for this competition, which was therefore postponed to 1874, and eventually cancelled. Why, then, did Jordan not compete for the 1871 Grand prix?

From 1867 to 1870, Jordan had developed his research in faithful accordance with the topics of the Grand prix, first on cubic surfaces in connection with the 1868 prize, and later on abelian functions with the intention of competing for the 1872 prize. But by 1871, Jordan was no longer in a position to compete for the Grand prix. As a matter of fact, he had followed the alternative strategy of earning academic recognition by winning the Poncelet prize. Founded in 1868, this prize was to be awarded for the entirety of one's mathematical works up to that point. The Grand prix, on the other hand, was awarded to solutions of specific problems. With this in mind, Jordan had started to work on a coherent treaties, presenting both new results and an original synthesis on a specific topic: the theory of substitutions. Interestingly, Mathieu, who had also failed to win the Grand prix, followed the same strategy with his *Traité de physique mathématique*, the first volume of which was published in 1873. While this strategy did not turn up to be a winning one for Mathieu, the Académie awarded Jordan the second Poncelet price in 1870.[28] Thanks to this academic recognition, Jordan was then in a position to start competing for entrance into the Académie, to which he would eventually be elected in 1882 after several attempts.

3.1 The 27 Lines and the *Traité* Des Substitutions

The preparation of the *Traité des substitutions* from 1868 to 1869 represents a shift in Jordan's career. As already noted, from 1860 to 1867 the mathematical issues he tackled had to a large extent been dictated by the prizes of the Académie. In his first thesis in 1860, and again in his application to the Académie in 1881, Jordan had claimed that these various issues were nevertheless connected one with another through a global and coherent framework, that of the 'theory of order' in the legacy of Louis Poinsot (Jordan 1881, p. 7–8).[29] This theory was characterized by Jordan

[28]The prize had been created in 1868 through the intervention of Jean Victor Poncelet's widow; it was first awarded in 1869 to Julius Robert von Mayer.

[29]From 1808 to 1844, Poinsot had highlighted the transversal role played by the notion of 'order' in the analogies encountered in various cyclic situations such as the investigations of cyclotomic equations, congruences, symmetries, polyhedrons, and mechanical motions. Poinsot had especially discussed the notion of order when commenting on Gauss's number-theoretic indexing of the roots of the cyclotomic equations in connection with the division of a circle. Later on, he also characterized the theory of order as having a relation to algebra analogous to the relations between Gauss's higher arithmetic and usual arithmetic, or that between analysis situs and geometry (Boucard 2011).

as the part of mathematics dealing with 'relations' between classes of objects, in contrast to classical concerns about quantities, magnitudes, or proportions. It was thus transversal to algebra, number theory, geometry, and kinematics. In this very general sense, the theory of order encompasses the various issues Jordan had tackled from 1860 to 1867: substitution groups, algebraic equations, higher congruences, motions of solid bodies, symmetries of polyhedra, crystallography, the analysis situs of deformations of surfaces (including Riemann surfaces), and the groups of monodromy of linear differential equations. As an example of the intertwining of these various topics, one may cite Jordan's second memoir in the *Annali* in 1868, which classified the groups of motions of solid bodies in the legacy of Bravais's approach to crystallography.

Yet when he was working on his *Traité* in 1868 and 1869, Jordan gradually switched his focus to substitution groups, and more precisely to the reduction of a substitution group into a chain of normal subgroups. The series of papers he published in 1868–1869 fit into a new thematic organization grounded in substitutions and equations, including the pieces on the equations of geometry he discussed with Brioschi and Cremona. As seen above, Jordan was already devoting all of his free time to the preparation of a monograph on substitutions in the early 1860s. As we have seen, however, Jordan had understood that substitution groups were hardly a topic that would attract a broad attention to his works, and that he therefore did not consider it a subject worthy of an entire monograph. For this reason, and in line with the French tradition that mathematical treaties should present both a general theory and its applications, Jordan worked on broadening the perspectives of his works on substitution groups to issues that were not limited to the Galois theory of algebraic equations.

As Jordan phrased it in the introduction of his *Traité*, his book aimed at achieving a "higher point of view on the transformation and classification of the irrationals" (Jordan 1870b, p. V). Jordan's 'irrationnelles' pointed to both quantities and functions. They implicitly referred to the works of authors such as Hermite, Kronecker, Betti, and Brioschi who had considered that the impossibility of solving general algebraic equations of degree greater than five raised the issue of identifying the nature of the 'orders of irrationalities' associated with general equations of higher degree. This program involved finding the 'most general' functions by which the roots of general equations of a given degree could be expressed, such as in the solutions Hermite, Kronecker, and Brioschi had given to the general quintic by using elliptic functions. For these mathematicians, the problem of expressing roots as algebraic functions of coefficients had to be replaced by the problem of associating adequate analytical expressions to each 'order of irrationalities' introduced by general equations of degrees higher than four.[30]

[30]In this context, most mathematicians rather focused on Galois' brief discussion of three modular equations generated by elliptic functions instead of Galois' approach to the general theory of equations (i.e. what we now refer to as Galois theory). In Bertrand's 1867 report on the progress of mathematics, Galois's works were clearly inscribed in a twofold collective dimension (Bertrand 1867, p. 3–17): first, the general theory of equations as it was treated in textbooks such as Serret's;

In the 1850s, these approaches reflected a specific research field that had developed interconnections between analysis, algebra, and arithmetic through the investigation of particular objects such as elliptic functions (Goldstein Schappacher 2007, p. 26). The minor role this research field attributed to geometrical aspects contrasts with Jordan's interest in geometrical equations. Actually, the unity of this research field had been torn apart in the 1860s, when geometric approaches to elliptic and abelian functions, such as Clebsch's, came to play an increasingly important role. It was in this context, that Jordan shifted perspectives in 1868–1869.

This evolution is exemplified by the 'Commentaires sur Galois' that Jordan published in Crelle's journal in 1869. This paper laid the ground for the opening chapter of the third section of the *Traité*: 'Théorie générale des irrationnelles' (i.e. Galois theory). Indeed, Jordan claimed that the interplay between Galois' adjunction of roots 'to an equation' and the resulting reduction of a group could be "very useful for the classification of algebraic irrationals" because the invariance of the number of links in the chain reduction of a group (known as the Jordan-Hölder theorem) "gives a very distinct definition of the degree of irrationality of the roots of a given equation" (Jordan 1866a, p. 1064).

In Jordan's *Traité*, the section 'Des irrationnelles' is actually mainly devoted to applications of the Galois theory. With the help of groups of substitutions, Jordan aimed at placing the problem of resolution with radicals of algebraic equations as "the first ring of a long chain of questions concerning the transformation of the irrationals and their classification."[31] But while the 'algebraic applications' presented classical results on the solvability of cyclotomic, abelian, and metacyclic equations, it was the geometrical and transcendental applications that constituted the actual core of this part of Jordan's treaties.

The section dealing with the 'Applications géométriques' of the theory of irrationals is divided into six parts, each dealing with a particular geometrical situation. These geometric applications exemplified the relevance of Jordan's method of chain reduction of a group. For instance, by successive adjunctions of roots, the group of the 28 double tangents to a quartic curve could be reduced successively to the groups of the 27 lines on a cubic surface and of the 16 lines on a quartic surface with a double conic section. The statement of the link between the 27 lines on a cubic and the trisection of abelian functions concluded the 'geometrical applications' of Livre III (Jordan 1870b, p. 333), which Jordan connected to the 'applications to transcendental functions' (Jordan 1870b, p. 365).

Most of the transcendental applications were devoted to the modular equations and to the equations of the division of the periods of elliptic functions; they presented an original synthesis of the works of Clebsch and Hermite on the special

(Footnote 30 continued)
second, the higher point of view of the nature of the algebraic and transcendental irrationals related to elliptic and abelian functions and their related special equations.

[31]In the *Traité*, all of the groups are composed of substitutions; even in the geometrical applications, while there is no group of transformations of space such as the ones Jordan investigated in his second paper in the *Annali*.

equations associated with elliptic/hyperelliptic functions. Jordan devoted particular attention to Hermite's solution of the general quintic through the modular equation of degree 5 (Goldstein 2011) within the framework of substitutions groups. The section 'Des irrationnelles' was then concluded by the proof that the general equations of degrees higher than five cannot be solved by elliptic functions and thus require investigations of hyperelliptic functions of multiple periods. It is from this perspective that Jordan investigated the reduction of the degree of the equation of the trisection of the periods of hyperelliptic functions of four periods by studying the corresponding algebraic equation which, as he managed to prove, has a reduced equation having the same group as the twenty-seven lines equation.

This connection between abelian functions and geometrical equations aimed at illustrating the efficiency of the theory of substitutions. As Jordan claimed in one of the papers he sent to Brioschi in 1867, the substitution groups attached to the special equations of geometry support "an investigation of the hidden properties of the given equation" (Jordan 1869b, p. 656).[32] In 1869, it was on the occasion of the publication of the series of papers he had sent to Brioschi that Jordan announced the forthcoming publication of his *Traité*.

In summary, the theorem connecting the 27 lines and hyperelliptic functions played a crucial role in Jordan's book, and represents one of the main original results of Livre III where it concluded the geometrical applications of Galois theory, and opened the discussion of the applications to transcendental functions. This theorem therefore played a key role in establishing the legitimacy of the *Traité* as a whole as well as in securing its early reception (Brechenmacher 2011). The *Traité*'s third section in particular argued for the relevance of substitution groups for developing a higher point of view on the 'order of irrationalities' related to the special equations of elliptic/abelian functions. In doing so, it situates the theory of substitutions within a framework completely different from that of the theory of order to which Jordan had appealed in the early 1860s. While no consideration of polyhedra, crystallography, analysis situs, or groups of motions appeared in the *Traité*, Jordan attributed an important role to Clebsch's geometric approach and to Hermite's works on modular equations.

[32]Such as the fact that the group of the 28 double tangents can be reduced successively to the groups of the 27 lines on a cubic and of the 16 straight lines on a quartic surface having a double conic. These groups were introduced by geometric permutations or symmetries in relation to the invariance of the algebraic form that defined the cubic surface. The reductions of the groups thus usually appealed to interplays between the consideration of fixed points (or lines) and factorizations of algebraic forms. But Jordan systematically associated equations with the substitution groups involved. A fixed point then corresponded to a factorization of the equation by the adjunction of a root and to the corresponding reduction of its Galois group.

3.2 Brioschi's Difficulties with Jordan's Group Theoretical Approach

It was in one of the papers he had sent to Brioschi in 1867, that Jordan had stated for the first time that the reduction of the 80th degree equation of the trisection of abelian functions led to a group he recognized as that of the equation of the 27 lines on a cubic surface. Yet in this preliminary version of his works on geometrical equations, this connection may not have been stated as clearly as in 1870 when it was presented as a key result of the *Traité*.[33] In his response to Jordan, Brioschi expressed his difficulties in identifying precisely the type of equation Jordan was dealing with:

> I have been deeply interested in your memoirs on the equations of geometry, and I have the intention to deal very soon with the reduction of the modular equation of the fortieth degree which corresponds to the transformation of the third order of the transcendental hyperelliptic of the third species. But I haven't clearly understood if you have obtained the reduction of the latter equation or rather of that corresponding to the triplication of hyperelliptic functions, even though the connections to which you are pointing between these equations and the cubic surfaces lead me to think that it is rather in this second equation that you have found a result so important. I doubt it is possible to establish a priori connections between these two questions, i.e. between the triplication and trisection of cubics; at least directly.[34]

Brioschi's difficulties highlight the important differences between the mathematical practices of the two authors. In his answer to Jordan, Brioschi struggled to give a formulation of the problem of the trisection that suited his own approach to abelian functions. This approach especially involved to writing down an explicit expression for the equation of the trisection, which Brioschi considered 'very simple'[35]:

[33]Recall that Jordan was at the time focusing on the Grand prix on algebraic surfaces while the one on elliptic functions had not been issued yet.

[34]Vos mémoires sur les équations de la géométrie m'ont vivement intéressé, et j'ai l'intention de m'occuper sous peu de l'abaissement de l'équation modulaire du quarantième degré qui correspond à la transformation du troisième ordre des transcendantes hyper-elliptiques de la première espèce. Mais je n'ai pas bien compris si vous avez obtenu l'abaissement de cette équation ou plutôt de celle qui dépend de la triplication des fonctions hyper-elliptiques, bien que les liens que vous signalez entre cette équation et les surfaces de troisième ordre me porte à penser que c'est sur cette seconde équation que vous avez obtenu un résultat si important. Je doute qu'on puisse trouver des liens a priori entre les deux questions, c'est-à-dire entre la triplication ou la trisection et les surfaces du troisième ordre; au moins directement.

[35]A, B, C, etc., are functions of the three parameters (i.e. modules) k, λ, μ which characterize a pair of abelian functions in two complex variables $u = (x_1, x_2)$ and $v = (y_1, y_2)$.

$$\begin{aligned}
&Ax_1^{\frac{3}{2}}x_2^{\frac{3}{2}}y_1^{\frac{1}{2}}y_1^{\frac{1}{2}} + B(1 - x_1)^{\frac{3}{2}}(1 - x_2)^{\frac{3}{2}}(1 - y_1)^{\frac{1}{2}}(1 - y_2)^{\frac{1}{2}} \\
&\quad + C(1 - k^2x_1)^{\frac{3}{2}}(1 - k^2x_2)^{\frac{3}{2}}(1 - k^2y_1)^{\frac{1}{2}}(1 - k^2y_2)^{\frac{1}{2}} \\
&\quad + D(1 - \lambda^2x_1)^{\frac{3}{2}}(1 - \lambda^2x_2)^{\frac{3}{2}}(1 - \lambda^2y_1)^{\frac{1}{2}}(1 - \lambda^2y_2)^{\frac{1}{2}} \\
&+ E(1 - \mu^2x_1)^{\frac{3}{2}}(1 - \mu^2x_2)^{\frac{3}{2}}(1 - \mu^2y_1)^{\frac{1}{2}}(1 - \mu^2y_2)^{\frac{1}{2}} + G = 0
\end{aligned}$$

For Brioschi, explicit formulations such as the one above were tightly connected to the geometrical interpretation he required for Jordan's connection between the 27 lines and abelian functions. Jordan, on the other hand, had a very different method for the geometrical equations. This specific and systematic method consisted of using the incidence relations which existed between the objects to generate an algebraic form having the same group of substitutions as the equation. Unlike Brioschi, Jordan dealt directly with the characterization of the group he was interested in, very often through the explicit analytical representation of its substitutions, and more generally by studying its order and its chains of normal sub-groups. Such characterizations allowed one to establish connections between the groups generated by different problems, such as those of the 27 lines on a cubic or the trisection of hyperelliptic functions. For Jordan, such group-theoretical characterizations touched the very 'essence' of the subject. He therefore did not bother to express the equations to which his substitution groups were associated (much less to express their roots), an issue at the core of Brioschi's approach to algebraic geometry through invariant theory.

Brioschi noted once again his difficulties with Jordan's works in another letter dated 7th January 1870. While expressing his thanks to Jordan for sending him his *Traité*, Brioschi focused again on the issue of "cubic surfaces and the equations of the trisection of the periods" of hyperelliptic functions. As in 1867, Brioschi argued that he was planning to tackle the problem of the trisection 'very soon.' But his letter focused on the more traditional issue of the equations generated by the bisection of the periods of hyperelliptic functions. Brioschi eventually stated that this equation of degree 2^{2p} can be reduced to an equation of degree p. When he archived this letter in his correspondence, Jordan added the following note: "the result indicated in this letter which M. Brioschi takes as a new one is already stated in my *Traité*." As a matter of fact, in his letter of January 1870 Brioschi had provided a new proof of a characterization already given by Jordan for the roots of the equation of the bisection of elliptic and hyperelliptic functions, one which Brioschi published a few months later in the *Comptes rendus* (Brioschi 1870).

3.3 Geometrical Interpretations of the 27 Lines Theorem

Cremona congratulated Jordan on his *Traité* in a letter dated 19th December 1869. This letter was entirely devoted the theorem on the 27 lines[36]:

> Until now, I could do no more than cut the pages [of the *Traité*] but the titles of the §§ already gave me an idea of the pleasure and the edification I will gain from reading your work, which I plan to do promptly.
>
> There is one question—among others—that excites my curiosity to the highest degree: that of the link between the research on the 27 lines on a cubic surface (which has been discovered by MM. Cayley and Salmon, before Steiner) and the trisection of elliptic functions. There is here a real puzzle to explain, especially from a geometrical point of view.[37]

As Brioschi did before him, Cremona required a geometrical interpretation of the equation of 40th degree whose group connected the 27 lines to the trisection of abelian functions. On January 11, Jordan answered to Cremona that it would indeed be interesting to find a 'definitive demonstration' of this connection, even though he had already provided a group theoretical proof of this issue in his *Traité*.[38]

Cremona's comments led Jordan to further develop his geometrical interpretation of the link between the equations of the 27 lines and those of the trisection of hyperelliptic functions. While most of his results on the geometrical equations had been deduced solely from the characterization of the associated substitution groups, in the case of the twenty-seven lines of a cubic surface Jordan deduced some resolution properties directly from the existence of geometrical objects.[39] In his first answer to Cremona, dated 11th January 1870, Jordan alluded briefly to an interpretation based on a system of nine planes with particular incidence properties as a certain configuration of lines that generates the same equation as the trisection of hyperelliptic functions. At this point, he wondered whether this interpretation was a

---

[36]The two letters written by Cremona to Jordan in 1869–1870 are found with the draft of the latter's response in the Archives of the École Polytechnique (ref. VI2A2(1855)9). These letters and their answers have been edited by Simonetta Di Sieno and Paola Testi Saltini in a volume currently in print (Israel 2015).

[37]Je n'ai pu jusqu'ici que tailler les pages mais les seuls titres des §§ ont suffi pour me donner l'idée du plaisir et de l'instruction que je pourrai tirer de la lecture de votre ouvrage: lecture que je commencerai sans délai. Entre autres, il y a une question qui excite au plus haut degré ma curiosité: celle du rapprochement de la recherche des 27 droites d'une surface cubique (qui ont été découvertes par MM. Cayley et Salmon, avant Steiner) avec la trisection des fonctions elliptiques. Surtout au point de vue géométrique, il y a là une véritable énigme à expliquer.

[38]Jordan also took into account the references to the works of Cayley and George Salmon that Cremona had indicated, by adding a note to the first volume of his *Traité* in 1870. This appendix was later merged with the introduction of the book when the full version of the *Traité* was published.

[39]Those objects are the forty-five planes containing the twenty-seven lines, the forty so-called 'triples of double Steiner's trihedrons' and the thirty-six double-sixes of Schläffli. A 'Steiner trihedron' is a trihedron made from some of 45 planes, having special incidence relations. These trihedrons can be grouped by pairs and then by triples of pairs, according to other incidence relations. The final objects are the 'triples of double Steiner's trihedrons.' See (Le 2015a).

new one, especially because he could not find it in Cremona's previous works on the 27 lines (Cremona 1868). In his answer dated 9th February 1870, Cremona was finally ready to send Jordan a more detailed comment about the issue of the trisection elliptic functions:

> I beg you a thousand times to forgive me for having taken a month to respond to your excellent letter [...]. I wish I could be as liberal as M. Clebsch in his communications; but to do so I would need a mind as admirably fecund as his. My situation is unfortunately very different and I do not even have the faculty so desirable of being able to tackle several different questions simultaneously. I thus beg you to forgive me if I reply to you almost empty handed.[40]

In this letter, Cremona developed a geometrical interpretation very similar to that of Jordan himself. The two mathematicians shared a very close reading of Clebsch's works. While working on his letter to Jordan, Cremona had actually studied a recently-published paper by Clebsch in *Mathematische annalen* (Clebsch 1870), in connection with a note that Jordan had published in the *Comptes rendus* (Jordan 1869a). But Cremona did not yet have access to Livre III of the *Traité*, or to the more recent note Jordan was about to publish (Jordan 1870b, p. 326), and in which he developed the geometrical interpretation he had briefly mentioned in his letter of January 11. This interpretation was detailed in a letter Jordan sent to Cremona on the 5th of April 1870, accompanied by three copies of his recent note for Cremona, Brioschi, and Casorati.

The important role played by Clebsch's approach to the resolution of the nine inflection points equation for both Jordan and Cremona is highlighted by the importance these three mathematicians attributed to the formation of geometrical objects encoding particular incidence relations, and especially to the groupings of lines as keys to understanding an equation (or links between equations). Clebsch's approach is also seen in their arguments in which algebraic results were directly deduced from the existence of geometrical objects.[41] This approach to algebraic surfaces had been especially influential in Cremona's pioneering works on what would lated be referred to as the 'Italian school of algebraic geometry.' We have already noted before Clebsch's influence on the elaboration of the chapter on the 'Applications géométriques' of Jordan's *Traité*. As a matter of fact, Clebsch had been the first to provide a geometrical understanding of Galois theory through a careful study of the groupings of geometrical objects (and not of roots), often

[40]Je vous demande mille fois pardon si j'ai attendu un mois à répondre à votre excellente lettre [...]. Je voudrais bien être si libéral que l'est M. Clebsch dans ses communications; mais il faudrait, pour cela, posséder son admirable fécondité d'esprit. Malheureusement, il en est bien autrement; et de plus, je n'ai pas la faculté si désirable de poursuivre simultanément plusieurs questions différentes. Je vous prie donc de m'être indulgent, si je viens devant vous, presque avec les mains vides.

[41]François Lê has characterized this approach as a specific mathematical practice linking the existence of certain geometrical objects and the understanding of a problem, and which can also be found around the link between the twenty-seven lines and the hyperelliptic functions. See (Lê 2015a).

expressed in tables.[42] In his letter to Jordan dated 9th February 1870, Cremona made extensive uses of such tables for characterizing the groups associated with various geometrical configurations such as the forty 'triples of double Steiner's trihedrons' and their decompositions. His alternative and more geometrically detailed approach to Jordan's statement on the 27 lines on a cubic was published a few months later in the *Rendiconti* of the Lombard Institute (Cremona 1870).

That same year, Jordan was named correspondant of the Lombard Institute for sciences and letters, to the 'great satisfaction' of Brioschi and Cremona. Yet Jordan would not further develop his work on these geometrical equations after 1870. There is therefore no direct evidence that he was inspired by Cremona's more geometrical interpretation of Clebsch's works. About a decade later, however, Jordan advised Poincaré and Léon Autonne, two young polytechnicians who were considering a mathematical career, to read both Clebsch's and Cremona's works. Autonne would later develop an original synthesis between the works of Jordan, Clebsch, and Cremona, and would especially devote a great number of studies to a geometrical approach to differential equations inspired by Cremona's contact transformations.

3.4 Giuseppe Janni and the Reception of Jordan's *Traité* in Italy

Jordan's *Traité* has often been described as the first autonomous presentation of both group theory and Galois theory. The book has also often been assumed to have had a very limited reception, either because of the novelty of its group theoretical approach, or, in contrast, because of its outdated focus on substitutions groups at a time when Klein and Lie were developing their research on transformation groups (Wussing 1984). Yet this reception was actually as multifaceted as Jordan's *Traité* itself—a 'book with seven seals,' as Klein designated it when recalling his time with Lie in Paris in 1870 (Klein 1921, p. 51).

Jordan's *Livre III* applications of Galois theory to the investigation of the geometrical equations and abelian functions had immediately attracted the attention of prominent mathematicians such as Brioschi, Cremona, Clebsch, Klein, and Lie. In contrast, *Livre IV*, which represented almost one third of the entire book, would have almost no reception until the end of the 19th century.[43] But Jordan's *Traité*

[42]As has been shown by François Lê, this practice of using successions of tables to represent groups of geometrical configurations and their decompositions into subgroups points to the legacy of Galois as developed by mathematicians such as Betti in his treatment of the modular equation of order 5 in 1853 (Lê 2015a).

[43]*Livre IV* presents a synthesis of the research Jordan had continuously devoted to solvable groups since his thesis in 1860. It is undoubtedly the most original section of the *Traité*. It aims at classifying maximal solvable subgroups of transitive groups through what Jean Dieudonné referred to as an 'enormous machinery' consisting of successive reductions of general groups to a chain of special ones (Dieudonne 1962, p. XXXIX– XLII), with a particular attention to the roles played by general linear groups $Gl_n(F_p)$ in the chain reduction of solvable transitive groups.

was not limited to the very original approaches and results of its *Livre III* and *Livre IV*. The two first sections of the book also provided a synthesis of previous works on higher congruences (finite fields, in modern terms) and substitution groups. For several of its early readers, the novelty of Jordan's *Traité* was actually more related to its synthetic presentation of previous works on substitutions than to the existence of a genuine new theory.

Such was the case of Giuseppe Janni in Naples, who devoted a series of papers to the theory of substitutions and Galois theory in Battaglini's journal from 1870 to 1874.[44] Shortly after the publication of Janni's first paper in 1870, Jordan complained that his works had not been properly cited in Janni's publication which, he claimed, was 'entirely deduced' from the two first sections of the *Traité*. Janni apologized to Jordan in a letter dated 13th May 1871, indicating that his 1870 publication was only the first of a series, and that he had set aside all bibliographic references for his final paper. Moreover, Janni claimed that his work was not only based on Jordan's book but also on the previous works of authors such as Cauchy, Betti, Hermite, Bertrand, Joseph-Alfred Serret, and Mathieu. This claim sheds light on an important form of reception of Jordan's *Traité* in which Jordan's book was appreciated not so much for its originality, but in continuity with previous works developed in France.[45]

At this occasion, Janni also asked for Jordan's advice about two specific statements, thereby showing his very close contact with Livre II of the *Traité*. One of his questions pointed to a specific part of Jordan's proof for the formula giving the order of the orthogonal group, which involved number-theoretical computations on congruences (Jordan 1870b, p. 166). The other concerned the main original result of Livre II, i.e. the statement of the theorem now designated as the Jordan canonical form theorem. This theorem provides the simplest form to which a linear substitution A (i.e. a matrix in $Gln(F_p)$) can be reduced through linear manipulations of its variables. This canonical form depends on the nature of the roots of the characteristic equation $det(A - SI) = 0$ (with S the unknown and I the identity). If this equation has n distinct roots, then A can be reduced to a diagonal form, but this may not be the case when multiple roots occur. Janni's question pointed to the main conceptual difficulty raised by this problem, which Jordan had solved for the first time in 1868, i.e. that a multiple root of the characteristic equation is not always associated with a number of linearly independent solutions equal to its multiplicity.[46]

Janni wrote once again to Jordan in 1873, sending him the final paper of his series for Battaglini's journal. On this occasion, he expressed his desire to further study the second, more original, volume of the *Traité* (i.e. Livre III and Livre IV).

---

[44]See (Janni 1871,1872, 1873, 1874).

[45]That the *Traité* has often been considered a continuation of previous works is also made clear by the series of papers in which Netto reformulated Jordan's approach in the legacies of Cauchy and Kronecker.

[46]In modern parlance, the dimension of the vector space of eigenvectors associated with the eigenvalue of a matrix is not always equal to the multiplicity of the eigenvalue.

But this never happened. As a matter of fact, Janni's involvement with substitutions came to a rather bitter conclusion. As Janni phrased it in his 1873 letter, Jordan's works on substitutions had 'unintentionally doomed' his career by 'ruining' his chances for appointment to the chair of algebra he had been competing for at the university of Turin. Opponents to Janni's nomination had indeed claimed that the latter's works on substitutions were no more than a mere translation into Italian of the beginning of Jordan's *Traité*. Furthermore, according to these opponents, Janni had demonstrated his incompetence by reproducing several errors in Jordan's book.

Janni also lamented that his application had suffered from the novelty and difficulty of substitution theory, in contrast with other candidates who had do little more than applying traditional works of analytical geometry to the investigation of conic sections. To be sure, Janni's focus on the synthesis provided by the two first sections of the *Traité* had narrowed the scope of the book. In contrast, we have seen that from 1868 to 1869 Jordan had struggled to legitimize the theory of substitutions by applying it not only to algebraic equations but also to geometry and analysis. Janni's sad experience with the theory of substitutions in Italy suggests retrospectively that Jordan had made the right choice in avoiding to presenting his works on substitutions as an autonomous theory, and therefore in refusing to publish these works in the *Annali* in the mid 1860s.

4 École Polytechnique and Italy

In addition to, but not independently from, the Académie des sciences, the École polytechnique is another important institutional framework through which Jordan developed relationships with Italian mathematicians. While the École may seem obviously French, we shall see that it had a fairly complex relationship with Italy in the 1860s and 1870s. The occupation of the North of Italy by the French empire at the beginning of the century had cast a long shadow in Italy. During this period several young men from Piedmont had been trained in mathematics in Paris in the same manner as regular 'French' polytechnicians.

4.1 Polytechnician Networks

In the first section of this paper, we have seen that it was through the action of Bertrand—their former professor at École polytechnique—that young polytechnicians such as Mathieu and Jordan ended up sending to the *Annali* the memoirs they had submitted to the Académie. Moreover, Jordan's first letter to Cremona in 1867 highlights the interplay between networks of polytechnicians and networks of Franco-Italian correspondants.

When he received Cremona's circular in January 1867, Jordan did not only feel that he had to respond personally to this call for papers but that he also had to suggest other potential contributors. Jordan therefore suggested to Cremona his "friend, M. Sarrau, who is one of our good geometers, and who may be the best in the methods of the calculus of residues [...] his portfolio must be furnished enough and, if he gives you his collaboration, I am certain he will be very useful to you." Émile Sarrau had entered École polytechnique 2 years after Jordan in 1857. Like Jordan, he started a career as an engineer and developed academic research in physics that eventually led him to become professor of mechanics at École polytechnique in 1884, and to be elected to the Académie in 1886. Between 1865 and 1867, Sarrau had published two memoirs on the propagation and the polarization of light in crystals, a topic that was connected to Jordan's works on polyhedra through the framework of Bravais's approach to crystallography (a former professor of both Sarrau and Jordan at École polytechnique) as well as through Poinsot's theory of order.[47] It is certainly because of this connection that Jordan introduced Sarrau to Cremona. As a consequence, Cremona sent to Jordan another copy of the circular for the attention of Sarrau. A few weeks later, Cremona congratulated Jordan for having "acquired [for the *Annali*] the collaboration of M. Sarrau."

Yet Sarrau did not eventually contribute to the *Annali*. His research in physics was actually hardly compatible with the turn of the journal in 1867 toward pure mathematics, in contrast with the editorial policy of the *Annali* under Tortolini's supervision. It is possible that, in 1867, Jordan was not yet fully aware of the changes taking place in the *Annali*. But it is far more likely that Jordan simply considered Sarrau's works on mechanics and crystallography as fully belonging to 'mathematics,' along the lines of the traditional definition of the mathematical sciences at École polytechnique.

This situation highlights a difference in perception of the very definition of mathematics between French and Italian mathematicians. In France, the 'mathematical sciences' were traditionally divided between analysis, geometry, and their applications. The last of the three potentially included all of the investigations in physics that were not limited to observations or experimentations, with topics from mechanics, physics, astronomy, navigation, heat theory, the theory of light, crystallography, electricity, etc. In this sense, the memoir on the stability of fluid motions that Jordan had submitted to the *Annali* was no more 'mathematical' than Sarrau's crystallographical approach to optics. In contrast, while after 1867 the *Annali* still occasionally published papers from applied mathematics (in coherence with its title), these papers were mostly limited to mechanics (both fluid and solid) and heat theory, i.e. topics very close to analysis. As for Jordan, he maintained his broad understanding of 'the mathematical sciences' throughout his career. Even though his own published works mainly contributed to pure mathematics, he supported the publication in his journal of contributions on a great variety of topics of physics. The physicist Pierre Duhem, for instance, was one of the most prolific contributors to the *Journal de mathématiques pures et appliquées* from 1885 to 1905.

---

[47]See (Sarrau 1867, 1868).

4.2 The Italian Connections of the École Polytechnique

Among the many congratulations that were sent to Jordan for publishing his *Traité*, one reveals the long shadow cast by an ancient connection between the École polytechnique and Italy.

While Jordan had mainly sent copies of his *Traité* to a few prominent European mathematicians, such as Clebsch, Carl Wilhelm Borchardt, Cayley, Cremona, and Brioschi, a copy was nevertheless sent to the general Luigi Federico Menabrea in Firenze. Menabrea had not only been an engineer, mathematician, and professor of mathematics at the Turin military school, but he had also started a military and political carrer at the time of the Risorgimento. Born in 1809 (15 years before Brioschi), Menabrea was, by his own account, "from a time when men were imbued from the teaching of the École polytechnique."[48] In Turin, Menabra had been a student of Cauchy, an illustrious polytechnician and professor of analysis at École polytechnique who had exiled himself to Italy in the 1830s after refusing to plead allegiance to the reign of Louis Philippe. On several occasions, Menabrea also claimed to follow the legacy of Lagrange, another illustrious professor of École polytechnique who actually originated from Turin.

During his military career, Menabrea had served under the general Augustin Jérôme Chiodo, a polytechnician, at the siege of Peschiera in 1848, and later succeeded Chiodo as the President of the committee of artillery and military engineering. Chiodo had been trained at École polytechnique during the time when Piedmont belonged to the French empire. In contrast with the 'external" foreign auditors who attended lectures at École polytechnique, a number of young 'Italians' such as Chiodo entered École polytechnique as regular students during the French occupation of Piedmont. Several of them returned to Italy afterward and made a career as military officiers, engineers, or scientists. Among them, Jean Antoine Amédée Plana and Dominique Joseph Botto were both professors at the university of Turin when Menabrea studied there (in mathematics and physics, respectively).

Menabrea had also developed a close relationship with Jean-Victor Poncelet, another polytechnician who was both a mathematician and a general. At the military school in Turin, the former had especially adapted the latter's lectures at the École d'application d'Artillerie et du Génie in Metz, "so that we may say that our technical education was penetrated by the polytechnician spirit of Paris."[49] Menabrea's scientific works were often presented to the Académie by its secrétaire perpétuel, namely Élie de Beaumont, another polytechnician who had often promoted the value of the 'Italians' of the École polytechnique such as Plana, whom he described as a "model of all the virtues that honor a scientist."[50] Moreover, Menabrea

[48]Les hommes de mon temps sont imbus des enseignements de l'École polytechnique.

[49]de sorte qu'on peut dire que notre enseignement technique était pénétré de l'esprit polytechnique de Paris.

[50]Beaumont also suggested that Plana's portrait should be placed at École polytechnique where he "had drawn his primary instruction."

benefited from the contribution of a polytechnician we have already mentioned several times, namely Bertrand, in the proof of one of his main theorem on the application of the principle of least work to an elastic truss.[51]

Menabrea wrote to Jordan for the first time in November 1872, shortly following his return to Florence after having served as the Italian ambassador in Vienna in 1870–1871. In addition to expressing his congratulations on the *Traité* and the Poncelet Price, Menabrea also sent to Jordan a few of his memoirs in mathematical physics (vibrations, heat propagation, wave equation, elasticity). These topics, he admitted, were "quite far from the horizon in analysis" that Jordan's *Traité* had opened. But Menabrea thought they could be interesting to Jordan for their connection to the works of Mathieu. The tone of the second letter sent by Menabra a month later is especially interesting, in the sense that Jordan's achievements were presented as the proof that the 'sacred fire of science' was still burning in France, despite the defeat of 1870:

> A few weeks ago, I had the honor of writing to you to thank you (quite late) for your precious Traité des substitutions, which you have had the kindness to send to me, and which, in my opinion, is the cornerstone of an important era in the history of the progress of the analytic science [...] Your noble example, Monsieur, proves that the cruel hardship suffered by our country did not extinguish the sacred fire of science and, that, by all means, France still has a great future.[52]

4.3 Jordan's *Cours d'Analyse* and Italy

A few years later, when Jordan became professor of analysis at École polytechnique, the connections between the École and Italy were quite different than those of the period Menabrea cherished in his recollections of Lagrange and Cauchy.

When the occupation of Piedmont ended with the fall of Napoleon, Italians could no longer join École polytechnique as regular students. From 1830 to 1850, 12 Italians nevertheless registered at École polytechnique as external auditors (among them Faà De Bruno), a status that allowed them to follow either all or a portion of the courses, but not to pass exactly the same examinations as regular students. The ranking of the external auditors therefore did not vary between their entrance and their release from the school, which is why they were called

[51]In 1868 Menabrea published a new demonstration of his principle of least work, which, although superior to the preceding one, still failed to note the independence of the variations of the internal forces and of the elongations of the members of the structure. This oversight was criticized by Sabbia, Genocchi, and Castigliano, giving rise to a controversy lasting until 1875. In 1870 Menabrea published jointly with Bertrand a note that advanced the first valid proof of his principle.

[52]Il y a quelques semaines, j'avais l'honneur de vous écrire pour vous remercier (un peu tardivement) de votre précieux Traité des substitutions que vous avez eu l'obligeance de m'envoyer et qui selon moi marque une époque importante dans l'histoire des progrès de la science analytique [...] Votre noble exemple, Monsieur, prouve que les cruelles épreuves qu'ont subi notre pays n'y ont point éteint le feu sacré de la science et que, sous tous les rapports, la France est toujours pleine d'avenir.

'constants' in the polytechnician slang. The number of Italians auditors decreased after the Risorgimento: only seven Italians attended lectures at polytechnique from 1850 to 1875, and the participation of Italian auditors stopped completely after 1875. As a consequence, no Italian ever actually attended Jordan's lectures at École polytechnique. Indirectly, however, Jordan's prestigious status guaranteed a diffusion of the printed version of his *Cours d'analyse* 'throughout the world,' as Charles de La Vallée Poussin phrased it in 1921 (de La Vallée Poussin 1921), and in particular throughout Italy.

Hints of the diffusion of the *Cours d'analyse* in Italy can be found in Jordan's correspondence. For instance, in 1892, Joseph Prisma, a student of mathematics at Sestri Levante near Genova, sent a letter to Jordan about a possible mistake in a formula given in the second edition of the *Cours d'analyse* for the computation of the difference between the volumes of two polyhedra. This letter highlights the very close reading of the *Cours* by this student, a reading that was more oriented toward specific problems and their actual resolutions, than toward the architecture that Jordan had given to the general organization of the second edition of his *Cours* for the sake of pursuing absolute clarity and perfect rigor.

The diffusion of the *Cours d'analyse* also benefited to that of the *Traité des substitutions* among Italian professors. When the second edition of the *Cours* was issued in 1892, Giovanni Moigano wrote a letter to Jordan seeking information about recent research in substitution theory, which he wanted to include in the course of analysis he was teaching at the university of Messina, Sicily.

On the other hand, the very design of the *Cours d'analyse* was inspired by Jordan's contacts with Italy. The second edition of Jordan's *Cours* in 1892 played a key role in raising concerns for the foundations of analysis in France in the early 1890s. It especially paved the way for the new approaches to real analysis that would be developed by Émile Borel, René Baire, and Henri Lebesgue at the turn of the 20th century (Gispert 1983). Since the 1860s, many remarks, definitions, and results relative to the elementary principles of analysis had been stated, mainly in Germany and Italy. Yet most of these statements had remained scattered in various periodical publications. The second edition of Jordan's *Cours* was the first textbook to provide a complete synthesis of this variety of new approaches, including Giuseppe Peano's reflections on continuity, Georg Cantor's set theory, Karl Weierstrass's theory of elliptic functions, as well as the emerging theory of real functions.

The three volumes of the first edition—published in 1882, 1883, and 1887 respectively—show the evolution of Jordan's concerns for the foundations of analysis. Jordan had become more and more preoccupied by foundational issues after he publishing the first volume of the first edition of his *Cours* in 1882. The first volume had followed a rather traditional approach to the teaching of analysis in France, with no explicit definition of limits, no proof of the properties of continuous functions, no construction of the real numbers, and no notion of set theory. Yet by 1887 Jordan had already began to reconsider this approach in a note he added to the third volume of the *Cours*, and in which he criticized the choices he had made in

1882. As we shall see, Peano played an important role in this shift of perspective. His correspondence with Jordan on Rolle's theorem indeed involved key properties of continuous functions.

In 1882, Jordan had stated the following theorem (Jordan 1882b, p. 22):

> Let $y = f(x)$ be a function of x with a determinate and finite derivative when x varies in a given interval. Let a and $a + h$ be two values of x taken in this interval. Then
>
> $$f(a+h) - f(a) = \mu h$$
>
> with μ designating an intermediary quantity between the highest and the lowest value of $f'(x)$ in the interval between a and $a + h$

Jordan had then deduced from this statement the formulation of the mean value theorem by making the following 'remark,' with no demonstration: "if $f'(x)$ is a continuous function from a to $a + h$, this function will pass by insensitive degrees from its maximum M to its minimum m" (Jordan 1882, p. 22). This proof was similar to those already given by Lagrange and Cauchy in their own respective *Cours d'analyse de l'École polytechnique*. It presupposes a double passage to the limit, and therefore the continuity of the derivative of f. This proof, therefore, is not satisfactory as had already been pointed on by Bonnet in the early 1860s. In his *Cours de calcul différentiel et intégral*, Serret had followed Bonnet's approach in proving the mean value theorem by using Rolle's theorem rather than the intermediate value theorem. Yet Serret had relied on the 'trivial' property that any continuous function which is not always equal to zero "has to either start to increase by taking positive values, or to decrease" (Serret 1868, p.17). In 1884, Peano provided a counter example to Serret's statement: $x sin(1/x)$ for $x = 0$ (Peano 1957, p. 45). He moreover criticized the proof given by Jordan of the mean value theorem. It was in this connection that Jordan wrote to Peano in 1884 with a request for the details of the proof of Rolle's theorem. Peano answered Jordan's request in a letter dated 16th February 1884, in which he detailed the proof that he and Angelo Genocchi were teaching in Turin, and that Ulisse Dini was teaching in Pisa, and which is still used today.

Jordan reproduced this proof in the note he added to the publication of the third volume of his *Cours* in 1887 (Jordan 1887, p. 581).[53] In this edition, Jordan also criticized his previous proof of 1882 which "made the useless assumption of the continuity of $f'(x)$." Peano's proof was reproduced once again in the first volume of the second edition of the *Cours* in 1892 in which Jordan fully developed his new perspectives on the foundations of analysis, with a presentation of set theory and a clarification of the elementary concepts of topology and real function theory.

While he was preparing the second volume of his second edition in 1894, Jordan requested once again Peano's advice on the "proof of the existence of the integral differential equations," i.e., what is now referred to as the Cauchy-Lipschitz theorem (or, alternatively, as the Picard-Lindelöf theorem), to which Peano had devoted

[53]Jordan actually made two minor modifications to Peano's proof. See (Gispert 1982).

a paper in 1890. In a letter dated 6th November 1894, Peano answered Jordan that his proof

> follows the usual procedure of integrating differential equation by passing to the limits on approximations by polygonal lines, as developed successively by Cauchy, Lipschitz, Gilbert, and others. But the conditions have to be reduced to their smaller number, and one has to explain the meaning of integrating by approximation as well as to analyse the passage to the limit. I have corrected all the gaps between the premises and the thesis, and reduced all the demonstration to a sequence of syllogisms and other known forms of reasoning. It is the first time that mathematical logic has been applied to a question of superior mathematics and this application is, in my opinion, the most important part of my work. But learning the symbols and operations of logic requires time, and my demonstration is not well known. M. Mie has cited it in Mathematische Annalen, Bd. 43 page 553. but several works have been published since then on the same topic with no new result and no reference to my works. I regret this situation, as I believe that mathematical logic will bring great progress in the analysis of difficult questions. I also regret this situation for more personal reasons; I have been teaching at the University for 14 years, and I still haven't been promoted to ordinary professor in contrast with younger colleagues, because my works is not known and appreciated enough here.[54]

On this occasion, Peano also sent to Jordan a copy of the mathematical 'formulario' that he had published as a supplement to his *Rivista di Matematica*, made a few comments about two points from the first volume of the second edition of Jordan's *Cours*.[55] and expressed his desire to "translate into the symbols of logics the entire section devoted to multiple integrals."[56]

[54]Je viens de terminer la longue suite des examens fatigants, et je puis enfin répondre à votre lettre du 12 octobre. Le chemin suivi dans ma démonstration de l'existence de l'intégrale des équations différentielles est bien simple et bien connu. On intègre les équations par approximation, au moyen d'une ligne polygonale et l'on passe à la limite. C'est l'idée de Cauchy, Lipschitz, Gilbert et d'autres. Mais il faut réduire les conditions au plus petit nombre, expliquer ce qu'on entend par intégration par approximation et analyser le passage à la limite. J'ai comblé toutes les lacunes entre les premisses et la thèse, et réduit toute la démonstration à une suite de syllogismes, et d'autres formes de raisonnement connues. C'est la première fois qu'on a appliqué la logique mathématique à l'analyse d'une question de mathématiques supérieures, et cette application est, selon moi, la chose plus importante de mon travail. Mais les symboles et les opérations de la logique exigent du temps pour être appris; et une démonstration est peu connue. M. Mie a publié un article explicatif dans les Mathematische Annalen, Bd. 43 page 553. Mais ensuite ont parus plusieurs travaux sur le même sujet, sans y ajouter rien de nouveau (sauf quelque inexactitude), et sans faire mention de mon travail.Je regrette cela, parce que je crois que la logique mathématique apportera des grands avantages dans l'analyse des questions difficiles. Je le regrette aussi par une question personnelle; il y a quatorze ans que je professe à l'Université, et je ne suis pas encore nommé ordinaire, à différence d'autres plus jeunes d'age et d'enseignement; car mes travaux sont ici peu connus et estimés.

[55]The first was a superfluous hypothesis that could be dropped in Jordan's proof about the derivability of a convergent series of function along the lines of a demonstration provided by Peano in his Lezioni di analisi infinitesimale. The second was a critic on a statement made by Jordan in the theory of enveloppes and according to which the equations of a figure defined as a limit of a sequence of figure were 'obviously' the limits of the equations of the sequence. Peano had given counter examples to this situation which he had discussed in details in his *Applicazioni geometriche del Calcolo infinitesimale*.

[56]En continuant l'étude de votre traité, je lui trouve une foule de nouvelles cognitions. J'ai désir de traduire en symboles de logique toute la partie qui regarde les intégrales multiples.

To be sure, Jordan was not familiar with (and not actually a supporter of) Peano's recent development in logics. He therefore eventually settled for not mentioning the Cauchy-Lipschitz theorem in the second volume of the second edition of his *Cours* in 1895, while Picard had included this theorem in his course at the Sorbonne in the mid 1880s. The limited inspiration Jordan was able to draw from Peano's approach to the foundations of analysis highlights, in turn, the limited modernization brought by the second edition of his *Cours d'analyse*. Even though it opened the worlds of set theory, topology, and real functions to a new generation of mathematicians, Jordan's *Cours* nevertheless did not take into account the new issues developed in logic in connection with the foundation of analysis.

5 A French Representative of Italian Mathematicians

The creation of the Circolo Mathematico di Palermo opened a new horizon to Jordan's relations with Italy. Founded in 1884 by Guccia, the Circolo aimed at gathering all Italian mathematicians in a learned society, and therefore to establish the first national mathematical society in Italy, in contrast with the various existing societies that were anchored in the academies of different cities. The society launched its own mathematical journal, the *Rendiconti del Circolo Matematico di Palermo*.[57] The first volume of which appeared in four parts from 1885 to 1887. The completed volume was praised by Bertrand as a publication of remarkably high quality at the meeting of the Académie on the 7th of November 1887.

A few months later, in February 1888, the Circolo passed a statute allowing foreign members and opened its *Rendiconti* to foreign contributors. By then, Guccia aimed at turning the Circolo into an international society with a leading mathematical publication. In 1888 and 1889, the foreign contributions to the second and third volumes of the *Rendiconti* were dominated by French polytechnicians.[58] Jordan was one of them,[59] along with Eugène Catalan, Halphen, Poincaré, Mannheim, Humbert and Georges Fouret.

In 1889, Guccia invited Jordan for a familial visit to Sicily on the occasion of the latter's election as a permanent, non-resident member of the Circolo. The election took place on the 9th of June. In Palermo, Jordan was introduced to the scientific community and more generally to Sicilian high society, including Guccia's own aristocratic family. This visit was the beginning of a close relationship between the two mathematicians and their families. In his letters to Jordan, Guccia systematically paid his respect to Jordan's wife and daughters, and frequently asked for news

[57]See (Brigaglia 2014).

[58]Actually, all of the foreign contributors were French polytechnicians, with the exception of Alexis Starkoff from Odessa (who was nevertheless a member of the French mathematical society).

[59]Jordan's paper, "Sur la marche du cavalier" (Jordan 1888), starts by dealing with a problem raised by the rules of chess and then develops a general method for obtaining a solutions (x_i) of a linear system with integer coefficients such as $\Sigma|x_i|$ is minimum.

about Jordan's sons, Edouard and Camille, who later paid visits to Guccia when they were travelling to Italy. As for Madame Jordan, she frequently shipped to Palermo flasks of quinine sulfate, a treatment for malaria.

The correspondence between the two mathematicians started on June 26, shortly after Jordan and his family had returned from their pleasant—though very hot—trip to Sicily. Jordan and Guccia set up an agreement with the publisher Gauthier-Villars for exchanging free issues of Jordan's *Journal* and Guccia's *Rendiconti* between Paris and Palermo. This type of exchange was particularly important for mathematicians such as Guccia who were located on the peripheries of European main mathematical libraries.[60] In return, Guccia offered to list all the papers published in Jordan's *Journal* in the 'Biblioteca matematica,' a section of bibliographic references of both periodical and non periodical publications that concluded each issue of the *Rendiconti*. The *Journal* was therefore one of the first periodical publications to benefit from such full coverage, which Guccia quickly extended to most European and American mathematical journals.

5.1 An Italian Delegate in Paris

Even though Jordan was one the first foreign members of the Circolo, he had nevertheless been preceded by ten other foreign mathematicians, starting with Catalan of Liège who had been elected in December 1886, and two other Parisians, Fouret and Humbert, who had both been elected in December 1887.[61] Yet shortly after his election in June 1889, Jordan was appointed by Guccia as the delegate of the Circolo in Paris. For the purpose of normalizing the bibliographical references provided by his 'Biblioteca matematica,' Guccia decided to follow the mathematical subjects classification that was to be set up at the Congrès bibliographique that took place in Paris from 16–19 July 1889 (Nabonnand et Rollet 2002). In a letter dated 30th June 1889, Guccia wrote to Jordan:

> Our Biblioteca Matematica will be more helpful to scientists when the mathematical classification will have been definitively fixed by the Paris bibliographic conference. On the matter of this important conference, I announce you that, in naming you as the only representative of the Circolo Matematico, our intention is to fully entrust you with all of the decisions that will be discussed and voted upon. You have therefore the complete freedom to push forward your ideas, either as M. Jordan, or as the delegate of the Mathematical Circle.

[60]As a matter of fact, we learn from a letter dated 25th July 1889, that Guccia did not have access to the issues of the *Comptes rendus* anterior to 1884. For this reason, Guccia had to ask Jordan to check his bibliographic references to papers published before 1884.

[61]The list of the foreign members who had been elected before Jordan is the following: Catalan (Liège, Dec. 1886), Hirst (London, Jan. 1887), M.N. and J.S. Vanecek (Bohemia, June 1887), Humbert and Fouret (Paris, Dec. 1887), and, in May 1888, Starkof (Odessa, April 1888), Le Paige (Liège), Schlegel (Hagen), and Mittag Leffler (Stokholm).

I believe that this situation should allow you two ballots in each vote. [...] Again, I assure you that all the ideas you may have concerning the lacuna that you may consider to exist in this Project will be shared and accepted by our Society, in advance and with no discussion.[62]

This delegation for representing the Circolo in Paris sheds light on Guccia's political involvement in the French mathematical scene. Shortly after he had completed his thesis "On a class of surfaces representable point by point on a plane" with Cremona in Rome in 1880, Guccia presented his main results at the Reims conference of the "Association française pour l'avancement des sciences" (Guccia 1881). That same year, Guccia had been elected as a member of the French mathematical society thanks to the support of Jordan, then the president of the society. In addition to the *Rendiconti*, Guccia's first mathematical works had been exclusively published in the *Comptes rendus* of the Académie. By 1889, Guccia was sufficiently familiar with the French mathematical scene so as to grasp the opposition between various groups of mathematicians and side with the small group of elite mathematicians, such as Jordan, Halphen Poincaré, Hermite, and Picard, who controlled the key institutions in Paris—the Académie, the *Journal de mathématiques*, Polytechnique, the É.N.S., the Sorbonne, the Collège de France, etc.—against what Guccia designated as the "league of the geometry of the triangle." This expression implicitly referred to another group of actors, such as Henri Brocard and Emile Lemoine, who were marginal figures with respect to the central institutions, but who had managed to develop the geometry of the triangle as a specific field of mathematics, particularly through their regular communications and publications at the Association française pour l'avancement des sciences, and through the medium of the *Nouvelles annales de mathématiques*.[63] As Guccia wrote to Jordan:

It would therefore be useful to exchange a few words with M. Poincaré[64] in order to agree upon a position with him before the first session (Tuesday July 16, 9 pm in the session hall of the Mathematical society) so that you will not be taken by surprise by the league of the geometry of the triangle and by the coterie of the association française (the one you know) which has many votes. If you consider it necessary, you can say that the Mathematical Circle of Palermo, in its act of membership of the congress, has allowed you to promise its collaboration to the Repertoire for all that concerns the mathematical production of Italy. This work will be shared between various members of our society, to which **all** Italian mathematicians belong.[65]

---

[62]Notre Biblioteca Matematica sera plus utile aux savants lorsque le congrès bibliographique de Paris aura définitivement fixé la classification de la matière. Au sujet de cet important congrès je vous préviens que le Circolo Matematico en vous nommant son unique représentant, il a eu l'intention de s'en remettre complètement à vous pour toutes les décisions qui seront discutées et votées. De sorte que vous avez pleine liberté de faire prévaloir vos idées, soit comme M. Jordan, soit comme délégué du Cercle Mathématique, ce qui, je crois, devrait vous permettre d'avoir deux voix dans les votations. [...] Encore une fois, je vous confirme que toutes vos idées au sujet des lacunes qui pourraient vous paraitre exister dans ce Projet, seront partagées, acceptées par notre société, d'avance et sans discussion.

[63]See (Romera-Lebret 2014).

[64]Poincaré was the President of the bibliographic conference.

[65]Il serait donc utile que vous échangiez quelques mots avec M. Poincaré pour vous entendre avant la 1ère séance (mardi 16 juillet à 9 h du soir, salle des séances de la Société Mathématique) dans le

Guccia's claim to represent all Italian mathematicians certainly played an important role in his nomination as one the members of the permanent international commission of mathematical classification that was created by the 1889 conference. This episode documents the interplay of the Sicilian and the Parisian scenes in Guccia's construction of the Circolo as both an Italian and an international mathematical society. Recall that the "international space of mathematics" was usually conceived at the time by elite mathematicians as a world of nations, with various powers, colonial possessions, and subtle alliances. To be sure, other mathematicians developed more universalist, or even socialist, views of internationalization, but such was not the case of actors such as Mittag-Leffler and Guccia who both aimed at developing elite international organizations and journals. Both Mittag-Leffler and Guccia used their position in the periphery of the main European capitals as a neutral position between the two main mathematical powers, i.e. France and Germany. Such a neutrality was a prerequisite to their claims of internationalism. Yet Guccia's neutrality was quite artificial; he was an objective ally of Parisian academicians.

In January 1892, Guccia sent another request to Jordan. The latter was to join forces with Fouret to have F. Caldarera, professor of mechanics and dean of the faculty of sciences of Palermo, elected as a non resident member of the French mathematical society.[66] Caldarera was indeed elected about a year later, when his application was presented by Jordan and Fouret on the 15th February 1893. A few years earlier, in 1890, Guccia had himself presented the application to the S.M.F. of two other professors in Palermo—Gebbia and Gerbaldi (with the help of Georges Humbert). As a result, Palermo had become in 1894 the most represented Italian city at the S.M.F., with four of the thirteen Italian members, surpassing Rome with its three representatives (Cremona, Valentino Cerruti and Baldassarre Boncompagni) and Milan with its two representatives (Brioschi and Giuseppe Jung).

5.2 The Death of Georges Halphen

Halphen's sudden death in may 1889, at age 44, was another event that served to strengthen the relationship between Guccia and Jordan. Halphen had been one of the leading experts in France on the type of approach to geometry that Guccia had

(Footnote 65 continued)
but de ne pas vous faire surprendre par la ligue de la géométrie du triangle et par la côterie de l'association française (celle que vous connaissez) qui disposent de beaucoup de voix. Si vous le jugez nécessaire vous pouvez dire que le Cercle mathématique de Palerme, en faisant acte d'adhésion au Congrès, vous a autorisé à promettre sa collaboration au Répertoire pour tout ce qui concerne la production mathématique de l'Italie. Ce travail serait distribué à différents membres de notre société, à laquelle appartiennent **tous** les mathématiciens italiens.

[66]Jordan and Fouret had both served as president of the French mathematical society, in 1880 and 1887, respectively. Both also served as perpetual societary of this society, and as members of its council.

developed in the legacy of Cremona. Shortly after his election as an academician in 1886, he had repeatedly presented Guccia's works to the Académie.

Guccia had been informed by Gauthier-Villars, the publisher of the *Journal*, that Jordan was working on an obituary of Halphen (who had been a member of the editorial board of his journal).[67] On July 11, Guccia wrote to Jordan about his project to establish a complete list of Halphen's mathematical works for the next issue of the *Rendiconti*, requesting advice concerning the relevant classification of two papers on continuous fractions, and expressing his desire to write a notice on Halphen's works in geometry:

> I have prepared the material for a short notice on Halphen's geometrical works but I have to postpone the writing of this notice, as I want to leave Palermo immediately after I'll have finished two short notes for our Rendiconti, because the heat is unbearable. I am not competent to appreciate all the progress in analysis due to Halphen, but in the case of geometry, I dare to claim that his works are of the foremost importance, which, in my opinion, has not been highlighted enough because these works deal with the most difficult questions that have ever been raised, and which are not already familiar to everyone, because of their difficulty.
>
> I will read with the greatest interest what you will write about this. Let it be said, and kept strictly between us: I am not at all satisfied by the discourses of MM. Hermite and Brioschi.[68] Am I wrong? I am convinced that by writing your notice you will pay a true favor to Science and to the memory of the eminent geometer who just died, especially because various publications on the history of modern geometry have, for some time, been pleased to create an unfair and illegitimate pedestal for M. Schubert on the theory of characteristics [in cubic space curves], and, on the classification of algebraic space curves, to make believe that the two memoirs of Halphen and Nöther complement each other, as if the two authors followed the same methods and helped each other by using the same procedures, which is false. Halphen never used Nöther's methods, and often worried about their frequent lack of rigor. In addition to this, Halphen has never left any unfinished work in any of the questions he tackled in geometry. The same cannot be said of Nöther. It often happens in the latter's works (which are also of the foremost importance) that what has not yet been rigorously proved has to be tacitly admitted. For instance, in everything that touches upon algebraic surfaces, M. Nöther seems to ignore that the public is not already convinced (even though M. Picard vaguely alludes to this in his last memoir) that each algebraic surface can be transformed, by a birational correspondence of space on a given surface with a single ordinary singularity!!!
>
> Etc. etc. etc. etc.
>
> I do not want to bother you by adding up other examples.
>
> Please, remember my best memories and express my best wishes to Madame Jordan and to all your children.[69]

---

[67]Jordan was also to participate in the edition of Halphen's collected work with his fellow academicians Picard and Poincaré.

[68]Guccia is referring to the discourses given by Hermite and Brioschi at Halphen's funeral and the Academie de Lincei respectively.

[69]J'avais préparé les matériaux pour une courte notice sur les travaux géométriques de Halphen mais je dois en remettre l'execution à plus tard, étant obligé en ce moment-ci d'achever deux petites notes pour nos Rendiconti et voulant tout-de-suite après (le mois prochain) quitter Palerme à cause de la chaleur qui est devenue insupportable. Il n'est pas de ma compétence d'apprécier tous les progrès dont l'analyse est redevable à Halphen, mais pour ce qui concerne la géométrie, je ne crains pas d'affirmer que ses travaux sont d'une importance capitale, dont, je crois, on ne se rend

Guccia's evaluations of Halphen's accomplishment would not have any influence on Jordan's obituary, which had actually already been written by the time Jordan received Guccia's letter. Jordan's notice actually made no mention of Hermann Schubert's works on the theory of characteristics of cubic space curves, and made no more than a brief allusion to the memoir by which Max Nöther had won, conjointly with Halphen, the Steiner price of the Academy of Berlin for "solving an important question in the theory of algebraic space cuves." As he expressed it in another letter in July 25, Guccia was very pleased by the succinct and synthetic nature of Jordan's notice, which left open the possibility of writing a larger piece on the geometrical aspect of Halphen's works:

> I am very grateful to you for communicating to me your notice on Halphen, which I find absolutely perfect in your moderation in giving a general overview of the principal discoveries of this eminent geometer. In each theory, you have found the right overtone, and there is nothing but truth, presented in the style both so simple and so elevated that you alone have the secret. Truly, entering into further details would require one to write the history of the progress of geometry for the past 20 years; a rough task given all the susceptibilities that would emerge from several illustrious geometers, among whom not even one had ever shown in his discoveries and demonstrations ideas as profound and a spirit as rigorous as our regretted Halphen.
>
> This will have to be said later, but it will have to be said!!!
>
> [...] I am sad to learn that Louis caught the scarlet fever. I hope that he will have gotten rid of it by the time this letter reaches you. Please, remember me to his good memory. Would you be so kind as to thank Madame Jordan for the kindness she had to send me the Quinine sulfate? My mother and my father ask me to send you all their best wishes. I pray

(Footnote 69 continued)
pas compte assez, attendu qu'il s'agit des questions les plus difficiles qu'on ait jamais abordées, et dont la plupart ne sont pas encore devenues familières à tout le monde, à cause de leur difficulté.Je lirai avec le plus grand intérêt ce que vous en écrirez. Soit dit absolument entre nous: les deux discours de MM. Hermite et Brioschi ne m'ont pas satisfait du tout. Ai-je tort? Je suis donc persuadé que vous rendrez un réel service à la Science et à la mémoire de l'éminent géomètre qui vient de disparatre, en écrivant votre notice, d'autant plus que, depuis quelque temps, en différentes publications sur l'histoire de la géométrie moderne, on s'est plu, pour ce qui regarde les caractéristiques, à créer un piédestal illégitime et injuste à M. Schubert, et pour ce qui regarde la classification des courbes gauches algébriques, à faire croire que les deux mémoires d'Halphen et de Nöther se complètent l'une l'autre comme si les deux auteurs suivaient les mêmes méthodes s'aidaient mutuellement en se servant des mêmes procédés, déjà comme par leurs publications antérieures. Ce qui est faux. Halphen ne s'est jamais servi des méthodes des Nöther, au sujet desquelles en bien des endroits il ne craignait pas d'avoir des scrupules au point de vue de la rigueur, ce qui se manifestait bien souvent. Ajoutez à cela que, pour ce qu'il a fait dans chaque question de géométrie, Halphen n'a rien laissé d'inachevé. On ne peut pas dire la même chose pour Nöther. Il arrive souvent pour les travaux de celui-ci (qui sont aussi de la plus grande importance) qu'il faut admettre tacitement ce qui n'est pas encore rigoureusement démontré. Ainsi, par exemple, dans tout ce qui touche aux surfaces algébriques, M. Nöther semble ignorer que le public n'est pas encore persuadé (quoique M. Picard en fasse vaguement mention dans son dernier mémoire) que toute surface algébrique peut être transformée, par une correspondance birationnelle de l'espace en une surface donnée uniquement de singularité ordinaire!!!Etc etc. etc. etc.Je ne veux pas ajouter d'autres exemples, car je crains de vous ennuyer.Veuillez me rappeler au bon souvenir de Madame Jordan et lui faire agréer mes hommages respectueux ainsi qu'à tous vos enfants.

you to kindly accept mine as well, and to pay my respectful homage to Madame Jordan and all your children; I shake your hand with great fondness.[70]

5.3 An Anonymous Reporter for the Circolo in Paris

After Jordan and Guccia's correspondence on Halphen's works in July 1889, the relationship between the two mathematician became even more intimate. Guccia came to visit Jordan's family during his trip to Paris in the fall, and made his return train trip to Turin with one of Jordan's son, Edouard. At the occasion of this visit to Paris, Guccia also shared with Jordan his fears about his application for the chair of professor of geometry at the University of Palermo, in which he was eventually ranked first (while the second, Pasquale Del Pezzo, eventually got the other position in Naples).[71]

A few weeks later, on December 13, Guccia requested, once again, that Jordan act as the representative of the Circolo in Paris on the occasion of Hermite's Jubilee:

> My dear friend,
>
> I have great favor to ask you! On December 24, a ceremony of an intimate character will gather at the Sorbonne, at 10 in the morning, the students, friends and admirers of the illustrious Hermite for celebrating his 70th birthday. The Mathematical Circle of Palermo, in its session of the 11th, has chosen you as its delegate for representing our Society (of which you are both one of the most illustrious members, and a perpetual member) to this solemnity of Science. […] I am begging you for
>
> 1° agreeing this delegation
>
> 2° agreeing to write in French an address that you will read on behalf of the Mathematical Circle of Palermo (which gathers the largest number of Italian mathematicians and friends of Science) during the ceremony of the 24th.
>
> In case, after discussing this matter with M. Darboux (the president of the ceremony), you would rather have the mathematical Circle send a telegram to you or M. Darboux on the afternoon of the 23rd, please inform me immediately. But you will understand that, rather than an inevitably laconic telegram, I would prefer an address that you would write!

[70]Je vous suis bien reconnaissant de m'avoir communiqué votre notice sur Halphen que je trouve parfaite dans la mesure que vous avez voulu garder, soit d'une aperçu général des principales découvertes de cet éminent géomètre. Dans chaque théorie vous avez trouvée la note juste, et il n'y a rien qui ne soit la vérité, exposée dans une forme si simple et si élevée en même temps dont vous seul avez le secret. Assurément, si l'on voudrait entrer dans des détails il faudrait écrire, pour ce qui regarde la géométrie, l'histoire des progrès accomplis par cette science depuis 20 ans; rude tâche, à cause surtout des susceptibilités qu'on verrait surgir de la part de plusieurs illustres géomètres, parmi lesquels il n'y en a pas un qui ait montré dans les découvertes et dans les démonstrations des idées profondes et de l'esprit de rigueur à un si haut degré que notre regretté Halphen.Il faudra dire cela plus tard, mais il faudra le dire!!![…] J'apprends avec tristesse que Louis a attrapé la scarlatine. J'espère qu'à l'arrivée de cette lettre il en sera complètement débarrassé. Veuillez me rappeler à son bon souvenir. Avez vous eu l'obligeance de remercier de ma part Madame Jordan de la bonté qu'elle a eu de m'envoyer le sulfate de Chinine? Ma mère et mon Père me chargent de vous faire agréer leurs bons souvenirs à vous tous. En vous priant de vouloir bien agréer les miens aussi, et de vous rendre interprète de mes respectueux hommages auprès de Madame Jordan et de tous vos enfants, je vous serre la main bien affectueusement.

[71]Guccia to Jordan, 16. Nov. 1889.

> and that I could afterward publish in one the forthcoming booklets of the 'Rendiconti,' with the details of the ceremony that you would have the kindness to communicate to me [...].
> How can we thank you for all the efforts we request from you? But we could not have found another person who could, better than you, make himself the spokesman of the feelings of the Italian mathematicians (and friends of mathematics) for this illustrious old man that every contemporary geometer considers as the grand master of Science.[72]

Guccia added to his letter the details of the subscription he had received from Italian mathematicians to participate in the Hermite medal. This subscription was raised in a very short time (the news of Hermite's Jubilee had only reached Guccia in late October) but "the smallest purses have immediately responded to my [Guccia's] invitation as soon they learned that it was to honor Ch Hermite!"[73] The subscription eventually amounted to 543 francs with the participation of 76 subscribers. The origins of this sum provide us with a kind of map of Guccia's mathematical Italy: Palermo (145 Fr.), Rome (85), Turin (53), Milan (35), Naples (30), Pisa (30), Padoue (25), Florence (20), Messine (20), Catane (19), Modica (19), Bologne (17), Gênes (15), Modène (10), Paris (5), Pavie (5), Venice (5). The city of Paris itself is listed as an Italian city, because one of the subscribers actually lived there (Bortolotti). More interestingly, Sicilian cities gathered almost 40 % of the whole subscription, of which Palermo itself accounted for about 27 %.

Jordan complied with Guccia's sollicitation. In a letter dated 29th December, Guccia congratulated Jordan on his 'masterpiece': "With no exaggeration, it would not have been possible to express better, and in such a few words, the feelings of

[72]Mon cher ami, J'ai à vous demander un grand service! Le 24 décembre, une cérémonie, d'un caractère tout intime, réunira, à 10 heures du matin, à la Sorbonne, les élèves, les amis et les admirateurs de l'illustre Hermite, pour fêter son 70e anniversaire. Le Cercle Mathématique de Palerme, dans la séance du 11, vous a choisi comme son délégué pour représenter notre Société (dont vous êtes une des plus illustres membres et en même temps membre perpétuel) à cette solennité de la Science. La délégation officielle vous parviendra presque en même temps que cette lettre. En attendant je vous demande:1 ° de vouloir bien accepter cette délégation2 ° de vouloir écrire en français une adresse de la part du Cercle mathématique de Palerme (qui réuni le plus grand nombre des mathématiciens et des amis de la Science, italiens) que vous liriez au nom de notre Société dans la cérémonie du 24.Dans le cas où, après avoir conféré avec M. Darboux (qui présidera la cérémonie) vous trouviez préférable, à l'adresse du Cercle mathématique, une dépêche de notre Société expédiée, dans l'après midi du 23, à vous, ou à M. Darboux directement, veuillez m'en prévenir tout-de-suite. Mais vous comprenez que je préfèrerais une adresse écrite par vous! et que je pourrais publier dans un des prochains fascicules des 'Rendicont' avec les détails de la cérémonie que vous auriez la bonté de me communiquer, à une dépêche qui ne peut qu'être nécessairement laconique. [...]Comment vous remercier pour toute la peine que nous vous donnons? C'est que nous n'aurions pas pu trouver une autre personne, qui aurait pu, mieux que vous, se rendre interprète des sentiments des mathématiciens (et des amis des mathématiques) italiens pour cet illustre vieillard, que tous les géomètres contemporains considère comme le grand matre de la Science.

[73]Les plus petites bourses ont tout-de-suite répondu à mon invitation dès qu'ils ont appris qu'il s'agissait d'honorer Ch. Hermite!

our society—as well of the Italian mathematicians who belong to our Circle—for M. Hermite." But Guccia was even more delighted by Jordan's chronicle of Hermite's Jubilee, which he had translated into Italian for his *Rendiconti*[74]:

> The chronicle of the celebration in Italian will naturally not be published under your name as it is a service of reportage that has to be attributed to the editors, and to the official representant of the Society. The public will assume that, in addition to our delegate, we also sent reporters who covered the ceremony for us in detail!!![75]

Guccia expressed his thanks to Jordan by having him—and Madame Jordan—delivered two boxes of Sicilian tangerines. Yet after this episode, the epistolary connection between Jordan and Guccia stopped almost completely. Only one letter after January 1892 is kept in the archives at École polytechnique. In this letter, dated 31st December 1894, Guccia expressed his thanks for the second edition of Jordan's *Cours d'analyse*, and apologized for not having written sooner. The tone of the letter remained very friendly and intimate, with Guccia asking for news of Jordan's wife, sons, daughters, and even daughters-in-law, and confiding the 'intrigues' and "absurd criticisms of MM. the Jews" against his recent promotion to ordinary professor.[76]

To be sure, the two mathematicians continued to see each other at the meetings of the Mathematical Circle of Palermo, and Jordan continued to suggest contributors to the *Rendiconti*, such as Autonne and Robert Montessus de Balore. But Guccia undoubtedly realized after 1892 that he could make use of other supports in Paris. First, Jordan himself had been delegating some of his responsibilities to Humbert, his former student at École polytechnique, whom he had always strongly supported (to the point of having him elected against Picard as professor at École polytechnique in 1895) and who would take charge of his lectures at Collège de France after 1904. Yet in the few letters he sent to Jordan from 1892 to 1894, Guccia complained several times about Humbert's lack of reactivity to his sollicitations.[77] Second, Hermite's jubilee was one the events that helped to highlight the new institutional importance of Poincaré—an early contributor to the *Rendiconti*—on the French and international mathematical scenes. The latter had indeed been in

[74]Jordan's address was published in 1892 (Jordan 1892a) in the *Rendiconti*, accompanied with his anonymous chronicle (Jordan 1892b).

[75]Il est bien entendu que le Compte rendu de la fête qui paraîtra en italien ne portera pas votre nom; car c'est un service de reportage qui doit être attribué à la rédaction et pas au représentant officiel de la Société. Le public supposera que, en dehors de notre délégué, nous avions aussi des reporters qui nous ont fourni les détails de la cérémonie!!!.

[76]Heureusement tout est fini bien au sujet de ma promotion à ordinaire. La nouvelle commission (octobre dernier) a été unanime pour reconnaître les critiques absurdes de MM. les Juifs et pour proposer à pleines voies ma promotion. après cela, ces mêmes MM. sont aller faire des intrigues partout, mais ils ne sont pas réussis, car le Ministre a fait signer le décret par le Roi le 29 novembre dernier. Tout est bien qui finit bien!.

[77]This lack of reactivity to Guccia began after Humbert lost his wife in 1882.

charge of the plenary lecture of the ceremony, one that Guccia would have been eager to publish in his *Rendiconti*.[78] Third, Guccia was well aware of the emerging generation of mathematicians trained at the École normale supérieure. Since the late 1880s, he had especially developed a close relationship with Picard, who became an important contributor to the *Rendiconti*. Both Picard and Poincaré began to play central roles in the French mathematical scene in the mid 1890s, at a time when Jordan was willingly stepping down.

6 Conclusions: Italian Mathematics or Sicilian Tangerines?

Our investigations have revealed scattered images of Italy in France from 1860 to 1900. During this whole time period, no more than one action taken by Camille Jordan may be understood as pointing to a global perception of the mathematical activities in Italy.

We have seen that when Jordan took over the direction of the *Journal de mathématiques pures et appliquées*, his eyes were turned toward a few mathematicians in Germany while no Italian would ever be solicited to contribute to this journal. We have seen also that this attitude may be understood in connection with both the national role played by the *Journal* and with the emergence of an international space of mathematics. The latter was usually conceived as a concert of nations, whose respective powers were reflected by their contributions to a few elite journals. In this context, France was struggling in the 1880s to demonstrate—especially to French people—its capacity to challenge the scientific leadership of Germany, which implied considering other nations, such as Italy, as secondary actors in the international space of science. Yet this context was not specific to mathematics. The German tropism of Jordan's *Journal* reflects, rather, the impact of both national and international politics on some institutional organizations of mathematics, more, than it involves a direct evaluation of 'German' or 'Italian mathematics' as a whole.

Jordan never actually expressed any global perspective on 'Italian mathematics.' His relationship with 'the mathematicians of Italy' points rather to a diversity of Italies, envisioned through different connections with the Parisian (rather than the French) mathematical scene. To be sure, this fragmentation is, to some extent, due to the very nature of our main source of investigation. A scientific correspondence does not constitute a homogeneous corpus. Moreover, it cannot be considered as providing an exhaustive account of one author's epistolary communications. Even though Jordan was carefully collecting his letters with a view on his own legacy—as highlighted by the note he added to Brioschi's 1870 letter—, he may have discarded a few letters that he considered unimportant or even compromising. Yet the diversity of Italies emerging from the investigation of Jordan's correspondence is not only the consequence of the fragmented and incomplete nature of correspondences as

[78]Guccia to Jordan, 29 December 1892.

historical sources. It also points to a variety of collective dimensions of mathematics, usually much smaller than, but always interlaced with, national frameworks. In order to discuss this issue further, two different scales of analysis have to be distinguished, the scale of institutions and the one of interpersonal correspondences.

We have seen in the present paper that Jordan established connections with Italian mathematicians through several institutional frameworks, such as the Académie des sciences de Paris, mathematical journals, the École polytechnique, and the various academies attached to Italian cities such as Milan, Turin, and Palermo. On this institutional scale, Jordan made clear distinctions between the various positions of his Italian correspondents. Italian mathematicians, in turn, often discussed with Jordan their specific positions, and their occasional grievances with other Italian actors, such as the cases of Janni's disappointment with the lack of recognition of his works on substitutions in contrast to the more traditional approaches of his competitors, Guccia's denunciation of his jewish colleagues, and with Peano's despair over his isolation in Turin.

National categories are even less relevant at the smaller, interpersonal, scale of Jordan's scientific correspondence with Italian mathematicians. We have seen that this correspondence focused on the connection between geometrical equations and abelian functions. This issue was neither French or Italian. On the contrary, the investigation of elliptic and hyperelliptic functions had been at the core of a specific research field that had involved prominent European mathematicians from the 1820s to the 1850s. The legacy of this research field was still vivid in the 1860s, as exemplified by Jordan's references to the works of Betti, Hermite, Brioschi and Kronecker. Yet the unity of this field eventually fractured into various lines of developments which involved small networks of mathematicians of various countries, as highlighted by the example of Clebsch's influence on both Jordan's and Cremona's approaches to the 27 lines equation. This example shows that the French Jordan and the Italian Cremona, both still quite young in the 1860s, independently identified Clebsch's research as crucial sources of inspiration.

In turn, Jordan's 27 Italies illuminate important aspects of the Parisian mathematical scene. Jordan himself did put on a variety of Italian costumes: correspondent of the academies of Milan, Palermo, and Turin, contributor to the *Annali* and the *Rendiconti*, 'scientific brother,' delegate of, and anonymous reporter for Palermo's Circolo, actor of the international legitimacy of Guccia, and even representative of Italian mathematicians in France.

Jordan's contributions to the *Annali* have allowed us to grasp the crucial role that the Académie was still playing in the 1860s for a young polytechnician willing to make a career in the mathematical sciences. More precisely, the riddle of Jordan's two Italian papers unveils the extremely strong impact of the system of academic prizes in the evolution of Jordan's mathematical works.

Moreover, our investigations have shed light on an important form of circulation of mathematics in the 1860s through the specific, and quite subtle, connection between the Académie and mathematical journals such as Liouville's and the *Annali*. In his correspondence with Cremona, Jordan expressed his conception of a 'mathematical journal.' In his opinion, journals of engineers, such as the *Annales des mines*, were not mathematical journals because they mainly published applied mathematics and had therefore a limited audience among 'mathematicians.' Jordan clearly had an elite conception of both 'mathematicians,' and 'mathematical journals,' which involved a close connection with the Académie's institutional organization of the 'mathematical sciences,' and involved both theoretical and applied original research. In contrast, Jordan did not only reject journals of engineers on the peripheries of mathematics, but also the publications targeting an audience of students and teachers, such as the *Nouvelles annales de mathématiques*, to which he never contributed. This elite, academic, conception of mathematics is also documented by Guccia's involvement in the opposition to the league of the geometry of the triangle and to the côterie of the Association française pour l'avancement des sciences.

Further, the correspondence between Cremona and Jordan reveals a differentiation between the types of mathematics that may, or may not, have been suitable for periodic publications. We have seen that Jordan's reluctance to publish his works on algebraic equations and substitution groups in the *Annali* did not result from a bad opinion of the Italian mathematical scene. On the contrary, Jordan was very much aware that, whether in France or in Italy, his research on substitutions should not be presented as an autonomous theory. As a consequence, he wanted to make sure that none of his publications on substitution groups would be isolated one from another, in coherence with the global organization he would later give to his *Traité* by distinguishing between the 'theory of irrationals' and its applications to algebra, geometry, and analysis. To be sure, this situation sheds some light on the status of group theory in the 1860s, a 'difficult' topic, as Janni phrased it, neither widely known nor highly considered, in contrast with cubic surfaces and abelian functions. But it also reveals that in the 1860s the *Annali* did not have in Paris a status comparable to Liouville's or Crelle's journals. It was indeed because he feared that a publication in the *Annali* would be isolated from his series of memoirs on substitutions that Jordan settled for sending two memoirs on much more traditional issues in mechanics.

The Italian reactions to Jordan's two main monographs, i.e. the 1870 *Traité* and the two editions of the *Cours d'analyse*, illuminate the diversity of perceptions of École polytechnique in Italy. General Menabrea not only cherished the École but even referred to France as 'our' nation. His letters to Jordan reveal the long shadow cast by Napoleon's Empire on Italian scientific activities, especially through the training at polytechnique of several young men from Piedmont. Yet by the time Jordan had become Professeur d'analyse, École polytechnique was no longer a beacon illuminating the world with the sacred fire of science. This evolution has often been commented upon in connection with the emergence of the École normale supérieure as the institution of reference for the training of elite French mathematicians. Yet Jordan's correspondence with Peano shows that the former was at least as aware of the innovations developed in Turin in the teaching of analysis as of those of prominent normaliens such as Picard.

Finally, although it was not the main purpose of the present paper, our investigations have documented various aspects of the mathematical activities in France viewed from Italy, as well as the role played by Jordan in the evolution of the persona of a 'French mathematician,' in the sense of the socio-cultural identities, values, and ethical dispositions expressed by the collective role taken on by individuals.[79]

In the 1860s, the young Jordan participated in forging the image of a promising French mathematician by distinguishing himself through a few beacons of French science with a strong visibility in Italy: the École polytechnique, the Académie, its system of prizes, and its the circle of collaborators with Italian scientific journals and societies, such as Bertrand. In this context, Jordan was spontaneously considered by Brioschi as a successor of Hermite, a prominent French member of an international network of mathematicians of the previous generation. Yet this 'French identity' that Brioschi attributed to Jordan did not help the former to understand the specific approach of the latter, which, as we have seen, was much closer to Clebsch than Hermite.

After the Franco-Prussian war of 1870, Jordan embodied the 'future' of French science from the perspective of several mathematicians in Italy, whether they had a proximity with his approach to substitution groups, such as with Cremona, or they considered Jordan in the continuity of the prestigious tradition of analysis at École polytechnique, such as with Menabrea. In any case, the *Traité des substitutions* was celebrated for its capacity to challenge issues very much valued at an international level, such as cubic surfaces and abelian functions, by opening the French tradition of research on substitutions to other approaches such as Clebsch's or Cremona's. In the 1870s, Jordan therefore participated in forging the new persona of a French mathematician very much open to the new algebraic approaches developed abroad, even though their very technical and specialized nature was often criticized at the time. This persona is still vivid in the retrospective testimonies of Klein and Dieudonné when they described Jordan as almost German or foreign to French mathematics.

But on the other hand, when Jordan came to occupy key positions of the Parisian mathematical scene in the 1880s, the interplay between Jordan and his Italian correspondents helped revive the persona of a French elite mathematician in continuity with the traditional role played by polytechnicians and academicians in the 19th century. Jordan's monographs were considered by several actors in Italy as a model to follow in the legacy of previous prestigious Professeurs d'analyse, such as Cauchy. We have seen, for example, that Janni considered Jordan's *Traité* to be as a synthesis of a French tradition of research on substitutions rather than as a groundbreaking new approach to group theory. We have also seen that students in Italy studied the *Cours d'analyse* for its formulation of traditional results rather than for the novelty of its organization.

[79]On the notion of persona and its uses in the history of science and mathematics, see (Mauss 1938); (Daston Sibum 2003); and (Brechenmacher 2015).

The persona of elite French mathematician that Jordan embodied was often considered in Italy as a natural ally for gaining recognition both on the international and the national scenes. Polytechnicians constituted the most important group of foreign contributors to the *Annali* from its creation to the early 1870s; in the late 1880s, they still constituted the most important group of foreign members of the Circolo. We have seen that Jordan's position was especially instrumental in Guccia's construction of the Circolo as both an Italian and an international mathematical society. But several episodes, such as the death of Halphen, or Guccia's disappointment with Humbert, also exemplify that the mathematicians of Italy were also able to use their contacts with Jordan to contrast the latter's persona of French mathematician with the other models of mathematical lives embodied by new actors emerging in the Parisian mathematical scene, such as Poincaré and Picard.

References

Bertrand, J. (1867). *Rapport sur les progrès les plus récents de l'analyse mathématique*. Ministère de l'instruction publique, Imprimerie impériale, Paris.

Billoux, C. (1985). La correspondance mathématique de C. Jordan dans les archives de l'ecole polytechnique. *Historia Mathematica*, 12:80–88.

Bordiga, G. (1882). La surface du sixième ordre avec six droites. *Comptes rendus hebdomadaires des séances de l'Académie des sciences*, 102:743–746.

Bordiga, G. (1886). Nouveaux groupes de surfaces de deux dimensions dans les espaces *n* dimensions. *Comptes rendus hebdomadaires des séances de l'Académie des sciences*, 102:1442–1446.

Boucard, J. (2011). Louis Poinsot et la théorie de l'ordre: un chaînon manquant entre Gauss et Galois? *Revue d'histoire des mathématiques*, 17(1):41–138.

Brechenmacher, F. (2007). La controverse de 1874 entre Camille Jordan et Leopold Kronecker. *Revue d'histoire des mathématiques*, 13:187–257.

Brechenmacher, F. (2009). Le 'journal de M. Liouville' sous la direction de Camille Jordan (1885–1922). *Bulletin de la Sabix*, 45:65–71.

Brechenmacher, F. (2011). Self-portraits with Évariste Galois (and the shadow of Camille Jordan). *Revue d'histoire des mathématiques*, 17(fasc. 2):271–369.

Brechenmacher, F. (2015). Récits de mathématiques: Galois et ses publics. *in Belles lettres, sciences et littérature*, volume 15, pages 135–161. Epistémocrique.

Brigaglia, A. (2014). The Annali di Matematica and the Rendiconti del Circolo Matematico di Palermo: two different steps in the dissemination and progress of mathematics in Italy. *In* Gerini, C., éditeur: *L'émergence de la presse mathématique en Europe au 19ème siècle*, pages 155–172. Milton Keynes: College Publications.

Brioschi, F. (1870). Sur la bissection des fonctions hyperelliptiques. *Comptes rendus hebdomadaires des séances de l'Académie des sciences*, 70:504–506.

Clebsch, A. (1870). Zur theorie der binären formen sechster ordnung und zur dreitheilung der hyperelliptischen functionen. *Mathematische Annalen*, 2:193–197.

Cremona, L. (1868). Memoire de géométrie pure sur les surfaces du troisième ordre. *Journal für die reine und angewandte Mathematik*, pages 1–133.

Cremona, L. (1870). Sulle 27 rette di una superficie del 3^0 ordine. *Rend. d. Ist. Lomb.*, (2^0 III).

Daston, L. et Sibum, O. H. (2003). Introduction. scientific personae and their histories. *Science in Context*, 16(1–2):1–8.

de la Vallée Poussin, C. (1921). Sur les fonctions à variation bornée et les questions qui s'y rattachent. *Comptes Rendus du Congrès International des mathématiciens*, 1:57–59.

Dieudonné, J. (1962). *Notes sur les travaux de Camille Jordan relatifs à l'algèbre linéaire et multilinéaire et la théorie des nombres*. in [Jordan Œuvres, 3, p. V–XX].
Galois, É. (1962). *Ecrits et mémoires mathématiques*. Gauthier-Villars, Paris.
Gergonne, J. (1810). Prospectus. *Annales de mathématiques pures et appliquées*, 1:I–IV.
Gispert, H. (1982). *Camille Jordan et les fondements de l'analyse*. Thèse de doctorat, Université Paris-Sud.
Gispert, H. (1983). Sur les fondements de l'analyse en France. *Archive for History of Exact Sciences*, 28:37–106.
Goldstein, C. (2011). Charles Hermite's Stroll through the Galois fields. *Revue d'histoire des mathématiques*, 17:135–152.
Goldstein, C. et Schappacher, N. (2007). A book in search of a discipline (1801–1860). *In* Goldstein, C., Schappacher, N. et Schwermer, J., éditeurs: *The Shaping of Arithmetics after C. F. Gauss's Disquisitiones Arithmeticae*, pages 2–65. Springer.
Guccia, G. (1881). Sur une classe de surfaces représentables points par points sur un plan. *Comptes rendus AFAS*, pages 191–200.
Hornsby, C., éditeur (2000). *The Impact of Italy: The Grand Tour and Beyond*. British School at Rome.
Israël, G., éditeur (2015). *Correspondence of Luigi Cremona (1830–1903) conserved in the Department of Mathematics, "Sapienza", University of Rome*. Brepols.
Janni, G. (1871). Esposizione della teorica delle sostituzioni. *Giornale di matematiche: ad uso degli studenti delle universita italiane*, 9:280–340.
Janni, G. (1872). Esposizione della teorica delle sostituzioni. *Giornale di matematiche: ad uso degli studenti delle universita italiane*, 10(98–206).
Janni, G. (1873). Esposizione della teorica delle sostituzione. parte terza. *Giornale di matematiche: ad uso degli studenti delle universita italiane*, 11(1–17; 71-86; 257–301).
Janni, G. (1874). Esposizione della teorica della risoluzione dell' equazioni di Galois. *Giornale di matematiche: ad uso degli studenti delle universita italiane*, 12:277–300.
Jordan, C. (1866a). Note sur les irrationnelles algébriques. *Comptes rendus hebdomadaires des séances de l'Académie des sciences*, 63:1063–1064.
Jordan, C. (1866b). Recherches sur les polyèdres. *Journal für die reine und angewandte Mathematik*, 66:22–85.
Jordan, C. (1866c). Résumé de recherches sur la symétrie des polyèdres non eulériens. *Journal für die reine und angewandte Mathematik*, 66:86–91.
Jordan, C. (1867). Mémoire sur la stabilité de l'équilibre des corps flottants. *Annali di Matematica Pura ed Applicata*, 1:170–221.
Jordan, C. (1868a). Mémoire sur les groupes des mouvements. *Annali di Matematica*, 2:167–215, 322–345.
Jordan, C. (1868b). Recherches sur les polyèdres (second mémoire). *Journal für die reine und angewandte Mathematik*, 68:297–349.
Jordan, C. (1869a). Sur la trisection des fonctions abéliennes et sur les vingt-sept droites des surfaces du troisième ordre. *Comptes rendus hebdomadaires des séances de l'Académie des sciences*, 68(865–869).
Jordan, C. (1869b). Sur les équations de la géométrie. *Comptes rendus hebdomadaires des séances de l'Académie des sciences*, 68:656–659.
Jordan, C. (1870a). Sur une nouvelle combinaison des vingt-sept droites d'une surface du troisième ordre. *Comptes rendus hebdomadaires des séances de l'Académie des sciences*, 70:326–328.
Jordan, C. (1870b). *Traité des substitutions et des équations algébriques*. Gauthier-Villars, Paris.
Jordan, C. (1875). Essai sur la géométrie à *n* dimensions. *Bulletin de la Société mathématique de France*, 3:103–174.
Jordan, C. (1881). *Notice sur les travaux de M. Camille Jordan à l'appui de sa candidature à l'Académie des sciences*. Gauthier-Villars, Paris.
Jordan, C. (1882). *Cours d'Analyse de l'Ecole Polytechnique*. Gauthier-Villars, Paris.

Jordan, C. (1885). Avertissement. *Journal de mathématiques pures et appliquées*, IV, 1:5.
Jordan, C. (1887). *Cours d'analyse de l'Ecole Polytechnique, volume 3*. Gauthier-Villars, Paris.
Jordan, C. (1888). Sur la marche du cavalier. *Rendiconti del circolo matematico del Palermo*, 2:59–68.
Jordan, C. (1892a). Adresse présentée, au nom du Cercle Mathématique de Palerme, à l'occasion du 70*ème* anniversaire de M. Ch. Hermite. *Rendiconti del circolo matematico del Palermo*, pages 161–164.
Jordan, C. (1892b). Pour le 70*ème* anniversaire de M. Ch. Hermite. *Rendiconti del circolo matematico del Palermo*, pages 249–256.
Jordan, C. (1961–1964). *Œuvres de Camille Jordan. Publiées sous la direction de M. Gaston Julia, par M. Jean Dieudonné*. Gauthier-Villars, Paris.
Klein, F. (1921). *Gesammelte mathematische Abhandlungen*. Springer, Berlin.
Lê, F. (2015a). Geometrical equations: Forgotten premises of Felix Klein's Erlanger Programm. *Historia Mathematica*.
Lê, F. (2015b). *Vingt-sept droites sur une surface cubique: rencontres entre groupes, équations et géométrie dans la deuxième moitié du XIXe siècle*. Thèse de doctorat, Université Paris 6.
Liouville, J. (1836). Avertissement. *Journal de mathématiques pures et appliquées*, 1:1–5.
Marie, M. (1873). Des residus relatifs aux asymptotes. *Comptes rendus hebdomadaires des séances de l'Académie des sciences*, 76:943–947.
Mauss, M. (1938). Une catégorie de l'esprit humain: la notion de personne, celle de "moi". *Journal of the Royal Anthropological Institute of Great Britain and Ireland*, pages 263–281.
Nabonnand, P. et Rollet, L. (2002). Une bibliographie mathématique idéale? Le répertoire des sciences mathématiques. *Gazette des mathématiciens*, 92: 11–26.
Peano, G. (1957). "Annotazioni" al trattato di calcolo de 1884. *InOpere scelte*, volume 1, pages 47–73. Cremonese, Roma.
Résal, H. (1875). Avertissement. *Journal de mathématiques pures et appliquées*, III, 1:5.
Romera-Lebret, P. (2014). La nouvelle géométrie du triangle à la fin du XIXe siècle: des revues mathématiques intermédiaires aux ouvrages d'enseignement. *Revue d'histoire des mathématiques*, 20(2):253–302.
Sarrau, É. (1867). Sur la propagation et la polarisation de la lumière dans les cristaux. *Journal de mathématiques pures et appliquées*, 12(1–46).
Sarrau, É. (1868). Sur la propagation et la polarisation de la lumière dans les cristaux. *Journal de mathématiques pures et appliquées*, 13:59–110.
Serret, J.-A. (1868). *Cours de calcul différentiel et intégral*. Gauthier-Villars.
Turner, L. E. (2011). *Cultivating Mathematics in an International Space: Roles of Gösta Mittag-Leffler in the Development and Internationalization of Mathematics in Sweden and Beyond, 1880–1920*. Thèse de doctorat, Aarhus University.
Verdier, N. (2009). *Le journal de Liouville et la presse de son temps: une entreprise d'édition et de circulation des mathématiques au XIXe siècle (1824–1885)*. Thèse de doctorat, Université Paris-Sud.
Wussing, H. (1984). *The Genesis of the Abstract Group Concept*. MIT Press, Cambridge: Mass.
Zerner, M. (1991). Le règne de Joseph Bertrand (1874–1900). *In* Gispert, H., éditeur: *La France Mathématique. La société Mathématique de France (1870–1914)*, pages 298–322. SMF, Paris.

Picard and the Italian Mathematicians: The History of Three *Prix Bordin*

Aldo Brigaglia

1 Introduction

It is usually said that in the transition period between 19th and 20th centuries, French scholars (mainly Picard and Humbert) as well as Italian scholars (mainly Castelnuovo, Enriques and Severi) were interested in the study of algebraic surfaces, though using different methods. The object of this chapter is to focus on the interrelations between these two groups. This question provides an interesting case–study of how different mathematical languages interact. The chapter is divided into two parts. The first one goes roughly from 1894, when the correspondence between Picard, Castelnuovo and Enriques began, to 1906 when Picard published the last volume of his treatise on algebraic functions in two variables. In this part, the main problem is the study of the irregularity of algebraic surfaces. Using different methods, Italian and French scholars had found invariants defining the irregularity. In a natural way, they looked for reciprocal links among those invariants, illustrating the difficulties met when translating one mathematical language into another. The second part relates to the discussions about the attribution of several prix Bordin by Paris Academy of Science after 1906. Enriques and Severi were awarded the first prize, Bagnera and De Franchis the second one, and Lefschetz the last one. We shall scrutinize how at that moment not only scholars played a leading role, but, above all, how the arguments involved were quite different. In this case, what was at stake was the completion of the classification of hyperelliptic surfaces and their generalization to abelian varieties. Let us emphasize that the two parts are closely connected. Lefschetz found the key to unify the two different (*French* and *Italian*)

A. Brigaglia (✉)
Dipartimento di Matematica, Università di Palermo, Via Archirafi 34, 90123 Palermo, Italy
e-mail: brig@dipmat.math.unipa.it

F. Brechenmacher et al. (eds.), *Images of Italian Mathematics in France*, Trends in the History of Science, DOI 10.1007/978-3-319-40082-2_5

mathematical languages and he expressed his discovery in a quite intriguing way as planting "the harpoon of algebraic topology in the whale body of algebraic geometry" (Lefschetz 1968, 854). Generally, historians of algebraic geometry tend to underrate the reciprocal connections between Picard, Castelnuovo and Enriques as well as the relations between Picard, Humbert and Scorza or between Scorza and Lefschetz.[1] On the contrary, we find large evidence about the very important links among the mathematical works of these scientists. We shall try to prove this fact in the sequel. Obviously, this chapter must be seen as a first approach of the problem and it would be necessary to go deeper in the examination of the specific, technical language used by the two groups to understand how these different mathematical languages, dealing with the same questions, could have great difficulties to properly understand each other.

2 Some Open Problems in the Theory of Algebraic Surfaces at the End of the 19th Century

On 14 May 1903 Enriques wrote to Castelnuovo: "I must communicate to you a seducing proposition from Picard. He wrote to me that he is completing his book about functions of two variables in which he is dealing only with his own works and more or less classical considerations about Abelian functions. However, at the end of the book he would like to add some complements and proposes that we write one or two notes (signed by us) «summarizing the current state of your research on two-variable algebraic functions…altogether would take up about forty pages…I think that this addition, which you have provided, will greatly enhance the value of my Treatise (Traité)…In your note or notes you could focus on the points in my text that your research completes when needed; and you could provide a bibliography of your most essential reports»".[2] (Bottazzini et al. 1996, 526–7).

Scientific relations between the two Italian mathematicians and Picard were clearly motivated by a common aim, which in many ways was central for the development of algebraic geometry. The problem was to extend Nöther methods, so successfully used to study curves, to surfaces, or, equivalently, to extend the study of complex algebraic functions in one variable to two variables. The study of

[1]For the relations between Picard and the Italian school, see for example (Dieudonné 1974, vol. 1) and, more recently, (Houzel 2002). In the preface to the edition of the correspondence between Castelnuovo and Enriques, U. Bottazzini, A. Conte and P. Gario provide the best dealing of such relations (Bottazzini et al. 1996); in this correspondence Picard is largely the most cited mathematician, apart Segre. One of the best survey on Lefschetz's work in algebraic geometry, the chapter by Phillip Griffths in (Griffiths et al. 1992) does not mention the influence of Scorza on the American scholar. However, a deep analysis of this influence is in (Hawkins 2013).

[2]Letter by Enriques to Castelnuovo. The letters of the two brothers in law are published in (Bottazzini et al. 1996). In the following, we shall quote many letters from this book, without repeating the reference. The same for the letters from Picard to Castelnuovo, which are published in the same volume.

algebraic curves had developed following Riemann considerations. Each curve, always considered in the complex field, may be viewed as a connected and orientable surface whose topological invariant (the *genus*) is defined as the number of his handles. The genus was a birational invariant. Riemann showed that this number is exactly the same as that appearing in Abel's theorem. It is the dimension of the vector space of Abelian first kind integrals (now called holomorphic differentials) of the form $\int_{z_0}^{z} R(x, y)$ where R is a rational function and y an algebraic function of x, without poles. Picard was really impressed by this result and wrote:

> it is really remarkable that two such different approaches, reduction of a sum of integrals and a question of geometry of position, lead to the same result (Picard 1894, p. 946).

Riemann also proved that the genus g may be used to find the greatest number r of linearly independent rational functions with given poles on the curve. He proves indeed that, if there are n poles, then $r \geq n - g + 1$, a theorem completed in 1864 by G. Roch, who interpreted the number $r - n + g - 1$ as the greatest number of holomorphic differentials (null in the poles) linearly independent (theorem of Riemann-Roch). To generalize the Riemann-Roch theorem for surfaces (and for higher dimensional varieties) will be one of the main tasks of algebraic geometry.

At the end of 1860s and in the 1870s, German scientists (mainly Clebsch, Nöther and Brill) found another important connection between different mathematical languages in the study of algebraic curves. They introduced an algebraic geometrical description instead of the analytic one. A typical result was, for instance, Clebsch's formula which interpreted genus of an irreducible algebraic curve of degree n from a geometric point of view, $g = \frac{1}{2}(n-1)(n-2) - k$ where k is the number of simple singularities (double points and cusps) of the curve. Moreover, it was quite easy to translate one mathematical languages into the other. The main point in Nöther's method was the consideration of linear series describing the sets of points cut on the curve by a linear family of hypersurfaces in a given projective hyperspace. There is indeed an intimate connection between the concept of rational maps in a projective space, and the existence of rational functions with given poles on the curve.

At the end of the 1870s, the formulation of the theory of algebraic curves in Brill and Nöther's algebraic geometrical language was accomplished. Corrado Segre, following the path indicated by his German colleagues, began to transpose this theory in the projective hyperspatial formulation that he had created with Veronese at the beginning of the 1880s. In a famous course of Higher Geometry held at University of Turin in the academic year 1890–1891,[3] and later, in a more complete way, in the paper (Segre 1894), he gave the final structure of curve theory in the "Italian" geometrical hyperspacial approach. As Segre wrote: "When, some years ago, I was dealing with research on ruled algebraic surfaces, I needed to use the properties of linear series studied in the Brill-Noether's paper, I realized how, by

[3]The "Quaderni" of lectures in higher geometry by Corrado Segre are available on the web site http://www.dm.unito.it/segre/quaderni.php (L. Giacardi, ed.). See also the introduction to the lectures held in 1890–1891 by A. Conte in the same web site.

representing these linear series by use of hyperspatial curves, one may recover (partly at least) these properties using simple geometrical means, avoiding thus algebraic calculations or functional considerations." (Segre 1894, Opere 199)

This kind of interrelations between several branches of mathematics for solving algebraic geometrical questions attracted a lot of studies in the last years of 19th century. The main task was to develop the methods that had been so successful with curves to surfaces, a task initiated by Clebsch and Nöther in a quite straightforward way by switching simple for double integrals of the first kind. It appeared however almost immediately that this was not at all an easy task. It was necessary to reformulate the whole theory. Many years later, Enriques expressed the fact in a famous quotation: "Fifty years ago began in Italy the study of these theories [of algebraic surfaces], hardly drawn by the genius of a precursor (Max Nöther); then, joking about the difficulties and the exceptions that emerged from everywhere, one used to say that, if algebraic curves (which formed already an harmonious theory) were created by God, the surfaces on the contrary are work of the Devil. Now, one says in fact that it pleased to God to create for the surfaces an order of more concealed order of harmony in which is hidden an amazing beauty" (Enriques 1949, 464).

To accomplish the difficult passage from curve to surfaces French and Italian mathematicians took different approaches: Picard (and later Poincaré) tried to generalize Riemann methods and developed what was then called transcendental way; the Italian geometers preferred to generalize Segre more geometrical approach. Therefore, Picard made the first important steps in this direction, examining the behaviour of algebraic integrals on a surface. Beginning in 1884, Picard studied in a very complete way the integrals defined on an algebraic surface: simple and double integrals which he divided (following Riemann steps) into different kinds.[4]

Some years later, Italian mathematicians such as Segre, Castelnuovo, Enriques, and later Severi, elaborated some specific geometric tools to study algebraic surfaces, extending the geometric methods which we can call of Nöther–Segre and which were so successful in studying curves, to algebraic surfaces. Among them were:

(1) the linear systems of curves (a direct generalization of linear series on a curve: the family of curves intersected on the surface by linear systems of hypersurfaces)
(2) continuous, nonlinear, systems of curves (which exist only on irregular surfaces); for example, on a ruled surface with non rational sections, a continuous system of curves is given by a system of straight lines.
(3) the characteristic series of a linear system (the linear series cut on a curve by the other curves of the system). This idea was introduced by Corrado Segre.

[4]For Picard's work and mathematical definitions of the terms used by the French scholar, see chapter 10 of (Houzel 2002). It is the better reference for Picard work in algebraic geometry.

Maybe the most original and effective idea introduced by the Italian school was the idea of virtual curve (now called divisor), that is formally a linear combination with integer coefficients of a finite number of curves on the surfaces. In this way we can speak of the sum and difference of curves.

The Italians had considered the maximum dimension of a continuous system of curves on S not linearly equivalent to each other. They proved it to be a birational invariant equal to the geometrical irregularity. However, at the end of the century, the relations between those various invariants and the nature itself of irregular surfaces were still very unclear. Picard and the Italians continued to follow their respective ways, but they hoped to clarify the mutual relations.

As already mentioned, perhaps the main difficulty arising from the passage to surfaces depended on the emergence of a new invariant, the irregularity, already discovered by Nöther and Cayley in the 1870s. Trying to extend the Clebsch formula to surfaces, in 1871 Cayley had found (Cayley 1871) a number, p_a, that Nöther proved to be an invariant, though it was sometimes different from the genus p_g of the surface, expressed as the dimension of the vector space of double integrals of the first kind on the surface. One has always $p_g \geq p_a$ and the difference $i = p_g - p_a$ is called the irregularity of the surface. When $i \neq 0$ the surface is irregular. Ruled surfaces with base curve of not zero genus are a well-known example of irregular surfaces. For a while, Nöther conjectured that these were the only irregular surfaces, but this statement is actually totally wrong. Other invariants were soon discovered. One of them was the dimension of the space of integrals of total differentials of the first kind,[5] which coincides in all the already known cases with the irregularity and may be called the analytical irregularity. Picard studied surfaces also from a topological point of view, proving the invariance of the first Betti number p_1 called the topological irregularity (Picard 1897).

3 Connections Between the Two Points of View. The Beginning: 1894–1897

In the 1890s and with these problems in sight, Picard, Castelnuovo and Enriques entered into a closer relationship. In the first months of 1894, Castelnuovo sent his work on the rationality of plane involutions (Castelnuovo 1894) to Picard. Picard's answer (7 July) was very kind. He wrote: "I find your results very remarkable, and I attach great importance to the demonstration of the properties of rational surfaces, which until now has been sought in vain." (Bottazzini et al. 1996, 659).

On that occasion, for the first time, Picard proposed Castelnuovo to work with him: "As I have a report to write on recent work in mathematics (for the *Revue Générale des Sciences*), this year I would like to talk a bit about work done in Italy, and specifically your work. In addition, I would like to ask if you could send me a short summary of work done in Italy in the past two or three years on the theory of

[5]See (Picard 1885).

algebraic surfaces (Castelnuovo, Enriques, etc. ….), because I do not want to leave out anything important." (Bottazzini et al. 1996, 659).

This was an exceptional moment for algebraic geometry in Italy. It was the beginning of the great geometrical construction of the theory of algebraic surfaces by Castelnuovo and Enriques, joined by Severi some years later. Let us besides note that, in January 1894, the two Italian mathematicians had only recently begun their work on algebraic surfaces. Castelnuovo wrote his first important paper on surfaces in 1891 (Castelnuovo 1891), and Enriques only in 1893 (Enriques 1893a, b) some months before the aforementioned exchange with Picard. They both were very young (Castelnuovo was 28 years old and Enriques, who did not hold yet any official academic position, only 22). Picard (and Nöther simultaneously) showed a really sharp mathematical insight in understanding almost immediately the scientific relevance of that work by the two young Italian scholars. In 1894, they were immediately recognized as two stars, at least in the restricted world of the mathematicians interested in extending the study of algebraic surfaces.

There was an obvious motivation for this early recognition. Castelnuovo and Enriques had immediately studied, and partially solved questions that gave the first answers to long awaited problems. Castelnuovo had constructed an irregular non-ruled surface, showing the wrongness of Nöther's conjecture and Enriques had introduced the main tools for the general study of surfaces in the "Italian style". In a sketchy way such questions can be described as the accurate study of the linear systems of curves on a surface, the definition of sums and differences of systems, the definition of adjoint system of a linear system, the definition of the Canonical systems, and that of *plurigenere*. Without entering further details, let us remember that the canonical system is the difference between a linear system and its adjoint. Its dimension is equal to the genus of the surface, while the new invariants, the k-plurigeneri (where k is a natural number), are the dimensions of k times the canonical system.

This constituted an impressive new language, the use of which provided numerous new results in an apparently easy way. Enriques identified the ultimate development of the geometry on a curve with such an interpretation using the hyperspatial methods through which Segre and Castelnuovo succeeded "in making the geometry of a curve independent from its representation on the plane and create a new theory of the geometry of a curve which merge simplicity and harmony of its basics with a power through which one may obtain important new results in this field." (Enriques 1893b, Memorie Scelte, 33).

Enriques exposed his ambitious program without hesitation: he wanted to bring the theory of algebraic surfaces to the same perfection as the theory of curves, using an extension of the same successful methods. He could therefore proudly state at the beginning of his paper:

> The geometry over a surface had not progressed in the same proportions as the geometry over a curve had, though one may assert that the latter has not yet entered the second phase of its development as the theory of linear systems of curves on a surface of any kind is not yet completed. The fundamental work still remains Nöther 1874–75 paper, in which the invariant functions belonging to a surface are studied in a deep way. One has then a work

by M.Picard where the surfaces admitting transformations into themselves are studied in particular, and two notes by M. Castelnuovo which contain important examples of particular classes of surfaces. On the contrary, the geometry of the plane [i.e. rational surfaces] entered its second phase with the remarkable work of M. Castelnuovo which contains original and important concepts that it seems possible to extend by applying them to the study of surfaces with positive genus (Enriques 1893b, Memorie Scelte, 33).

It was not only a declaration about future studies. As already mentioned, the paper contains many tools necessary to develop this program. It is remarkable that Picard, even if Enriques had quoted him only marginally, was perfectly able to understand the importance and the deepness of this new framework for the study of surfaces on which he put most of his energy in the previous 10 years. In December 1894, Picard wrote his paper for the *Revue* (Picard 1894) where the only contemporary works cited are those coming from the young Italian scholars:

These are primarily the continuous systems of curves on a surface, which have recently been the focus of a significant amount of work, particularly in Italy... The linear systems of curves are particularly interesting. This is the name for a system of curves drawn on F by the surfaces of a linear system ... Mr. Enriques noted... (Picard 1894, 949).

Picard was interested in the study of surfaces through their plane sections. He had proved (Picard 1887) that if the plane sections of a surface were rational the surface was a ruled rational or a Steiner surface. He was therefore impressed by the papers of Castelnuovo and Enriques which went much beyond his own results by proving that if the sections were of genus one, they are rational or ruled (Castelnuovo 1890) and the same result is true if the sections are hyperelliptic (Enriques 1893c, 1894a, b).

It is interesting to have a look at Enriques' paper as it is related to the first concrete occurence of a connection with Picard works. In fact, in his first paper, Enriques knew only vaguely Picard result and became aware of it only some months later, when he asked Castelnuovo "if there is something similar in the works by Halphen or Picard on the surfaces with rational sections" (Bottazzini et al. 1996, 62). Within a few days, Enriques got acquainted with Picard work and began to cite it.

Until this moment, both Italians had not yet well perceived the importance of Picard's research on the integrals of total differentials.[6] It was only in October that Enriques began to study this kind of questions:

This morning I was in the library and I read the notes to the Comptes-Rendus by Picard and Poincaré (1884–1886) on integrals of total differentials. Nothing new: however it is necessary to consult Picard's paper in the Journal of Liouville (84–86).[7] (Bottazzini et al. 1996, 137).

The interest of Italian geometers in the extension of Picard's results gradually increased. The "resumé" sent by Castelnuovo to Picard and the publication of his review[8] was an important step in this direction. Let us now present the chronicle of

[6]Indeed, Enriques never refers to them in his paper on surfaces of the same year.

[7]He referred to (Picard 1885) and (Picard 1886).

[8]We remark that this paper was published also in the *Rendiconti del CircoloMatematico di Palermo*, 9, 1895, pp. 159–166.

this progressive rapprochement as we can infer it from their correspondence. Enriques' aim was to translate Picard results into geometrical language. In November 1894, he wrote to Picard:

> I wrote today to Picard. ...I clearly submitted to him the problem of geometrical interpretation of his integrals of the first kind.[9]

One month later, he received Picard's answer and wrote to his colleague:

> I received from Picard a very kind answer with apologies for the delay. He told me that he thinks it interesting to find a geometrical interpretation of his integrals of the first kind: «I sent Mr. Castelnuovo a card to thank him for the very interesting analysis that he sent me; ...Both his work and yours completely captured my interest, and I was very glad to have a summary of it, which I found helpful in writing a review article in which I had the pleasure of citing your work....The statements that you sent me on integrals of the first kind are worthy of publication; some of them were not unfamiliar to me, but I have abandoned my research along these lines. It is important to distinguish between cases where the integrals are distinct and those with two or more that are functions of each other....».
>
> When I answer him I shall tell him about our new research on surfaces.[10]

On 20 January 1895, Enriques read Picard's resumé and wrote:

> I could have a look to Picard's paper. The last part of the paper dealing with us is very brief. We can be satisfied with it but it gives me the impression that he had not really read in depth the Italian works and hardly used your relation.[11]

It is clear enough that Picard understood very well the results obtained by the Italian mathematicians, which fit his own in a beautiful way. He had however great difficulty adopting their geometrical language thoroughly. We shall see later that this aspect characterized the tight but somewhat "strange" collaboration between Picard (and Humbert) on the one hand, and Enriques and Castelnuovo on the other hand. Both groups would make large use of the results of the other, though they would continue to speak different languages and were never really able to translate their methods from their own language into the other.[12] However, after this moment, scientific relations, along with reciprocal esteem and friendship, will never stop between them. A short time after the publication of his resumé, Picard wrote to Castelnuovo:

[9]Letter of Enriques to Castelnuovo, 3 November 1894 (Bottazzini et al. 1996, 144).

[10]Letter of Enriques to Castelnuovo, no dated but surely written between 3 and 11 of December 1894. (Bottazzini et al. 1996, 157–8).

[11]Letter of Enriques to Castelnuovo, 20 January1895 (Bottazzini et al. 1996, 166).

[12]Here I do not mean the difficulties to understand their respective kind of reasoning. Castelnuovo and Enriques were certainly able to follow a proof written in the "transcendental" language. The problem consisted in *translating* from one language into the other one. One of their aims was, in fact, to give a methodologically homogeneous treatment of the study of algebraic surfaces and not merely to give a set of results.

> I read your letter with great interest; I see that our work has much in common and I will benefit greatly from yours.…Surfaces that admit birational transformation with arbitrary parameters also continue to occupy me.[13]

The study of surfaces invariant through birational transformations was one of the main topic in the correspondence between Castelnuovo and Enriques during the next 6 months. As Picard had remarked, his work had many connections with the results of the young Italian scholars. In March, Picard returned to this kind of question:

> Regarding surfaces that admit a group of birational transformations…without changing anything about my demonstration method, I have discovered a theorem which appears to have much in common with some of the results that you told me about.[14]

Picard published a note on this theme in the *Comptes-Rendus* (Picard 1895a) immediately after this letter and the corresponding paper in the *Rendiconti di Palermo* in June (Picard 1895b). In the last lines of the latter, Picard observed that his last theorem was incomplete and wrote:

> I have every reason to believe that the recent work from Mr. Castelnuovo and Mr. Enriques on the theory of surfaces will yield more precise details on the nature of the surface[15] (Picard 1895b, 255).

Castelnuovo read this paper in July and immediately wrote to Enriques:

> I have also received Picard's note and I have interpreted the last words as a kind invitation for us to work on the problem. It is therefore indicated that we should immediately begin to write our work.[16]

The paper was ready in a few days and, ten days after the aforementioned letter, Picard received it and decided to include it in the *ComptesRendus*:

> I am delighted to be presenting your interesting research to the Academy;…For the surfaces that you are further studying, it follows naturally from the theorem…from my paper in the *Rendiconti* that the coordinates are expressed by doubly periodic functions…of two parameters. I considered conducting a more extensive study using this as a starting point, but as you wrote to tell me that you were studying this subject, I didn't want to spend any more time on it, and I simply wrote up a more complete account of my previous research for the *Rendiconti* of Palermo.[17]

Enriques and Castelnuovo note was immediately published on 29 July. One may observe that this was the very first paper with both Castelnuovo and Enriques as authors even if their correspondence shows that almost every article written by the two scholars since 1893 was the result of a tight collaboration. In the next 2 years, Castelnuovo became mainly engaged in his research on what is now called the

---

[13]Letter of Picard to Castelnuovo, 21 January 2 1895 (Bottazzini et al. 1996, 659–660).

[14]Letter of Picard to Castelnuovo, 22 March 22 1895 (Bottazzini et al. 1996, 660).

[15]For the contributions of Picard on this kind of surfaces see (Houzel 2002, 217–218).

[16]Letter of Castelnuovo to Enriques, 14 July 1895 (Bottazzini et al. 1996, 200-1).

[17]Letter of Picard to Castelnuovo, 24 July 1895 (Bottazzini et al. 1996, 662).

Castelnuovo rationality criterion,[18] while Enriques wrote a new, deeply modified, version of his *Ricerche* of 1893, (Enriques 1896) significantly published in the same issue of the same journal as Castelnuovo's paper. On a request by Max Nöther, they also wrote a complete *Bericht* (Castelnuovo and Enriques 1896)[19] of their work on algebraic surfaces which may be seen as the conclusion of the first period of their work, defining tools and mathematical language necessary to complete what Enriques had called the "2° momento" ("second period") of the theory of surfaces.

It is curious to compare Enriques' *Ricerche* with this *Bericht*. From the very introduction, there are sharp differences between the two texts. The Italians now had a clearer vision of the situation and had assimilated Picard work, exploiting a relevant part of it. The introduction of the article clearly presents the status of the research before new and profound developments were obtained in both the geometric and the transcendent directions.

"The Geometry of general algebraic surfaces…has been enriched by new results recently, thanks to the work of French and Italian geometers. In France we have considered primarily transcendental functions associated with algebraic surface equations. Mr. Noether had already introduced to the study of surfaces certain double integrals (with algebraic differentials) that are always finite and play the same role as Abelian integrals of the first kind relative to algebraic curves. A few years later, Mr. Picard provided additional momentum to this theory by considering,…, certain simple integrals of total differentials that are also worth noting. In particular, he studied simple integrals that are always finite. The most well-known surfaces do not truly possess these integrals; but Mr. Picard identified some classes of surfaces with notable properties that depend precisely on the presence of such transcendental functions.…But the field of research that Mr. Picard began is far from exhausted; because even the question of determining,…whether the surface possesses such simple integrals has not yet been resolved, in a general manner.

The Italian geometers, however, have dedicated themselves to the study of linear systems of algebraic curves that belong to an algebraic surface or, …to the study of rational functions of three quantities x, y, z, linked by an algebraic relationship f(x, y, z) = 0. This type of research came to their attention naturally; because two auxiliary theories that serve as the basis for new developments have been cultivated in Italy in recent years; namely, the geometry of an algebraic curve, and the theory of linear systems of plane curves." (Castelnuovo and Enriques 1896, 355–6).

Castelnuovo and Enriques had now a quite complete understanding of the state of art in the study of algebraic surfaces. They knew there was still a long way to go for the full understanding of the problems and concluded their introduction in the typical poetical style of their works:

---

[18]See (Castelnuovo 1896). For more details on the elaboration of Castelnuovo criterion, see (Conte 1994). Here it is possible to get the flavour of this intellectual adventure, which has the pathos of a novel.

[19]This paper had been elaborated for a very long time—from 1894 when it was requested by Nöther until March, 1896. During this time the two mathematicians hardly worked on solving the question.

> The concept that inspires this reasoning may be simple...but developing it requires envisioning so many particular cases and overcoming so many objections that the reader tires of the piecemeal demonstration and loses sight of the final results. This is a common fault of theories that are still being developed; and very often progress consists in introducing broader concepts, more appropriate expressions, which allow a large number of seemingly distinct cases to be understood at a single glance. The time has not yet come to introduce such refinements to the theory of surfaces. At the moment, what is important is to overcome the obstacles met along the way, to arrive at the shining summit before us....
>
> This field of research is far from being exhausted, and the area that remains to be explored is vast and may lead to brilliant discoveries. Even so, it is important to remember that such a large horizon cannot be seen from a single view point....It is only by the precise application of all the resources that geometry and analysis put at our disposal today that we can hope to develop the theory of surfaces and enrich it with new discoveries (Castelnuovo and Enriques 1896, 358).

Youthful enthusiasm was mixed with mature consciousness. Both were convinced to be the successors of a bright tradition and to be in position to accomplish a great task. They therefore could promise to lead their readers to discover "with interest the description of the landscape that can be seen from how far we have already come." (Castelnuovo and Enriques 1896, 358).

The next 3 or 4 years were probably the most fruitful on the scientific ground, especially for Enriques. Picard had moreover slightly slowed down his work on algebraic surfaces, absorbed by the completion of his influential *Traité d' Analyse*.[20] During that time Enriques read and studied eagerly everything published by French geometers, particularly Picard, but also Painlevé, Humbert or Poincaré:

> I had a look to Poincaré long paper on Analysis Situs. Reading this paper seems to me very difficult.[21]

4 The Crucial Years: 1897–1904

In 1897, Picard completed his treatise on algebraic surfaces, (Picard and Simart 1897) and started a new research program on this topic, which was more and more intertwined with the studies of his Italians colleagues. We will not discuss the treatise here, but let us however emphasize several references to the works of the Italians and expressions of high esteem for them it contains. For instance, in Chap. 7, a whole paragraph (Digression sur les systèmes linéaires de courbes tracées sur une surface) is inspired by Enriques' papers: "The essential points from

[20](Picard 1891 – 1896) Enriques read attentively this book and was enthusiastic of it: «Sto leggendo i fascicoli usciti del 3° vol del "Traité d'Analyse" di Pic. Ci sono delle cose interessantissime! In particolare sono entusiasta della teoria delle equazioni algebrico—diff del 1° ord, e della sua integrazione pel caso che l'integrale non abbia punti critici mobili: questo problema si riduce a quello delle trasf. biraz d'una curva alg in sé.» (Letter of Enriques to Castelnuovo, 21 January 1896, Bottazzini et al. 1996, 241–2).

[21]Letter of Enriques to Castelnuovo, 30 November 1896 (Bottazzini et al. 1996, 294).

this section were borrowed from two notable reports by Mr. Enriques, which we will come back to later" (Picard and Simart 1897, 197).

Picard refers here to Enriques' *Ricerche* and *Introduzione*. Picard's volume is entirely devoted to the study of regular surfaces. Only in the end does he mention the arithmetic genus and irregularity. It is in this context that he returns to Enriques' work:

"Mr. Zeuthen demonstration, like that of Mr. Noether, …is very long. We are not following the same path. In his recent work, Mr. Enriques studied from a new point of view, extremely fruitful, the arithmetic genus of a surface; this elegant research will be described in Volume II of this Work." (Picard and Simart 1897, 242). As may be seen, the passion for simple reasoning instead of tedious calculations had conquered the great analyst. Already in the introduction Picard highly prizes the importance of a full understanding of geometric language:

> In volume II…we also hope to describe the main results obtained recently by Mr. Castelnuovo and Mr. Enriques, results which have renewed an entire part of the Theory of Surfaces (Picard and Simart 1897, VI).

Before the publication of the volume, in March 1897, Picard had written enthusiastically to Castelnuovo about his paper in the *Annalen*:

> Your remarkable summary of Math Annalen is extremely useful to me, and I can now admire your body of research, which is quite splendid, with a full understanding. The two of you have produced an enormous body of work (Bottazzini et al. 1996, 662–3).

But it was not only to congratulate him that Picard wrote to Castelnuovo. He also wanted to begin a real collaboration:

> You know that I greatly admire the transcendental point of view, being above all an analyst. I would like to be able to define the arithmetic genus p_n from this point of view; do you have any results along these lines? I would also like to hear your opinion on the two invariants that I discussed earlier in a Note in the Reports (*ComptesRendus*)…; do you think it would be worth pursuing the study of these invariants?[22]

Enriques' comments followed almost immediately:

> I am very glad with what you wrote to me about Picard. It would be a good thing if Picard could tell us the analytical meaning of p_n, which should be easier for him than for us.[23]

Both Enriques and Picard clearly saw the significance of translating one language (topological, analytical, geometrical) into another but also considered that each mathematician had his own speciality. Picard was an analytical "native speaker" as they naturally considered themselves as geometrical ones. Enriques wrote some time later:

[22]Letter of Picard to Castelnuovo, 23 March 1897 (Bottazzini et al. 1996, 663).

[23]Letter of Enriques to Castelnuovo, 27 March 1897 (Bottazzini et al. 1996, 326).

I saw Picard's note in the CR. It seems to me that the result is already present in Poincaré's Analysis Situs. Anyway it is an interesting thing. We shall see what Picard could give us by following the same path. Probably $p_g$is the other order of connexion of the manifold. But the p_n?[24]

And again some days later:

I am not surprised that Picard stopped, as understanding the correspondence between the other orders of connexion of a Riemannian manifold and Noether's double integrals should be much more difficult.[25]

In his paper Picard proved that the number of linearly independent total differential of the second kind is equal to the first Betti number.[26] One can be surprised to read in this very short paper another enthusiastic comment on the works of the Italian scholars:

From the point of view of Algebra and Analytical geometry, in the past few years the theory of algebraic surfaces has been the subject of considerable work on the part of two extremely distinguished young Italian geometers, Mr. Castelnuovo and Mr. Enriques (Picard 1897a, 533).

In the following years, Picard efforts were mainly devoted to complete his Traité. At the same time, Castelnuovo and Enriques encountered increasing difficulties to have their language and methods fully understood by the mathematical community, and they tried new ways of diffusion. A proposal arrived from MittagLeffler:

[Mittag-Leffler] told me that in France he has heard many positive opinions about our work, but unfortunately our language, Italian and geometrical, make the situation quite difficult; and our Bericht, even if it is written in French is nevertheless also too much geometrical for him. However, Mittag-Leffler would like to publish in Acta Mathematica a survey of the recent progresses of the theory of surfaces where one could find the results presented in a purely analytical form (Bottazzini et al. 1996, 408).

Another opportunity came from Teubner who proposed to the pair to write a volume on algebraic surfaces. MittagLeffler warmly supported this idea:

I spoke with MittagLeffler about Teubner's offer, and he told me that the publication of a book would be extremely useful; and that Picard also finds great difficulty in presenting our theorems in the way he would like to.[27]

Finally, in 1900, Picard and Simart published the first part of the second volume of their treatise.[28] Enriques received it in March 1900. He immediately wrote to his brother in law:

[24]Letter of Enriques to Castelnuovo, without date, but written after 27 March and before 9 April 1897 (Bottazzini et al. 1996, 327).

[25]Letter of Enriques to Castelnuovo, without date, but written after 9 April and before 22 April 1897 (Bottazzini et al. 1996, 328).

[26]Picard denotes p_1 the first number of Betti plus one.

[27]Letter of Enriques to Castelnuovo, 14 April 1899 (Bottazzini et al. 1996, 409).

[28]This second tome has been published in three issues, 1900, 1904 and 1906.

> I see that he deeply entered into our considerations which are also valid for the double integrals of second kind but then a conclusion is missing which would connect the determination of these integrals, or better their number, to the known invariant characters of the surface. Therefore I do not see effective progress in the sense that we would like. Anyway, the issue will certainly provide a great advertisement for us.[29]

The volume contains a great deal of material taken from the main results of the Italian geometers, and there are evident traces of Picard's efforts to grasp in depth their results and methods. Enriques was without any doubt right. Picard had gone inside their works and the volume was a great advertisement for their methods. Let us scrutinize some of the points where the French mathematician compares his methods with the Italian ones. The second chapter (*La Géométrie sur une courbe algébrique*) is entirely based[30] on recent papers by Segre and Bertini (Segre 1894; Bertini 1894). That was not at all surprising: in the case of the geometry on a curve, the comparison between transcendental and geometrical methods had already been realized in a satisfying way.

But when Picard dealt with surfaces, the situation was quite different. Picard made efforts to provide a convincing picture of the similarities and differences between the two languages. Almost the whole Chap. 6 (*Système adjoint à un système linéaire de courbes*) is devoted to this task. Picard exposed in it a major problem of the theory of algebraic surfaces, namely the study of the irregularity of surfaces and its characterization through other known invariants. In a famous paper of 1896,[31] Castelnuovo had proposed an open problem:

"which relations do exist between the difference $p_g - p_n$ of the two genera and the number of simple integrals of the first kind which are mutually linearly independent?" (Castelnuovo 1896) and Picard tried to approach it with his transcendental methods in Chap. 6. He hoped to provide an analytical proof of a theorem of Enriques (Enriques 1896), who had introduced sub-adjoint curves and their defect, proving their sum to be an invariant equal to the arithmetic genus. Picard wrote that «he based his work on "considerations of a transcendental nature", and on "Mr. Enriques' basic theorem." (Picard and Simart 1906, 129). However, the French mathematician was not completely satisfied. He wanted to be sure his results, were the same as the Italians' ones, which seemed to be more general:

> Mr. Enriques followed a completely different path than that which we used in the previous section to define adjoint curves. This purely algebraic and geometric method, which can be applied in all cases…is based on the preliminary study of sub-adjoint curves….We will describe the main points of this method, primarily focusing on the case where the surface has a null geometric genus. We will then show that, regardless of the procedure used,

[29] Letter of Enriques to Castelnuovo, March 15, 1900 (Bottazzini et al. 1996, 451).

[30] Picard writes (p. 21): «Nous avons utilisé aussi dans notre rédaction une Mémoire de M. Segre … et une Mémoire de M. Bertini.» Picard is always very precise in his quotation and never fails to recognize the work of other scholars.

[31] Castelnuovo and Enriques collaborated strictly—their paper appeared in the same journal and complemented each other. The problems linked with the definition and the calculation of irregularity were one of the most important topic in algebraic geometry till recent times. For a throughout discussion of it see (Bardelli 1994).

> whether algebraic or transcendental, we arrive at the same canonical system, and *we will therefore have established a link between these two different methods of approaching the study of algebraic surfaces* (Picard and Simart 1906, 129–130; Picard's emphasis).

Another aspect taken by Picard from the works of the Italians was the notion of bigenus ("a very interesting idea, which Mr. Enriques is the first to have drawn attention to" (p. 146)) and he stressed the use of Castelnuovo's rationality criterion: "the significance of the bigenus P emerges clearly from an extremely remarkable theorem by Mr. Casteluovo...For a surface to be unicursal, it is necessary and sufficient that $p_n = 0$, $P = 0$... It is very interesting that, for a surface to be unicursal, it is sufficient that two invariant numbers relative to this surface be equal to zero" (Picard and Simart 1906, 148).

To finish with the many references to the work of the young Italians in the *Traité,* I want to cite the use of Picard of a result by Castelnuovo on the system of curves traced on a plane by a complete system of algebraic surfaces ("This research was made possible due to Mr. Castelnuovo's general theorem of linear systems of surfaces, p. 180").

The French and Italian research of these years met also on a topic that had been treated some years earlier: the surfaces invariant for a group of birational transformations. On 5 November 1900, Picard presented to the Academy a paper by Castelnuovo and Enriques, (1900) where, under the form of a corollary, they proved that the surfaces, admitting a continuous system of birational transformations which does not form a group, are necessarily rational or are equivalent to cylinders. This result extended Picard research published in 1889 and struck him (Picard 1889). He wrote to Castelnuovo on 5 November 1900:

> You have my warmest compliments on your results, which I found extremely striking. Very specifically, the results regarding surfaces that permit a continuous series of transformations without permitting groups of transformations is truly remarkable; I have often attempted to demonstrate this, but in vain. You were cleverer than I (Bottazzini et al. 1996, 663).

Another hint of the growing interest of Picard in the results of the Italian scholars is expressed in another letter from Enriques to Castelnuovo a few months later (31 January 1901):

> Picard took the opportunity to thank us for the note we had sent to him. "I read with the outmost interest what you have just published with Castelnuovo about the double plane. How your study on the bigenus prove to be fruitful! It is amazing (Bottazzini et al. 1996, 470).

In March 1902, Enriques was in Paris and met Picard and other leading French mathematicians, in particular Humbert, Appell and Poincaré ("I shall tell you now that in Paris I had an interesting discussion with Poincaré on the cubic variety of S_4").[32]

During his stay in Paris, Enriques' relationship with Picard seems to have evolved from purely scientific to friendly:

---

[32]Letter of Enriques to Castelnuovo of 11 March 1902 (Bottazzini et al. 1996, 492).

I spent yesterday with Picard an unforgettable day.[33]

We believe that the story we have narrated reasonably explains Picard's proposal to Enriques and Castelnuovo in 1903 to write a note to be added to his Traité. In the previous 10 years or so, the scientific agenda of the French and Italian scholars became quite close and they made a large use of their mutual results. However, many relations between invariants found through transcendental methods and those found by geometrical ones were still often unclear. Language and methods remained very distant, and Picard's efforts to describe the geometrical methods by analytical tools were rather unsuccessful. As a proof, the second part of the second tome is entirely written in the transcendental language. Preparing the third and last part, Picard was forced to obtain the help of his Italian colleagues and their contribution was the only part of the *Traité* not written by Picard or Simart.

5 1904–1906: A New Entry, Francesco Severi

From 1904 to 1906 a strict interrelation occurred between the results of Italian geometers and Picard's research project. In that period, there was a major innovation: the young Francesco Severi joined Enriques and Castelnuovo in order to further develop the theory of algebraic surfaces. Severi gave an important contribution to the study of some relevant problems on which Picard had hardly worked, namely to the links between the number of the linearly independent Picard integrals and algebraic surfaces and their irregularities. We shall illustrate some of the results deduced by Italian and French mathematicians in a very short time.

September 1904: Severi published a paper (Severi 1904) in which he proved that any algebraic surface admittingPicard integrals of second kind is irregular.

14 November 1904: Severi completed (Severi 1905a) a more detailed paper containing the same results as the first one. It appeared in the second half of the next year.

December 1904: Enriques proved (Enriques 1905a) the inverse of Severi theorem. He showed that on any irregular surface a non linear continuous family of curves having dimension equal to the irregularity exists. If we use some previous result as well as some theorems of Humbert and Severi, such a theorem means that Picard integrals of the first kind on any algebraic surface exist if and only if that surface is irregular. In a few days, at the end of December, the problem of the links between Picard integrals and irregularity seems completely solved. Before publishing their papers, Enriques, Castelnuovo and Picard exchanged their respective results in several letters, and Picard elaborated a strategy in order to publish all their works in the *ComptesRendus*.

About December 20th Enriques communicated his theorem to Picard, who proposed him to publish a paper in *ComptesRendus*:

[33]Letter of Enriques to Castelnuovo of 7 March 1902 (Bottazzini et al. 1996, 490).

When I answered to Pic. I communicated my theorem. Moreover Picard replied in a very polite way and asked me to write a note for the CR and to give him some explanation (particularly if there is a relation between $p_g - p_a$ and the number of integrals of the first kind).… I received his letter yesterday and I will answer him as soon as possible by promising a note for the CR for which I take some time.[34]

Before the end of 1904, Castelnuovo proved the main equalities: $q = i$; $r = 2i$ (where r is the dimension of the linear space of the integrals of second kind and therefore it is equal to the first number of Betti, q is the number of the integrals of the first kind and i is the irregularity). Immediately afterwards, he sent his results to Picard.

4 January 1905: Picard wrote back to Castelnuovo: "Having been laid low by the flu, which has put my brain in a very poor state, I have not yet been able to respond to your very interesting letter. The beautiful result that you told me about has been suspected for quite a while, as all of the known examples satisfy this statement. For some time I searched for a demonstration using my usual methods, but I was stymied by a circumstance related to my linear differential equations that seemed improbable to me. Sure of the result now, thanks to you, I will take this question up again when I have the time. As for your demonstration, it belongs to a class of ideas that I would never have dreamed of." (Bottazzini et al. 1996, 666). Castelnuovo results inspired Picard with new ideas. He immediately communicated them to Castelnuovo.

5 January: Picard wrote: "I have reflected a bit since yesterday on the question that occupies us. While still remaining true to my point of view, I have managed to demonstrate with one stroke of the pen that, if a surface is regular, it does not have integrals of the second kind (Severi theorem). I also establish, in two lines, the relationship $r = q + i$". Picard explained "the question that occupies us", as a member of a group involved in the same research project, and in fact he actually felt himself as a member of that same "circle".

16 January: Picard published his paper (Picard 1905a). He stressed that he obtained the same result as Severi, "from a completely different point of view." (Picard 1905a, 119).

16 January: In the meantime, Enriques completed the paper (Enriques 1905b) he published in the same issue of the *ComptesRendus*. In this paper, Enriques asked some questions that appear as an actual research program: "Can this be taken further? Can a numeric relationship be found between the difference $p_g - p_a$ and the number of integrals of the first kind belonging to a surface?" (Enriques 1905b, 135).

The problem was very intriguing for Picard, and Enriques exposed its status by referring to the research of the Italians:

"Mr. Severi…has told me about the following result: $p_g - p_a$ equals the difference between the number of Picard's integrals of the first kind and of the second kind belonging to the surface. Mr. Castelnuovo thinks that the relationship between $p_g - p_a$ and the number of integrals of the first kind could be established by studying non-linear systems of curves on a

[34] Letter of Enriques to Castelnuovo without date, but dated by the editors between 20 and 24 December (Bottazzini et al. 1996, 568).

surface." (Enriques 1905b, 135) By reading Enriques paper, Picard found out that Severi got the same results as he had already deduced a few days before. When he received Castelnuovo paper, he wrote back to him.

21 January: Picard wrote to Castelnuovo: "You told me (and Mr. Enriques confirms in his note) that Mr. Severi sent you this formula…I very much admire your demonstration of the formula q = i. Personally, I hope one day to demonstrate, using my own methods, the relationship r = 2i, which amounts to the same thing; at the moment, I can glimpse the next step, but it will require some more serious thought….If Mr. Severi would like to present his demonstration of the formula in the *Comptes-Rendus*, I would be happy to include it." (Bottazzini et al. 1996, 667).

23 January: Castelnuovo published his paper (Castelnuovo 1905a). The problem of the irregularity and Picard integrals is completely solved. It is a little bit surprising that he introduced in this paper (which "belongs to a class of ideas that I [Picard] would never have dreamed of") ani-dimensional variety, which parameterizes the continuous non linear systems of curves on an irregular surface he called (and that is still now called) Picard variety. This variety admitted a transitive and abelian group G_i of birational transformations in itself.[35] So Castelnuovo could state: "Considering the extensive research that Mr. Picard has conducted into surfaces that admit a group of birational transformations, I propose that the variety (and group) be called the Picard variety (or group) of the surface" (Castelnuovo 1905a, 221).

6 February: Severi proved the results already stated by Enriques (Severi 1905b).

3 April: Picard published a paper (Picard 1905b) in which he proved Castelnuovo theorem in a different way, and wrote: "In the current issue of the *ComptesRendus*, … Mr. Severi delivers a new demonstration of the same theorem [by Castelnuovo]; these two theorems are based on the properties of linear systems of curves on algebraic surfaces. I would like,…, to describe in a few lines the principles of another demonstration of the same theorem that I have had for some time, and which employs the same type of ideas that I have happily used in my research on the theory of two variable algebraic functions." (Picard 1905b, 915). Picard tried to claim the priority of his studies. Even if he was extremely fair with the Italian scholars, his version is not exact. As he said only a few months before, though he had some ideas about the proof of Castelnuovo theorem by transcendental methods, he had not yet found out its complete proof. In the same issue of the *ComptesRendus*, Severi published his results (Severi 1905c).

16 July: In the *Annalen* Severi published a large account of his results of September containing a short reference to the last results of the group (Severi 1905d). In less than 1 year, the understanding of links between Picard integrals and irregularity developed dramatically![36]

[35]This group, called the Picard group, must not be confused with the Picard group in number theory consisting in the subgroup of the Möbius group made by transformations with coefficients Gaussian integers.

[36]It is necessary to note that the proof of the theorem of Enriques of 1904 was, many years later, considered erroneous. Luckily, in the meantime, Poincaré had found a new demonstration using transcendental methods. For the entire story I refer to the quoted paper by Bardelli.

It is evident that the three scholars advanced maintaining strict relationships. It could be interesting to have a close look at their discussion the importance of which is also shown by the huge number of letters exchanged between Castelnuovo and Enriques (32 letters from Enriques in the period from 13 December 1904 to 30 January 1905) and of papers about this subject published in the *ComptesRendus* (4 by Picard, 1 by Castelnuovo, 2 by Enriques, 2 by Severi. Overall, we count 9 papers in some months).

The period of intense collaboration between Picard and the Italian geometers ended in 1906, when Castelnuovo and Enriques published a joint paper in the journal directed by Picard (Castelnuovo and Enriques 1906)—which is the last joint paper of the two brothers in law; perhaps it may be useful to recall that also their first joint paper had been proposed by Picard. Another, even more important, final episode of this collaboration is the publication of the third and last issue of Picard-Simart in 1906. As already mentioned, Picard asked Castelnuovo and Enriques for a contribution, which was then published as an Appendix to the volume—the only Appendix by scholars different from the authors.

In such a volume, Picard devoted Chap. 13 (*Sur le nombre des intégrales de différentielles totales de première et seconde espèced'une surface*), and particularly the third paragraph, to problems studied in 1905 and gave an appreciative account of the results of the Italian mathematicians. When he tried to give a new version of their results by using transcendental methods, he realized the incompleteness of their results. So he now was not so sure that his proof of Severi's theorem ($i = r - s$) was right, though on January 5th, 1905 he had communicated to Castelnuovo he was able to "demonstrate with one stroke of the pen"; he, indeed, wrote: "But as this demonstration must be completed, we will not stop (p. 432)." Now Picard was not longer sure of what he had written in his paper of April 3rd (namely that he got the proof of Castelnuovo theorem). He remarked that for his proof he had to use a relation that "we have not succeeded in demonstrating directly (p. 434)". He frankly admitted in a footnote: "Mr. Picard thought he had obtained such a demonstration by carefully applying ideas borrowed from the Analysis Situs of four-dimensional space, but we now believe that this demonstration may be somewhat in doubt, and therefore simply call attention to the gap which remains to be filled". (Picard and Simart 1906, 434).

The last chapter of the *Traité* is devoted to the hyperelliptic surfaces. That is the subject of a new interaction between the two different schools (even if indirect).

In his introduction, dated 15 January, 1906, Picard wrote: "The theory of algebraic functions of two variable, the broad strokes of which are known today, have recently, and particularly in Italy, been the focus of significant research. The functional point of view and the geometric point of view come together at multiple points in the theory; several examples can be found in the elegant work by Mr. Humbert, Mr. Castelnuovo and Mr. Enriques, and Mr. Severi." (Picard and Simart 1906, V–VI) Castelnuovo and Enriques wrote the Appendix 5 consisting of a beautiful account of methods and results of the Italian school, that does not contain new results. We stress that the inextricable connections between the two different languages emerges also in this section. I mention only one example, without going into technicalities: the proof that the defect is equal to irregularity for every linear system is still obtained only using transcendental means only ("We have not succeeded in demonstrating this theorem using geometric methods") (Picard and Simart 1906, 489). We conclude this short account with a reference to two recent papers of Enriques and Castelnuovo in which they proved that every surface with arithmetic genus strictly smaller than—1 can be transformed birationally in a ruled surface (Enriques 1905c; Castelnuovo 1905b). Essentially in the same way as before, they pointed out: "These conditions can only be expressed in a transcendental manner. Essentially…it involves recognizing that the surface has $p > 1$ simple integrals of the first kind but no double integrals of the first kind " (Picard and Simart 1906, 517).

In the introduction Picard remarked: "For subjects that are so new, it is not surprising to encounter many questions that have only just begun to be answered; among these I will point out…the question of the integrals of total differentials of the third kind, where I have introduced an integer ρ, which, having recently been considered by Mr. Severi from a new point of view, will require a more complete study." (Picard and Simart 1906, V).

Picard was again a good prophet. On this subject, he will indirectly interact again with the Italian school.

6 A New Field of Mutual Interactions: The Two Prix Bordin and Hyperelliptic Surfaces

It was Picard (or Humbert) who suggested to the Académiedes Sciences the subject of the Prix Bordin for the year 1907: "Recognizing generally if the coordinates of the points of an algebraic surface can be expressed as Abelian functions of two parameters, such that to every point on the surface is associated more than one system of values for the parameters (up to the periods). Specific study of the case in which the equation of the surface would be $z^2 = f(x, y)$, f being a polynomial, and giving explicit examples of such surfaces".[37] This subject is directly connected to the problems left open in the *Traité*, and Picard was aware that only some of the Italian geometers could answer this question.

Only the work by Enriques and Severi was communicated to the Academy. The history of their work is tortuous, and may partially be reconstructed[38] through both the huge correspondences Enriques–Castelnuovo and De Franchis–Guccia.

The deadline for delivering the work was the end of December, 1906. Enriques and Severi worked hard during all 1906; however, 1 month before the deadline they were still in a mess:

[37] *Comptes Rendus de l'Académie des Sciences*, séance du 2 décembre, 1907, pp. 981–983.

[38] On this history, see for instance (Ciliberto and Sernesi 1991).

> Several times I mentioned my research on surfaces in collaboration with Severi. Such research has gone through a hard time. The proof that I had developed that "the invariants (of genus r) are associated to finite transformation groups of the surface" is not valid, and neither is the demonstration Severi gave of the theorem that "each transformation changes in itself a system ∞^2 of curves of genus 2". But now we hope we have redone those demonstrations…. The proof that I did is, in truth, a bit daring, so that I hesitated to use it without first excluding a certain case; but Severi drove me to do it… I do state you briefly… What do you think about it? And particularly the point concerning the Analysis situs is delicate.[39]

In the meantime, De Franchis and Bagnera were also working on the same subject. Michele De Franchis was an algebraic geometer from Palermo, who was professor at the University of Parma and in 1909 moved to the University of Catania. Giuseppe Bagnera was a specialist of group theory and professor of calculus in Messina. De Franchis was used to research on subjects close to those studied by Castelnuovo, Enriques and Picard. In recent times he had published at least two papers closely related to this kind of problem (De Franchis 1904, 1905) According to a letter he sent to Giovan Battista Guccia, the editor of the *Rendiconti del Circolo Matematico di Palermo*, he and Bagnera decided not to send their work to the *Académie* when they knew that Enriques and Severi had ended their paper. De Franchis wrote: "We had not completed the research and we were informed that it [the research] was instead was made comprehensively by Enriques and Severi. So we gave up the competition."[40] According to De Franchis, in February they met Enriques and Severi and realised that their work was wrong and that "they were forced to deal with the easiest cases, … and return[ed] [the work] after a long time." Indeed, in the Archives of the Académie there are two different versions of their paper: the first one consists of 15 pages and the second of 400! This episode confirms the strict relations between Picard and Enriques.

They finally published the paper in the final form in 1909; it was an important, even if not complete, contribution to the classification of hyperelliptic surfaces (Enriques and Severi 1909). As Humbert pointed out on behalf of the commission[41]: "The problem posed by the Academy can therefore be considered resolved for irregular surfaces. Regarding regular surfaces, the authors confine themselves to studying those that correspond to an involution formed…by ordinary transformations." (Humbert 1907, 982).

It was an argument very near to the studies of De Franchis and Bagnera.

In the letter mentioned above, De Franchis wrote to Guccia: "You now understand my mood to see that, for an excess of scruples, our almost complete and correct work is condemned, and that it had not even the honor of a discussion. I am worried that, since the synthetic reasonings are not usual in France, it could happen that the Academy do not notice certain errors in the work of E. and S.… I wish that

[39]Letter of Enriques to Castelnuovo 2 December, 1906 (Bottazzini et al. 1996, 646–8).

[40]Letter of De Franchis to Guccia October 21, 1907 in the Archive of the Circolo Matematico di Palermo.

[41]Besides the "rapporteur" Georges Humbert, the members of the commission were: Jordan, Poincaré, Émile Picard, Appell, Painlevé, Maurice Levy, Darboux, Bousinnesque.

you, who have many friends among the members of the Academy, could work to make the Academy now accept our work or, at worst, … renew the competition."[42]

Guccia obtained two different things from Picard: first of all, the paper by De Franchis and Bagnera sketching the results obtained by the two Palermitans, was immediately published in the *Comptes Rendus*; secondly (and more important) the Académie proposed the next Prix Bordin on a subject very close to the previous one and close to the research of Bagnera and De Franchis.

The paper on the *ComptesRendus* was published on November 4th, just 1 month before the publication of the commission report and only two weeks after De Franchis' letter.[43] (Bagnera and De Franchis 1907b) It is strange that, contrary to any other paper in *ComptesRendus*, no member of the Académie communicated the work. After those preliminary publications, Bagnera and De Franchis published an important paper containing all their main results (Bagnera and De Franchis 1908). Although the work was published after getting the Bordin prize, it appeared before Enriques and Severi's one. The classification of the hyperelliptic surfaces seems to be really more complete than that of Enriques and Severi, as it also contains the regular surfaces. But, the main difference between the two papers is the method used by the scholars. The Sicilian mathematicians make a detailed study of the group of automorphisms of the surface and "their point of view is of great modernity and more general than Enriques–Severi one."[44] The methods used by Bagnera and De Franchis are much closer to those of the French school (above all of Picard and Humbert) and they make large use of what they call "tables of periods" (Riemann matrices).

Also De Franchis' second request was fulfilled: for the Prix Bordin of the year 1909 the Académie proposed a subject strictly connected to that of the 1907, namely: "The absolute invariant, which represents the number of distinct double integrals of the second kind of an algebraic surface, depends on a relative invariant that plays an important role in the theory of integrals of total differentials of the third kind and in the theory of algebraic curves on a surface. We propose to conduct an extensive study of this invariant and to specifically explore how to determine its exact value, at least for large categories of surfaces."[45]

Even in such a case the two scholars were unlucky. The commission report explains: "This work was not received by the Secretariat of the Institute until 15 January, 1909. Even though the required period had passed, we did not think to create a precedent by including it in the competition for the *Prix Bordin*. It was actually sent to Palermo on the evening of 27 December, 1908, and the horrible

[42]Letter of De Franchis to Guccia October 21, 1907 in the Archive of the Circolo Matematico di Palermo.

[43]Bagnera and De Franchis had already published on 7 and 21 April a paper on the same argument in the Accademia dei Lincei, see (Bagnera and De Franchis 1907a). This time the paper was presented directly by Enriques.

[44]See (Ciliberto and Sernesi 1991).

[45]*Comptes rendus de l'Académie des Sciences*, Séance du 20 Décembre 1909, 149, pp. 1185–1188.

catastrophe that destroyed Messina occurred the following night. The mail wagon containing the *Mémoire* that we refer to was at the Messina train station when the earthquake occurred, and more than three weeks elapsed before the objects retrieved from under the rubble could be sent on." (Picard 1909, 1185).

The members of the commission of the prize Bordin were: Camille Jordan, Henri Poincaré, Paul Appell, Georges Humbert, Maurice Levy, Gaston Darboux, Jacques Bousinnesq, Emile Picard (rapporteur).

Picard motivated the prize by confirming the continuity between the two prizes, and wrote: "Two years ago, the Academy awarded a prize to a *Mémoire* written by Mr. Enriques and Mr. Severi that established the theorem that dominates surface theory, and which included a complete enumeration of irregular surfaces; the case of regular surfaces was dealt with but with the omission of certain unusual types. Around the same time, Mr. Bagnera and Mr. de Franchis applied themselves to the same research based on the fundamental theorem stated earlier by Mr. Enriques and Mr. Severi. Their classification was complete, and the *Mémoire* on this subject was published last year; this forms the basis of the current work." (Picard 1909, 1186).

Bagnera and De Franchis published the paper containing their new results later on, and, indeed, it was a very innovative paper (Bagnera and De Franchis 1910). The Picard number ρ may be considered as the maximum number of algebraically independent curves on the surface. All the mathematical methods used in the proof were deduced from the table of periods, and could be extended to higher dimensions. Some years later Gaetano Scorza was able to do this further step.

7 The Contributions of Gaetano Scorza

Scorza, who graduated in Pisa under the supervision of Luigi Bianchi and worked for 1 year in Turin with Corrado Segre, went to Palermo as a high school teacher in 1907. Here he worked in the Circolo Matematico, becoming a member of the editorial board of the *Rendiconti* in 1909 and being acquainted with Bagnera and De Franchis, particularly with the latter. In 1912 he became professor at the university of Cagliari and then in Parma (1913), Catania (1916), Naples (1921) and Rome (1934).

When he went to Palermo, Scorza completely abandoned his works on algebraic geometry (he had published his last paper on this subject in 1901). During his stay in Palermo and Catania, he gradually developed a research program aiming to generalize many results deduced by Humbert (and Bagnera and De Franchis) to abelian varieties on hyperelliptic surfaces. He realized that the fundamental results deduced in the awarded paper by the two Palermitan scholars were the best starting point for getting that generalisation. He wrote:

> Starting from the existence theorem for hyperelliptic functions in the flexible and handy form due to Mr. Bagnera and Mr. De Franchis, it was easy for me to give to it a geometric

> interpretation under which the theorem of Mr. Humbert was reduced to a very elementary proposition; and then I immediately realized that, in order to get the desired extension, it was necessary to use the geometric interpretation of the existence theorem of Abelian functions for any number of variables, and that a first step towards this interpretation was already the theorem by which, 2 years ago, I generalized the correct result of Mr. Bagnera and Mr. De Franchis (Scorza 1913, 387).

In 1916, he published his fundamental work on the subject (Scorza 1916). In it, he extended the main results on hyperelliptic surfaces to abelian varieties of any dimension, by using as a fundamental tool the table of periods, which he interpreted as a matrix and called Riemann Matrix.[46] It is not only a matter of choosing the right name; Scorza largely used the theory of matrices in order to find the properties of Abelian varieties. In this way, he did not limit himself to extending the methods used for surfaces to n-dimensional varieties, but he also established a much more general and abstract point of view. The concept of Riemann matrix is not at all restricted to the "period table", but it is an abstract complex matrix Ω, px2p, such that an alternate form f such that (a) $\Omega f \Omega^t = 0$; (b) $i\Omega f \overline{\Omega}^t$ is an hermitian matrix exists. The dimension of the linear space of the forms satisfying such properties is an invariant, k. The first part of the paper is devoted to the investigation of an abstract Riemann matrix, and only in the second one Scorza applied the general results to the case when the elements of the matrix represent the periods of an Abelian function.

He wrote: "The theory of Abelian functions… and some of the highest theories of algebraic geometry have so frequent analogies and similarities that anyone who analyzes them carefully is induced, almost spontaneously, to assume that all of them must be based on the same general theory and this fact is the best explanation for the several and close connections between them. And, since the crucial point of each research on these subjects always concerns… a certain "period table", … one is forced immediately to think that precisely the properties of the matrix should have in those theories an essential role… Through more or less recent works… they [the properties of the matrix] have become so many and so prominent that it seemed to me it is necessary… to write an organic and systematic treatment that is independent of the various concrete interpretations of which they are susceptible." (Scorza 1916, 263).

A very audacious abstraction indeed, which poses the question of studying the problem of the connections between geometric and analytic language for varieties of higher order. We shall not go further in the examination of this paper.[47] Anyway, we remark that 5 years later Scorza did another step in the direction of abstraction, by linking the study of a Riemann matrix to the study of the structural properties of the algebra of its endomorphisms (Scorza 1921).

[46]This name has been universally adopted and is still used.

[47]A detailed examination of this paper of Scorza is in (Hawkins 2013).

8 The Third Prix Bordin—Solomon Lefschetz

The work of Scorza was one of the starting points of Solomon Lefschetz, who was awarded the Prix Bordin of 1919 on abelian functions: his paper constitutes a real bridge between two different mathematical languages (Lefschetz 1921a). Lefschetz wrote in the Introduction:

> In the development of the theory of algebraic functions of one variable the introduction by Riemann of the surfaces that bear his name has played a well-known part. Owing to the partial failure of space intuition with the increase in dimensionality, the introduction of similar ideas into the field of algebraic functions of several variables has been of necessity slow. It was first done by Emile Picard, whose work along this line will remain a classic. A little later came the capital writings of Poincare in which he laid down the foundations of Analysis Situs, thus providing the tools needed to obviate the failure of space intuition. Meanwhile, other phases of the theory were investigated in Italy and to some extent also in France, receiving an especially powerful impetus at the hands of Castelnuovo, Enriques and Severi. It is, however, a rather remarkable fact that in their work topological considerations are all but absent, practically never going beyond the study of linear cycles, and this on the whole by a very indirect method (Lefschetz 1921a, 328).

We just mention one of the many results of Lefschetz, i.e. the so-called invariant k introduced by Scorza is equal to $\rho - 1$ (where ρ denotes, as usual, the Picard number). Lefschetz often quoted Scorza's results in his papers, as well as some other results obtained by Bagnera and De Franchis. For instance, he wrote (p. 39): "We propose now to recall some concepts and definitions incipient in the works of various authors, but formally introduced and only fully developed in recent writings of Scorza and also Rosati. The nomenclature which we shall use is Scorza."[48]

Lefschetz was a student of Picard and Appell, and graduated from the University of Paris in 1905. Afterwards, he went to the U.S.A. where he got his Ph.D. from Clark University and then became professor at the University of Kansas. While in Clark University, he "took advantage of the library to learn about a number of interesting new fields, notably about the superb Italian school of algebraic geometry." (Lefschetz 1970, 201). In the following years "the guidelines in my [his] research were: Picard–Simart: Fonctions algébriques de deux variables (two volumes, mostly Picard); Poincaré's papers on topology (= analysis situs) and on algebraic surfaces; Severi's two papers on the theory of the base; Scorza's major paper (dated 1915 [1916]) in Circolo di Palermo on Riemann matrices". In my opinion, Lefschetz was the actual follower (through Scorza) of the research program of Picard, Castelnuovo and Enriques. In his paper which won an award, Lefschetz found relevant development of Poincaré topological methods. So, he was able to achieve his main goal—"to plant the harpoon of algebraic topology into the body of the whale of algebraic geometry" (Lefschetz 1968, 854). Immediately after

[48] I will not develop here the contributions by Carlo Rosati, which are really important and principally oriented towards the study of correspondences between curves. Also, many other Italian mathematicians were involved in the study of Abelian varieties. I cite here only Annibale Comessatti and Fabio Conforto. To have a deep examination of the work of Lefschetz in algebraic geometry, see the chapter by Phillip Griffiths in (Griffiths et al. 1992).

receiving the Prix Bordin, Lefschetz was in France and Italy in the period 1920–1921, "half in Paris and half in Rome" (Lefschetz 1970, 204). When he came to Italy, Lefschetz immediately got in touch with Italian algebraic geometers—in particular with De Franchis and Scorza in Sicily, and Castelnuovo in Rome. On 18 November, 1920 he wrote to De Franchis (then director of the *Rendiconti del CircoloMatematico di Palermo*): "I am actually on sabbatical for a year and I plan to spend several weeks in Italy and, more particularly, in Sicily in order to get acquainted with your school of algebraic geometry that has contributed so much to the advancement of science."[49]

In December, Lefschetz met Scorza (probably in Catania), and immediately after their meeting read Scorza's paper of 1921 and remarked several common points with his own work: in the translation of his awarded paper, he wrote: "(Added in 1922)The results of this note have since been extensively exposed by Scorza in the Pa1ermo Rendiconti, … Not a few contacts with this part of our work are of course to be expected." (Lefschetz 1921a, 395). Perhaps the most striking evidence of the tight links between the methods of the two mathematicians emerges in Lefschetz's lectures held in Paris. In Spring 1921, Lefschetz gave some lectures in the Séminaire Hadamard, where he also talked about Scorza recent works. Lefschetz himself referred to this meeting in a paper published 2 years later.[50]

Lefschetz clearly expressed his aims: "The theory of Abelian functions is without a doubt one of the most important theories that geometers have addressed. It is even more remarkable that our understanding of period matrices per se has remained mostly fragmentary until quite recently. Several fundamental theorems existed of course for the genus p, and the late Georges Humbert, especially, studied the case of p = 2 extensively. But in this case direct calculation, while not simple, is at the very least possible; in addition, his methods hardly lend themselves to a simple extension. It is primarily due to Mr. Scorza that the veil has been lifted. His work would no doubt have attracted more attention, if it had not been published during wartime, and if he had used a different method. Taken from projective geometry and applied on this occasion to a very significant question for analysts, this method requires special training that we can hardly expect analysts to have received. It therefore seems highly desirable to one day demonstrate Mr. Scorza results by an analytical route. I must note that his method, whose elegance cannot be doubted, has been of considerable help to me in the course of my own research. Therefore, I believe myself to be engaged in a useful task by summarizing here the broad strokes of his work." (Lefschetz 1923, 120).

With uncommon modesty, Lefschetz referred to the "considerable help" of Scorza's works. In particular, he mentioned some points:

[49]Letter of S. Lefschetz to M. De Franchis, Archivio del CircoloMatematico di Palermo.

[50]Now I always quote from the paper (Lefschetz 1923), which is evidently a "resumé" of his lectures, written probably to clarify what Lebesgue (and some others, I guess) had not well understood during the talks.

1. One of the main innovation of Scorza method is the general study of the properties of abstract Riemann Matrices before applying them to the study of Abelian varieties. This is precisely the point of view Scorza expressed in his paper of 1921.
2. Humbert's works were very important for the further research on Abelian varieties, but he used direct methods, which were unsuitable in the passage from surfaces to higher dimensional manifolds.
3. French analysts were not able to understand the full meaning of geometrical methods. For 20 years the situation had been almost unchanged: the translation from one language to the other was still necessary. That was Lefschetz aim in his lectures.

Apparently, Lefschetz did not simply aim to apply Scorza's methods, but he also innovated them deeply by planting into them the harpoon of algebraic topology; this point of view was necessary in order to give a relevant contribution to the interrelations between different languages.

In an intriguing footnote, the American scholar speaks about a misunderstanding of Lebesgue during one of his talks. In Lefschetz's words: "This was the sense that I intended to convey in a remark made at the end of a talk (Mr. Hadamard's course at the Collège de France, in the spring of 1921) in which I analyzed Mr. Scorza's work. I must have expressed myself very poorly for a scholar as astute as Mr. Lebesgue to have interpreted my comments as he did in his inaugural lecture to the Collège." (Lefschetz 1923, 120).

Lebesgue did not agree at all with—or better did not understand—the point of view expressed by Lefschetz. Lebesgue's words we report here clearly showed the difficulty for an analyst as he was to understand the importance of the translation from an analytical language to a geometrical one."His [Humbert's] *Mémoires* on this subject are comprehensive and, like everything he writes, easy to read, elegant, clear and concise. It is perhaps this final quality that has lead some distinguished gentlemen to publish work that they believe extends Humbert's results, when they really have done no more than to comment on them. In summarizing one of these works during a meeting that Mr. Hadamard organized at the Collège de France to analyze the *Mémoires*, one of Humbert's most brilliant students, Mr. Lefschetz, very fairly concluded in approximately the following words: «This work provided the geometric translation of Humbert's results using a representation that Humbert himself described in a few words. It adds essentially nothing new to what Humbert has already taught us about Abelian functions.»" (Lebesgue 1922, 220–221).

It is evident that Lefschetz was disappointed by reading such a summary of his talk, completely contrary to the principles he had tried to explain. In Lebesgue's opinion, hence:

1. Scorza's work is not original at all, and is only a commentary (maybe a plagiarism?) of Humbert work.
2. Lefschetz had expressed the same opinion with words that he certainly never pronounced.

3. The geometrical translation of the analytical results into geometry is basically useless.

A very different point of view from Picard's one! In order to clarify Scorza' papers, Lefschetz did not hide his enthusiasm towards methods that have "lifted the veil" on the study of Abelian functions. Concerning Scorza's proof of the theorem of Picard–Poincaré, Lefschetz pointed out: "One must read Mr. Scorza's proof, which is hardly half a page long, to understand the power of his method, which can be characterized as follows: Once the beginning is accepted, the rest follows naturally" (Lefschetz 1923, 126).

Here Lefschetz referred to the Humbert–Bagnera and De Franchis–Scorza–Lefschetz tradition.

Finally, in 1924, Lefschetz published his famous book on the use of topological methods in algebraic geometry that concludes the period in which algebraic geometry was the main subject of his studies (Lefschetz 1924).

9 The Last Contributions of Castelnuovo to Algebraic Geometry

Lefschetz was in Rome during the winter 1920/21 and the Italian mathematical milieu impressed him a lot. "You should have gone to school in Rome in the twenties. That was the Princeton of its time!", he said to Giancarlo Rota many years later (Rota 1988). During his stay in Rome, Lefschetz often discussed with Castelnuovo, who was struck by the American mathematician.

As it is well known, Castelnuovo did not work on algebraic geometry since 1906, when he and Enriques published the Appendix to Picard's treatise. To know his motivation would be important in order to understand the evolution of Italian methods, but it is unfortunately unclear to me. Anyway, Castelnuovo's long silence was broken after his meetings with Lefschetz. In 1921, Castelnuovo published one of his last papers on algebraic geometry.[51] (Castelnuovo 1921) Significantly, this paper appeared in the same volume of the *Rendiconti della R. Accademia dei Lincei* where Lefschetz published one of his notes, (Lefschetz 1921b) which was "written at the very kind invitation of Mr. Castelnuovo".

In 1937 in his *Selected Papers*, Castelnuovo explained the reasons why he wrote his work: "Frequent conversations with Lefschetz in the winter of 1921, when he went to Rome, led me to publish a part of an extensive research on Abelian functions I had begun several years ago in order to clarify and extend the good

[51]Between 1906 and 1921, apart from the two chapters in Klein Encyclopaedia, Castelnuovo wrote only one paper on Algebraic Geometry, in a volume dedicated to Enrico D'Ovidio, his former professor. After (Castelnuovo 1921) he published only one work on algebraic geometry, after 28 years, (Castelnuovo 1949).

results of G. Humbert on singular hyperelliptic functions. The remaining part of this research stopped." (Castelnuovo 1937, 549).

Castelnuovo began his paper with a survey of his large research program strictly linked with the works of Scorza and Lefschetz, which he had developed in the last few years: "Prof. Lefschetz, in the still unpublished memoir, which won the Bordin Prize of the Academy of Paris, settled some important results on Abelian functions, with a very significant method based on considerations of Analysis Situs and that is mentioned in the Note published before my own. Some time ago I had got similar results by following a purely analytical way that was, in truth, already shown by Frobenius, but here it is stripped from accessory parts, greatly improved, and highlighted by making use of the most brilliant points of view by Scorza and of the language of algebraic geometry." (Castelnuovo 1921, 50).

Therefore, it is possible to have an idea of why Castelnuovo abandoned the active research on this subject in the last 15 years. After 1906 and Severi's basis theorems, Enriques went on his program of classification of algebraic surfaces, while Severi, whose interests were much closer to those of the French school, wanted to show the supremacy of the "Italian" methods and their independence of the new ones coming from topology and abstract algebra. In the twenties, from the French side, Picard, Humbert, Appell and the others were not engaged in algebraic geometry any more. By reading his 1921 paper, indeed, we may infer that Castelnuovo was mainly interested in the development and the clarification of the mutual relations between analytical and topological methods and algebraic-geometrical ones. In some sense we can say that the works of Picard and Humbert from one side, and of Castelnuovo, Enriques, Severi, Bagnera and De Franchis from the other side, had sufficiently enlightened the situation for surfaces (with some points to develop and many points to prove in a more rigorous way), but they appeared almost completely obscure in the case of n-dimensional varieties.

In our opinion, the discover of the new results obtained by Scorza and Lefschetz—as well as the works of Rosati and Torelli—interested Castelnuovo a lot, and led him to contribute to the development of algebraic geometry in the spirit of Segre,[52] i.e. by finding strict links between geometric and analytical (and now also topological) methods.

Castelnuovo's enthusiasm towards this new development did not end in 1921. We can find it again in Castelnuovo's suggestions to Zariski in 1927—when Zariski left the United States in order "to explore the work by Solomon Lefschetz".[53] In the International Congress of Mathematicians of Bologna of 1928, Castelnuovo held the general speech "La geometria algebrica e la scuola italiana". On 23 September, 1927, one year before the Congress, Castelnuovo wrote to Lefschetz: "The Organizational Committee of the International Conference of Mathematicians, which will take place in Bologna, has invited me to hold a general conference on

[52]Segre expressed this point of view many times, particularly in his celebrated talk during the Heidelberg International Congress in 1904; see (Segre 1905).

[53]Cited in (Parikh 2009). Zariski, coming from Russia, had studied in Rome with Enriques and Castelnuovo.

"Algebraic Geometry and the Italian school". Naturally, I will also speak about your work, which continues the work of the Italian and French schools brilliantly. I therefore re-read your recent publications, particularly your Mon. I would like to ask you several questions that arose from my reading."[54]

On September 1928, Castelnuovo held his lecture at the International Congress; (Castelnuovo 1928) after the general survey where he greatly summarized the typical methods of the Italians, he devoted about six pages to the transcendental methods, by speaking of the *idee geniali* of Lefschetz and, very extensively, of the topological point of view in algebraic geometry. This seems to me a new image of Italians' contribution to algebraic geometry. As Castelnuovo had already written to Lefschetz, he was looking for new ideas coming from the American mathematician as the "natural" continuation of Italian (and French) school. In fact, in his talk he remarked: "The interest of Poincaré's memoir also concerns the fact [...] it inspired the most recent research on algebraic surfaces. These were not carried out in Italy or in France, but they are closely linked to the results of the two schools. They are due to Lefschetz, a brilliant geometer of Russian origin, now a professor at the American University of Princeton. He set out to investigate how far, with only topological considerations, we can proceed in the study of algebraic surfaces. If, as expected, concerning the various issues, topology asks the help of the theory of integrals already mentioned, the attempt deserves great interest and arouses, together with the results achieved, a number of questions to be answered, to which I would now like to draw your attention." (Castelnuovo 1928, 196).

Then Castelnuovo considered the same open questions he had already asked Lefschetz in his private discussions with him in the quoted letter of the preceding year.

This was a point of view completely opposed to Severi who, during the next International Congress of Mathematicians held in Zurich in 1932, wrote: "I do not agree with either Mr. Van der Waerden or Mr. Lefschetz on this subject. I believe that I have already demonstrated how highly I esteem both topology and its prophets; but sometimes health can be found even outside of the field of topology. For some time now, geometry has had the means of rigorously evaluating the intersection multiplicity of any two varieties or algebraic cycles on an algebraic variety; and the topological method that has been used is only the transposition—conscious in the first case and perhaps unconscious in the second—of these methods." (Severi 1934, 213).

In Italy, Castelnuovo did not continue this kind of research. Scorza was now mainly interested in problems of pure algebra, Torelli was killed in the First World War, and Rosati died in 1929. Therefore, no Italian mathematician was able to or interested in the research program begun many many years before by Scorza. In the meantime, a new mathematical star appeared: in 1930 William Hodge published his first fundamental paper in algebraic geometry, typically using Lefschetz methods for solving a problem stated by Severi (Hodge 1930). That marks the beginning of a new story.

---

[54]The draft of this letter by Castelnuovo is in the Archive of the Accademia dei Lincei. Now in the web, edited by P. Gario, http://operedigitali.lincei.it/Castelnuovo/Lettere_E_Quaderni/menu.htm.

10 Conclusions

The interrelations between different languages in mathematics is a main topic for its history and for epistemological studies. For example, historians have extensively studied synthetic versus algebraic methods in the geometry of the 19th century, or Newtonian versus Leibnitzian approaches to the calculus. Specialists of algebraic geometry generally know the links between geometrical and transcendental methods fairly well, but, in my opinion, historians have not paid enough attention to their connections and to the development of their mutual interactions. However, such an investigation is one of the main issues that allow us to fully understand the birth and the development of mathematical ideas in the last century. The present paper aims at showing how historical research on such a matter can be fruitful.

As a conclusion, we report the following quotation from André Weil, which illustrates well the links between new and old languages in algebraic geometry:

> Nor should one forget, when discussing such subjects as algebraic geometry, and in particular the work of the Italian school, that the so-called "intuition" of earlier mathematicians, reckless as their use of it may sometimes appear to us, often rested on a painstaking study of numerous special examples, from which they gained an insight not always found among modern exponents of the axiomatic creed. … Our chief object here must be to conserve and complete the edifice bequeathed to us by our predecessors. "From the Paradise created for us by Cantor, no one shall drive us forth" was the motto of Hilbert work on foundations of mathematics. Similarly, however grateful we algebraic geometers should be to the modern algebraic school for lending us temporary accommodation, makeshift constructions full of rings, ideals and valuations, in which some of us feel in constant danger of getting lost, our wish and aim must be to return at the earliest possible moment to the palaces which are ours by birthright, to consolidate shaky foundations, to provide roofs where they are missing, to finish, in harmony with the portions already existing, what has been left undone (Weil 1946, VII–VIII).

Therefore, the interaction between Italian and French different mathematical languages is an interesting example that allows us to give a unitary interpretation of a fascinating branch of mathematics.

References

G. Bagnera, M. De Franchis, Sopra le superficie algebriche che hanno le coordinate del punto generico esprimibili con funzioni meromorfe quadruplamente periodiche di due parametri, *Rendiconti della R. Accademi dei Lincei*, (5), XVI, 1907, pp. 492–498 and 596–603; 1907a.

G. Bagnera, M. De Franchis, Sur les surfaces hyperelliptiques, *Comptes Rendus de l'Académie des Sciences*, 145, 1907, pp. 747–749; 1907b.

G. Bagnera, M. De Franchis, Le superficie algebriche le quali ammettono una rappresentazione parametrica mediante funzioni iperellittiche di due argomenti, *Memorie della Società Italiana delle Scienze detta dei XL*, (3) 15, 1908, pp. 251–343.

G. Bagnera, M. De Franchis, Le nombre ρ de Picard pour les surfaces hyperelliptiques et pour les surfaces irrégulières de genre zéro, *Rendiconti del Circolo Matematico di Palermo*, 30, 1910, pp. 185–238.

F. Bardelli, On the origins of the concept of irregularity of an algebraic surface, in (Brigaglia, Ciliberto, Sernesi, 1994), pp. 11–26.

E. Bertini, La geometria delle serie lineari secondo il metodo algebrico, *Annali di matematica pura e applicata*, (2), 22, 1894, pp. 1–40.
U. Bottazzini, A. Conte, P. Gario, *Riposte Armonie*, Boringhieri, 1996.
A. Brigaglia, C. Ciliberto, E. Sernesi (eds), *Algebra e geometria: il contributo italiano*, Supplemento ai Rendiconti del Circolo Matematico di Palermo, (2), 36, 1994.
A. Brigaglia, G. Masotto, *Il Circolo Matematico di Palermo*, Dedalo, 1982.
G. Castelnuovo, Sulle superficie algebriche le cui sezioni piane sono curve iperellittiche, *Rendiconti del Circolo Matematico di Palermo*, 4, 1890, pp. 73–88.
G. Castelnuovo, Osservazioni intorno alla geometria sopra una superficie algebrica, *Rendiconti dell'Istituto Lombardo*, (2), 24, 1891, pp. 127–137 and 307–318.
G. Castelnuovo, Sulla Razionalità delle Involuzioni Piane, *Rendiconti della R. Accademia dei Lincei*, (5), 2, 1894, pp.205–209.
G. Castelnuovo, Sulle superficie di genere zero, *Memorie della Società Italiana delle Scienze detta dei XL*, (3), 10, 1896, pp. 103–123.
G. Castelnuovo, F. Enriques, Sur une classe de surfaces algébriques, *Comptes Rendus de l'Académie des Sciences de Paris*, 131, 1900, pp. 739–742.
G. Castelnuovo, Sur les intégrales de différentielles totales appartenant à une surface irrégulier, *Comptes Rendu de l'Académie des Sciences de Paris*, 140, 1905, pp. 220–222; 1905a.
G. Castelnuovo, Sulle superficie aventi genere aritmetico negativo, *Rendiconti del Circolo Matematico di Palermo*, 20, 1905, pp. 55–60.
G. Castelnuovo, Sulle funzioni abeliane, *Rendiconti della R. Accademia Nazionale dei Lincei*, (5), 30, 1921, pp. 50–55; 99–103; 195–200; 355–359.
G. Castelnuovo, La geometria algebrica e la scuola italiana, *Atti del Congresso Internazionale dei Matematici*, v. I, Zanichelli, 1928, pp. 191–201.
G. Castelnuovo, *Memorie Scelte*, Zanichelli, 1937.
G. Castelnuovo, Sul numero di moduli di una superficie irregolare, *Rendiconti della Accademia Nazionale dei Lincei*, (8), 7, 1949, pp. 3–7; 8–11.
G. Castelnuovo, F. Enriques, Sur quelques récents résultats dans la théorie des surfaces algébriques, *Mathematische Annalen*, 48, 1896, pp. 241–316.
G. Castelnuovo, F. Enriques, Sur les intégrales simples de première espèce d'une surface ou d'une variété algébrique à plusieurs dimensions, *Annales de l'Ecole Normale Supérieur*, 23, 1906, pp. 339–366.
A. Cayley, On the deficiency of certain surfaces, *Mathematische Annalen*, 3, 1871, pp. 526–529.
C. Ciliberto, E. Sernesi, Some aspects of the scientificactivity of Michele De Franchis, in M. De Franchis, *Opere*, Supplemento ai Rendiconti del Circolo Matematico di Palermo, (2), 27, 1991, pp. 3–36.
A. Conte, La scoperta e i primi tentativi di classificazione delle superficie di Enriques nella corrispondenza inedita di Federigo Enriques a Guido Castelnuovo, in (Brigaglia, Ciliberto, Sernesi, 1994), pp. 103–116.
M. De Franchis, I piani doppi dotati di due o più differenziali totali di prima specie, *Rendiconti della R. Accademia Nazionale dei Lincei*, (5), 13, 1904, 688–695.
M. De Franchis, Sugli integrali di Picard relativi ad una superficie doppia, *Rendiconti del Circolo Matematico di Palermo*, 20, 1905, 331–334.
J. Dieudonné, *Cours de géométrie algébrique*, (2 vols.) Presses Universitaires de France, 1974.
F. Enriques, Una questione sulla linearità dei sistemi di curve appartenenti ad una superficie algebrica, *Rendiconti dell'Accademia dei Lincei*, (5), 2, 1893, pp. 3–8; 1893a.
F. Enriques, Ricerche di geometria sulle superficie algebriche, *Memorie dell'Accademia delle Scienze di Torino*, (2), 44, 1893, pp. 171–232; 1893b (Memorie Scelte, pp. 31–106).
F. Enriques, Sui sistemi lineari di superficie algebriche le cui intersezioni variabili sono curve iperellittiche, *Rendiconti della Reale Accademia dei Lincei*, (5), 2, 1893, pp. 281–287; 1893c.
F. Enriques, Sui sistemi lineari di superficie algebriche le cui intersezioni variabili sono curve ellittiche, *Rendiconti della R. Accademia dei Lincei*, (5), 3, 1894, pp. 481–487; 1894a.

F. Enriques, Ancora sui sistemi lineari di superficie algebriche le cui intersezioni variabili sono curve ellittiche, *Rendiconti della R. Accademia dei Lincei*, (5), 3, 1894, pp. 536–543; 1894b.

F. Enriques, Introduzione alla geometria sopra le superficie algebriche, *Memorie della Società Italiana delle Scienze detta dei XL*, (3), 10, 1896, pp. 1–81.

F. Enriques, Sulla proprietà caratteristica delle superficie algebriche irregolari, *Rendiconti della R. Accademia delle Scienze di Bologna*, 9, 1905, pp. 5–13; 1905.

F. Enriques, Sur les surfaces algébriques irrégulières, *Comptes Rendu de l'Académie des Sciences de Paris*, 140, 1905, pp. 133–136; 1905b.

F. Enriques, Sulle superficie algebriche di genere geometrico zero, *Rendiconti del Circolo Matematico di Palermo*, 20, 1905, pp. 1–33; 1905c.

F. Enriques, *Le superficie algebriche*, Zanichelli, 1949.

F. Enriques, *Memorie scelte di Geometria*, Zanichelli, 1956.

F. Enriques, F. Severi, Mémoire sur les surfaces hyperelliptiques, *Acta Mathematica*, 32, 1909, pp. 283–292.

P. Griffiths, D. Spencer, G. Whitehead, *Solomon Lefschetz*, National Academy of Sciences, 1992.

T. Hawkins, *The Mathematic of Frobenius in Context*, Springer, 2013.

W. Hodge, On Multiple Integrals Attached to an Algebraic Variety, *Journal of the London Mathematical Society*, 5, 1930, pp. 283–290.

C. Houzel, *La Géométrie Algébrique*, Blanchard, 2002.

G. Humbert, Rapport sur le prix Bordin, *Comptes Rendus de l'Academie des Sciences*, 145, 1907, second semestre, pp. 981–983.

H. Lebesgue, L'oeuvre mathématique de Georges Humbert, *Bulletin des Sciences Mathématiques*, 46, 1922, pp. 220–232.

S. Lefschetz, On certain numerical invariants of algebraic varieties with application to abelian varieties, *Transactions of the American Mathematical Society*, 22, 1921, pp. 327–482; 1921a.

S. Lefschetz, Sur le théoréme d'existence des fonctions abéliennes, *Rendiconti della R. Accademia Nazionale dei Lincei*, (5), 30, 1921, pp. 48–50; 1921b.

S. Lefschetz, *L'analysis situs and the Géométrie Algébrique*, Gauthier Villars, 1924.

S. Lefschetz, Progrès récents de la théorie des fonctions abéliennes, *Bulletin des Sciences Mathématiques*, 47, 1923, pp. 120–128.

S. Lefschetz, A page of mathematical autobiography, *Bulletin of the American Mathematical Society*, 74, 1968, pp. 854–879.

S. Lefschetz, Reminiscences of a mathematical immigrant, *American Mathematical Monthly*, 77, 1970, pp. 201–207.

C. Parikh, *The unreal life of Oskar Zariski*, Springer, 2009.

E. Picard, Sur les intégrales de différentielles totales algébriques de première espèce, *Journal de Mathématiques pures et appliquées*, (4), 1, 1885, pp. 281–346.

E. Picard, Sur les intégrales de différentielles totales de 2e espèce, *Journal de Mathématiques pures et appliquées*, (4), 2, 1886, pp. 329–372.

E. Picard, Sur les surfaces algébriques dont tous les sections planes sont unicursales, *Journal für reine und angewandteMathematik*, 100, 1887, pp. 71–78.

E. Picard, Mémoire sur la théorie des fonctions algébriques de deux variables, *Journal de Mathématiques pures et appliquées*, (4), 5, 1889, pp. 135–319.

E. Picard, *Traité d'Analyse*, Gauthier–Villars, 1891–1896.

E. Picard, Sur la théorie des surfaces algébriques, *Revue Générale des Sciences pures et appliquées*, V, 24, 1894, pp. 945–949.

E. Picard, Sur la théorie des surfaces et des groupes algébriques, *Comptes Rendus de l'Academie des Sciences*, 20, 1895, pp. 658–660; 1895a.

E. Picard, Sur la théorie des groupes et des surfaces algébriques, *Rendiconti del Circolo matematico di Palermo*, 9, 1895, pp. 244–255; 1895b.

E. Picard, Sur la théorie des surfaces algébriques du point de vue de la Géométrie de situation et sur lesintégrales de différentielles totales, *Comptes Rendus de l'Académie des Sciences*, 124, 1897, (premier semestre), pp. 532–8; 1897a.

E. Picard, Sur les intégrales doubles de seconde espèce dans la théorie des surfaces algébriques, *Comptes Rendus de l'Académie des Sciences*, 125, 1897, pp. 909–910; 1897b.

E. Picard, Sur quelques théorèmes relatifs aux surfaces algébriques de connexion linéaire supérieur à l'unité, *Comptes Rendus de l'Académie des Sciences*, 140, 1905 (premier semestre), pp. 117–122; 1905a.

E. Picard, Sur la dépendance entre les intégrales de différentielles totales de première et de seconde espèce d'une surface algébrique, *Comptes Rendus de l'Académie des Sciences*, 140, 1905 (premier semestre), pp. 915–917; 1905b.

E. Picard, Rapport sur le Prix Bordin, *Comptes Rendus de l'Academie des Sciences*, 149, 1909, (second semestre), pp. 1185–1188.

E. Picard, G. Simart, *Théorie des fonctions algébriques de deux variables*, 1e vol., Gauthier-Villars, Paris, 1897.

E. Picard, G. Simart, *Théorie des fonctions algébriques de deux variables*, 2e vol., Gauthier-Villars, Paris, 1906.

G. Rota, Fine Hall in the Golden Age, in *A century of Mathematics in America*, American Mathematical Society, 1988, pp. 223–236.

G. Scorza, Sul teorema di esistenza delle funzioni abeliane, *Rendiconti del Circolo Matematico di Palermo*, 36, 1913, pp. 386–395.

G. Scorza, Intorno alla teoria generale delle matrici di Riemann e ad alcune sue applicazioni, *Rendiconti del Circolo Matematico di Palermo*, 41, 1916, pp. 263–380.

G. Scorza, Le algebre di ordine qualunque e le matrici di Riemann, *Rendiconti del Circolo Matematico di Palermo*, 45, 1921, pp. 1–204.

C. Segre, Introduzione alla geometria sopra un ente algebrico semplicemente infinito, *Annali di matematica pura e applicata,* (2), 22, 1894, pp. 41–142 (Opere, v. I, pp. 198–304).

C. Segre, La geometria d'oggidì e i suoi legami con l'analisi, *Verhandlungendes 3en intern. Math-Kongress in Heidelberg*, 1904, pp. 109–120; and *Rendiconti del Circolo Matematico di Palermo*, 19, 1905, pp. 81–83.

F. Severi, Sulle superficie algebriche che posseggono integrali di Picard di 2a specie, *Rendiconti della R. Accademia dei Lincei*, (5), 13, 1904, pp. 253–258.

F. Severi, Sulle superficie algebriche che posseggono integrali di Picard di 2a specie, *Matematische Annalen*, 61, 1905, pp. 20–49; 1905a.

F. Severi, Sulla differenza fra i numeri degli integrali di Picard della prima e della seconda specie appartenenti ad una superficie irregolare, *Atti della R. Accademia delle Scienze di Torino*, 40, 1905, pp. 254–262; 1905b.

F. Severi, Le théorème d'Abel sur les surfaces algébriques, *Comptes rendus de l'Académie des Sciences*, 140, 1905, pp. 926–928; 1905c.

F. Severi, Sulle superfici algebriche che posseggono integrali di Picard della 2a specie, *MathematischeAnnalen*, 61, 1905, pp. 20–49; 1905d.

F. Severi, Le rôle de la géométrie algébrique dans les mathématiques, *Verhandlungen des Internationalen Mathematiker Kongresses*, I, 1934, pp. 209–220.

A. Weil, *Foundations of Algebraic Geome*try, AMS, 1946.

Lines on the Horizon

Hadamard and Fréchet, Readers of Volterra

Angelo Guerraggio, Frédéric Jaëck and Laurent Mazliak

1 Introduction

This present chapter concerns a very specific aspect of the relationship between French and Italian mathematicians at the turn of the 19th and 20th centuries, and is centred on the work of Vito Volterra (1860–1940). This emblematic figure of the Italian scientific milieu has already been extensively studied, including two major biographies (Goodstein 2007; Guerraggio and Paoloni 2013). The omnipresence of this mathematician in the historical works that examine the scientific, and even more generally intellectual, life in Italy in the decades following Italian unification shows the fundamental role played by Volterra. Born at the very moment that the Italian peninsula achieved political unity, Vito Volterra may properly be considered as embodying the aspirations and transformations of the new nation, at a time when she wished to claim her place in the world of European science. Meteoric progress had been made in this direction since the Risorgimento; in the years following the

A. Guerraggio
Dipartimento di Economia, Università degli Studi dell'Insubria, Varese, Italy

F. Jaëck
Laboratoire SPHERE, Université Paris Diderot, Paris, France

L. Mazliak (✉)
Laboratoire de Probabilités et Modèles Aléatoires,
Université Pierre et Marie Curie, Paris, France
e-mail: Laurent.mazliak@upmc.fr

F. Brechenmacher et al. (eds.), *Images of Italian Mathematics in France*,
Trends in the History of Science, DOI 10.1007/978-3-319-40082-2_6

French defeat in 1870 and the rise of Germany, when a concerned Darboux wrote to Houël that if things continued in that way, the Italians would also surpass the French on the mathematical scene (see Gispert (1987)), the prediction did not seem unrealistic. Under the leadership of eminent mathematicians such as Enrico Betti (1823–1892) and Ulisse Dini (1845–1918), various institutions rose in prestige in the country, such as the Scuola Normale Superiore in Pisa, where the young Volterra was trained and where he was able to learn the latest developments in mathematics and physics from all horizons, Germany in the first place. Dini had worked a great deal following works of Dirichlet, Riemann and their successors on the properties of functions of a real variable. His masterful treatise (Dini 1878) inaugurated a new phase in the trend which tended to focus increasingly on functions as the object of studies in their own right. Thanks to the works of Dini, certain families of functions were characterized by properties related to the existence of limits, to their Riemannian integrability, to different interpretations of differentiability, and so forth.

Volterra, reprising his master Dini, thus had at his disposal a sufficiently stable definition of the notion of function, and a rough idea of classes grouping these objects according to affinity. These classes, which prefigured certain algebraic structures, allowed him to consider functions as variable elements on which to work particularly with infinitesimal methods. These ideas were useful to the young Volterra for rethinking issues of mathematical physics and mechanics, which naturally appear as outgrowths of the infinity of parameters composing a function. In the first part of this chapter, we outline the main stages of the story. In particular, we will describe how the young scientist played along with a mathematical construction that followed its own developmental process, and was more concerned with the issues of applications that it inspired. Within a few months Volterra published a number articles introducing two concepts, functions depending on functions, and a sort of geometric twin of that, the functions of lines. He also developed a variety of techniques for manipulating these objects and formulated equations to which they were the solutions. The birth of functional calculus, as we will refer to it hereinafter, was a decisive step in the transferral to more sophisticated situations of infinitesimal methods that had appeared at first to be restricted to the original case of numerical functions of a real variable.

What primarily concerns us here is how two French mathematicians, Jacques Hadamard (1865–1963) and Maurice Fréchet (1878–1973), came into contact with the work of Volterra, took it on themselves to develop it, and, in the case of Fréchet, to criticize and to some extent go beyond it on their own terms. The two figures belong to two successive generations, the older one to Volterra's generation. Their trajectories are interesting to us, among others reasons, because of the uniqueness of the tandem they formed with their Italian colleague. Hadamard and Fréchet, as we will show, were in a relationship of master and disciple; this meant that each declined their reading of Volterra's work in a different way. Hadamard's reading of

Volterra is the subject of the second part of the chapter. He took part in the extensive research program on partial differential equations (referred to as PDEs in what follows) that Hadamard began in 1890 on the occasion of his arrival in Bordeaux, especially after his meeting with Pierre Duhem (1861–1916).

This is by far not the first time that the Volterra-Hadamard duo has been studied to stress its importance in the study of PDEs that overtook physics in the second half of the nineteenth century, especially following Riemann's work on shock waves. We can cite in particular the voluminous scientific biography by Maz'ya and Shaposhnikova (1998) and above all the chapter by Gray (2001), which presents a detailed study of the roles of Poincaré, Volterra and Hadamard in this history. Our objective is much more modest than these panoramic overviews, and focuses on the passing of the baton from Volterra to Hadamard and Fréchet. Expressing it lightly, Gray says that in reprising Volterra's work on line functions, *his friend Hadamard* (Gray 2001, p. 127) introduced the term *functional.* However, this word *friend* is really too static to describe the growing, evolving relationship between the two men; this deserves greater attention, because it is highly correlated with Hadamard's research program in mathematical physics. This seems a convincing illustration of the hypothesis that Italian progress in mathematics was so valuable that the French mathematicians had to watch more closely the work that was being carried out on the other side of the Alps. We will therefore provide a fairly detailed account of the construction of Hadamard's program and how its progressive development put him in touch with Volterra. In fact, as we shall see, it was more broadly speaking the work of many Italian mathematicians that would at that time nourish French research on these topics. Volterra's considerations on line functions and differential calculus were only recognised in a later moment by Hadamard as possibly providing tools to solve problems of PDEs where the initial conditions undergo deformation over time, as in some wave situations. In the second section of the paper, we will examine the original circumstances of the encounter between the two mathematicians, who came to know and gradually appreciate each other between one international conference held in Zurich in 1897 and another in Heidelberg in 1904, when their personal relationship proper began.

In the case of Fréchet, the situation is very different. When he came into contact with Volterra, he was a young student fresh out of the École Normale Supérieure. Under the advisement of his mentor Hadamard and supported by émile Borel (1871–1956), a great friend of Volterra's since the Congress of Zurich, Fréchet sought a thesis topic with the Italian mathematician. The correspondence that began at that time shows that for a few months the young mathematician tried to follow in the footsteps of the older one. We will show how the young man quickly and markedly deviated from this initial orientation to follow his own path of radical originality, a path whose modernity somewhat baffled many of his contemporaries. Exploding the framework imposed by Volterra on classes of functions, Fréchet introduced a topological view of abstract spaces and general differential calculus that became the natural framework for further developments in functional analysis, relegating Volterra to the rank of pioneer, although not without some clashes, as we shall see.

Reflections on the birth of functional analysis and Volterra's role are certainly not new too. In the 1980s Reinhardt Siegmund-Schultze's comprehensive thesis (Siegmund-Schultze 1990) provided a thorough study of the subject. More recently, in another direction, Archibald and Tazzioli (2014) published an important contribution on the relationship between functional analysis, equations integrals and nuclei introduced by Fredholm operators as they were studied in France and Italy at the period we are concerned with, and Volterra also plays a central role here. Our research, however, makes a number of additions to these studies, emphasizing the reading of Volterra by his two French colleagues, but also highlighting a network of Italian and French contributions to the debate that the perspective chosen by other historians has not necessarily shed light on. For example, it is interesting to note that the story described in Archibald and Tazzioli (2014) has close affinities with the topics we will discuss, but also includes aspects that do intersect them at all (and vice versa), revealing how the bubbling research on functional spaces at the beginning of the twentieth century may have resulted in surprisingly independent lines of exploration. The lines we describe here appear to us to be those where the direct influence of Volterra's work on French mathematicians was the strongest and most decisive.

2 Vito Volterra: The Tools of Analysis at the Service of Mathematical Physics

2.1 Mathematical Physics and Rigorous Analysis

Vito Volterra was just over 25 years old when he published a note (Volterra 1887c)—presented by Betti—in the *Rendiconti dell'Accademia dei Lincei* that represents the beginning of his research in functional analysis and introduces a concept that is central to his work: the notion of *a function that depends on other functions*. This publication can rightly be thought of as marking a singular moment in the history of nineteenth-century mathematics. It is also often identified today, and was even by the protagonists of the period, as the founding act of a new science. Volterra wrote his degree thesis on hydrodynamics under the advisement of Betti at the Scuola Normale Superiore in Pisa. It was this period that witnessed the large movement to transfer the rigour of German mathematical analysis into Italian mathematics, particularly through the translation of the works of Riemann, under the impetus of Betti and Dini. Betti mainly focused on issues related to complex variables, while Dini primarily worked in real analysis. During the 2 years that he spent at the Scuola Normale, the young Volterra was seduced by Dini's lectures—Dini had published in 1878 his fundamental book on the theory of the functions of a real variable (Cooke 1984). Volterra's first publications clearly show Dini's influence. In 1881, when he was still a student at the Scuola Normale, Volterra published two articles (Volterra 1881a, b) with evocative titles—*Alcune osservazioni sulle funzioni punteggiate discontinue* (Some observations on pointwise discontinuous functions) and *Sui principi del calcolo integrale* (On the principles of integral calculus)—in the *Giornale di Matematica*.

He described in detail what is meant by the smallness of a set and gives an example of a subset of $\mathbb{R}$ which is nowhere dense yet has a non-null measure (in today's words). He further exhibited differentiable functions whose derivative, albeit limited, is not Riemann integrable, demonstrating that in general integration and differentiation are not inverse operations of each other. As we can see, the brilliant student Volterra showed that he had grasped the nature of various central issues of nineteenth-century analysis, modernised by Dirichlet, Riemann and their heirs beginning from their studies on trigonometric series.[1]

However after these notes of 1881, Volterra did not again intervene in these questions concerning the basis of real analysis, preferring instead to turn to the study of functions of a complex variable and differential equations. In the years following his degree thesis, Volterra published many articles in mathematical physics that reflect the presence and influence of Betti in his choice of research topics. The detailed study of the theory of elasticity by Gabriel Lamé (1795–1870) (JaËck 2010), which will be discussed in the next section, gave Volterra the opportunity to publish his best-known product of this era (Volterra 1892). Lamé treated the propagation of light in birefringent media assuming that the incident beam is split into two polarized rays that vibrate in perpendicular planes. Twenty years later, Sofia Kovalevskaya (1850–1891) took this up, criticising Lamé's method in an article published in *Acta Mathematica* in 1886 (Kowalevski 1885). Shortly after Kovalevskaya's death, Volterra noted an error in her approach, in that she had never taken into account the discontinuity of one of the parameters. He corrected the error and provided in his article the righteous differential system for light waves (Cooke 1984, p. 173 ff).[2]

2.2 Differential Equations and Functions of Lines

The two years 1887 and 1888 were particularly fruitful for Volterra. They actually saw the publication of several sets of work, each comprising a series of articles, which show the originality of his thinking and his growing independence from the ideas of his teachers, Betti and Dini. Especially noteworthy are the first three articles (Volterra 1887a, b, 1888) that mark the development of a proper theory of

[1]For the history of this very important chapter in the development of integration, see the fine work by Hawkins (2001).

[2]This publication provided the occasion for Volterra to begin an important correspondence with Gösta Mittag-Leffler (1846–1927), who had founded the journal *Acta Mathematica* in 1882 and directed it energetically for almost half a century. A study of this correspondences (see Nastasi and del Colombo 2015), which was most frequent in the years 1888–1892, offers a very vivid picture of the young Volterra's boundless activity.

linear differential equations. Here we should note Volterra's clear orientation towards mathematical physics. The subjects of these three articles are no longer based on the examination of concrete physical situations, but on a strictly mathematical analysis. More to the point for our purposes, in a single year, 1887, Volterra published three notes (Volterra 1887c, d, e) in which he introduced the concept of a function that depends on other functions in order to study the quantities that all depend on the values that one or more functions of one variable can assume in a given interval. Volterra uses the expression *function that depends on other functions*, clearly stating in his text that he is not dealing here with functions of functions, that is, a function obtained by composition of functions in the sense of Dirichlet. Volterra insists on the essential fact that the concept of function that depends on other functions attributes two different roles to the elements involved. Such a function has functions as variables, the latter maintaining a sort of general and indeterminate character proper to the nature of being a variable. For simplicity's sake, we will use the term *functional* (which will be introduced by Hadamard, as we shall see in the next section) to refer to the concept of Volterra. Volterra's insistence on indicating the variable x in his expressions involving the variable function φ—thus systematically written $\varphi(x)$—is a good illustration of the fact that this concept of variable function was still in its infancy for him. This also creates difficulties for the reader accustomed to modern presentation. These fluctuations in language and notation remind us that at that time functional analysis was not yet born as a field of research, and show how the abstract form it received in the twentieth century helped to synthesise the concepts introduced by Volterra and others; the third part of our article, devoted to Fréchet, also concerns the first steps in this spectacular synthesis. For Volterra, the domain of functionals, that is to say, the set of elements on which the operations is carried out, is not a general set but is systematically made up of the class of functions of one variable which are all continuous in an interval $[A, B]$. The notion of uniform metric on an abstract space is not yet clarified (it will be later, again by Fréchet) and the distance between two functions is exclusively given by a property of upper boundary property inherited from Cauchy and Weierstrass. Thus, a functional y is said to be continuous if[3]:

> making vary $\varphi(x)$ of a variation $\psi(x)$ such that the absolute value $\psi(x)$ is always less than ε, the corresponding variation in y can be made smaller than an arbitrarily small δ[4] (Volterra 1887c, p. 296).

The main objective of the three notes of 1887 on functions of functions is to extend to functionals the concept of derivative as well as that of differential. Specifically, the central section of the first note (Volterra 1887c) is based on the

[3]We must emphasize here the scope of this statement, which does not produce a generality that will only be arrived later. Volterra's terms are intended to define the continuity of the functional in the case of continuous functions in [ab]. In other words, the set of continuous functions should not be thought of here as an archetype of a more general abstract space.

[4]diremo che y è continua se, data a $\varphi(x)$ una variazione $\psi(x)$ tale che invalore assoluto $\psi(x)$ sia sempre inferiore ad ϵ, la variazione corrispondente di y può rendersi inferiore a δ, piccolo ad arbitrio.

notion of variation of a function that depends on another function and designates with the symbol σ the first-order term of the variation Δ of y for a small variation in the variable (which here is a function). It is significant that this approach takes place within the general development of analysis at a time when mathematicians were attempting to define a differential calculus for functions of several real variables that has the greatest possible analogy with what was known at the time for functions defined on $\mathbb{R}$. In keeping with the work being done in these last decades of the nineteenth century, Volterra did not seek to specifically form a notion of a differential or of a differentiable function, but to generalize the classic formulas $df(x) = f'(x)dx$ and $df(x_1, x_2, \ldots, x_n) = \sum f'_{x_i} dx_i$ that were known for the functions of one or several variables.

One of his main results is that of having shown that, under reasonable conditions on the functional y defined on the space of continuous functions in the interval $[A, B]$, its variation can be expressed as an integral of the type[5]

$$\delta y[\varphi(x)] = \int_A^B y'[\varphi(x), t]\delta\varphi(x)dt$$

in which can be specifically discerned the analogy with the corresponding formulas in real analysis and how to generalize them. The operation of integration generalizes the sum used in the expression of the differential referred to above, δy represents the variation of the functional y engendered by the variation $\delta\varphi$ of the independent variable, and y' denotes what Volterra called the *functional derivative* of y. The ad hoc conditions proposed by Volterra to permit the proof of his integral formula are quite clearly inspired by the analytical-geometric methods of the calculus of variations, consisting in the creation of a localised perturbation. Let us consider, Volterra says, a subinterval $[m, n]$ of $[A, B]$ and a variation θ of the variable function φ such that $0 \le \theta(x) \le \varepsilon$ for all $x \in [m, n]$. We set $\int_m^n \theta(x)dx = \sigma$, which thus represents the area comprised within the graph of φ and that of $\varphi + \theta$. Finally, δy represents the variation $y(\varphi + \theta) - y(\varphi)$. Volterra then formulates four hypotheses (Volterra 1887c, p. 296):

(1) $\frac{\delta y}{\epsilon h}$ is always less than a constant M.
(2) If ϵ and h tend to 0, such that the interval $[m, n]$ always contains the point t, the ratio $\frac{\delta y}{\sigma}$ tends to a finite limit, denoted $y'[\varphi(x), t]$, and called *functional derivative* of y.
(3) This limit is uniform for the possible choices of φ and t.
(4) $\varphi \mapsto y'[\varphi(x), t]$ and $t \mapsto y'[\varphi(x), t]$ are continuous.

[5]In 1887 Volterra introduced the notation $y|[\varphi(x), t]|_B^A$, which we simplify as $y[\varphi(x), t]$ in our commentary.

The publication of article (Volterra 1887c) is undoubtedly a milestone in the development of functional analysis. Beginning with a new mathematical concept, Volterra constructed for the first time a new differential calculus that allowed him to envision higher order derivatives and arrive at a Taylor formula in this general framework. Volterra revisited his system of assumptions again and again, thus the framework presented in Volterra (1887c) was still provisional. Aside from the ambiguities in the choice of symbolism and terminology that we have underlined, we see that it is necessary that the increase $\theta(x)$, to which the function φ is submitted, always have a constant sign, a condition that will disappear during the course of Volterra's successive publications. Similarly, the fourth hypothesis will evolve into the strongest condition, but naturally verified in some typical situations, that the derivative y' is uniformly continuous, making it possible to give a simpler proof of the result. This property will also become the hypothesis canonically proposed to guarantee the existence of the first derivative of a functional. In their 1936 work that became a point of reference (Volterra and Pérès 1936), Pérès[6] and Volterra also remarked that they prove the result of representation under conditions that are obviously not as general as possible. Another observation that should be noted is that Volterra is not concerned about the independence of the first derivative of the functional with respect to the particular choice of θ. As mentioned, the main purpose in Volterra (1887c) is not to study the formal properties of differentials or to specify the class of differentiable functions but rather to seek a representation theorem of the variation and it is on this occasion that a notion of differentiability is introduced. This will be greatly extended in subsequent work by Fréchet, which we examine below.[7]

We mentioned that the representation theorem was valid under certain assumptions. Volterra's second note on the functions that depend on other functions (Volterra 1887d) seeks to relax these constraints and focuses particularly on functionals for which the conditions of the first article are not all met at exceptional points t. Thus Volterra notes that there may be some cases where around certain points of the segment $[A, B]$, hypothesis 1 of the previous note, which required the variation of the functional to be an infinitesimal of an order higher or equal to εh, cannot be verified. The three cases examined by Volterra, where this assumption is not satisfied, make it possible to confirm that the representation formula for the differential established from the beginning remains valid, with the possible addition of terms that depend on exceptional points. In the third section we will return to this

[6]On Joseph Pérès (1890–1962), see Mazliak and Tazzioli (2009).

[7]These representational issues also arise in connection with specific works in the context of functional calculus, as in the case of the studies by René Gateaux (1889–1914), as well as those by Paul Lévy (1886–1971) in potential theory, introducing a notion of differential engendered by a variation of the independent variable in a given direction ψ. On this exciting chapter linking functional calculus and probability theory, see Mazliak (2015a) and Barbut (2013). We see in Barbut (2013) that when Levy began a correspondence with Fréchet after World War I, he never misses a chance to voice strong criticism of the narrow framework proposed by Volterra for the derivation of higher-order functionals. Younger by almost 30 years, Levy was probably not in position to fully comprehend how radically novel the older mathematician's approach was at the time it appeared.

point, which gave rise to a dispute between Volterra and Fréchet regarding the generality of the theory of functionals.

The third note (Volterra 1887e) focuses on particular issues, where the functional or its derivative has a specific form which is defined by its dependence on the relation to the variable φ or its derivatives at given points. Volterra especially examines the case where there is a function F such that $y'|[\varphi(x), t]| = F(\varphi(t))$ or where $y|[\varphi(x)]| = \int_A^B \int_A^B F(\varphi(t), \varphi(t_1))dt_1$. He also studies the case in which an ordinary differential equation

$$f\left(y, \frac{dy}{dx}, \frac{d^2y}{dx^2}, \ldots, \frac{d^ny}{dx^n}, \varphi(x), \varphi'(x), \ldots, \varphi^{(m)}(x)\right) = 0$$

is given, where the function φ is also given, as are the initial conditions

$$y(A), y'(A), \ldots, y^{(n-1)}(A).$$

One thus considers the functional Y which to φ associates the value in B of the solution associated to φ. As Volterra himself observes, that value depends on that of φ in the entire interval $[A, B]$. This then raises the question of the derivative of that functional, which Volterra obtains by means of the solution of an auxiliary equation. The consideration of particular functionals in the form of integrals given as solutions to differential equations provides an initial glimpse of the reasons that led Volterra to introduce the concept of functional. His initial motivations were in fact those of a mathematical physicist. However, behind the new mathematical notion that he introduced and the calculus that it allowed him to develop, he discovered tools essential for posing other problems of analysis. Volterra also seemed surprised to see how the idea that is the source of the concept of functional made it possible to revisit some classic chapters of mathematics and was already present in some elementary observations and experiments of physics in which one seeks identify the dependence of certain continuous parameters. His first note therefore opens with an optimistic vision, with Volterra saying that he will introduce.

> some considerations that will serve to illuminate concepts that I think are necessary to introduce for an extension of Riemann's theory on functions of a complex variable, and which I think lend themselves to uses in various other research areas.[8] (Volterra 1887c, p. 294).

A few lines further, we read:

> in fact in many questions of physics and mechanics, as well as in the integration of partial differential equations, there may be a need to consider quantities that depend on all the

[8]Alcune considerazioni le quali servono a chiarire dei concetti che credo necessari introdurre per una estensione della teoria di Riemann sulle funzioni di variabili complesse, e che penso possano tornar giovevoli anche in varie altre ricerche.

values that one or more functions of one variable assume in given intervals For example, the temperature at a point of a conductive blade depends on all the values that the temperature takes on the edge...[9]

The prediction of possible applications of the notion of functional, which Volterra presents in his first note (Volterra 1887c), effectively materialises in some of the research he carried out in the years that followed. In general there is no direct reference to functions that depend on other functions, but to their geometric versions—the *functions of lines*—that Volterra examined in two notes published in that same year, 1887, in the *Rendiconti dell'Accademia dei Lincei* (Volterra 1887f, g). Here the motivation behind the study of functions of lines is similar to that which generally tends to associate with each concept of analysis a geometric representation.[10] As Volterra wrote:

> The usefulness of geometric representation in the domain of variability of a function is well known.... One can obtain a geometric image of the same kind for the functions that depend on another function. (Volterra 1887f, p. 315)

The objectives are first of all identical to those set forth earlier in the study of the functions that depend on other functions:

> Such an idea is familiar to physicists: it presents itself spontaneously when one thinks of certain electrical phenomena ... For certain studies that I hope to be able to communicate in a very short time, it is beneficial to consider functions of lines of a three-dimensional field.[11] (Volterra 1887f, p. 315)

The study of functions of lines in fact permitted giving the representation theory of a form even more visible than what was known for the differential of a function of several variables. For the functional $\varphi[L]$—here φ indicates the functional and no longer the function on which it operates!—Volterra provides the following notation of its variation:

$$\delta\varphi = \int_L (X\delta x + Y\delta y + Z\delta z)ds$$

[9]Infatti in molte questioni di Fisica e di Meccanica, e nella integrazione di equazioni differenziali alle derivate parziali, capita di dover considerare delle quantità che dipendono da tutti i valori che una o più funzioni di una variabile prendono in dati intervalli Cos per esempio la temperatura in un punto di una lamina conduttrice dipende da tutti i valori che la temperatura ha al contorno ... (Volterra 1887c, p. 294).

[10]The notion of *line* covers several aspects that will be evoked via Volterra's texts in the what follows. Although at the time *line* could be interpreted as a function with values in the plane, with varying properties and regularity, Volterra makes very little appeal to this parametric version. For him the line is primarily manipulated as a geometric, almost physical object, for which he will define an operation of addition (the piecing together of two lines) and conceive a notion of neighborhood (word belonging to the vocabulary of an observer) that does not rely on parameterisation.

[11]una tale idea è familiare ai fisici; essa si presenta spontaneamente quando si pensa a certi fenomeni elettrici (...) Per alcuni studi, che spero di poter comunicare quanto prima giova considerare le funzioni delle linee di un campo a tre dimensioni.

where X denotes the derivative of φ with respect to x, defined as the limit of the ratio $\frac{\Delta_x \varphi}{\varepsilon h}$ when ε and h tend to 0, ε and h having the sense given in Volterra (1887c), while $\Delta_x \varphi$ is the variation of the functional that corresponds to the variation of the line L with respect to the x axis. The quantities Y and Z are defined in an analogous way. The three partial derivatives of φ corresponding to x, y, z are not independent of each other but are linked by relationships described in the form of equalities. These ideas are then extended to the second derivatives and developed more particularly in the second note on the functions of lines to study *simple functions*, which verify the equality $\varphi[L_1 + L_2] = \varphi[L_1] + \varphi[L_2]$ in which L_1 and L_2 are two paths with the restrictions described in the first note: the lines are closed or terminate at the boundary of the area in which they are considered, there are a finite number of singular points and elsewhere there is a tangent, and finally, there are no knots. The sum is defined, with a suitable parameterization, as the line obtained by removing the common part of L_1 and L_2. These properties of linearity drawn from problems of functionals emerged in various forms devised by many mathematicians in that period. We can observe a strong resemblance to the notion of distributive function introduced by Peano at almost the same time (Peano 1887).[12] Moreover, we will see in the next section that Volterra's *simple functions* fall within the scope of Hadamard's *additive functions*. We can also recall the work of Emmanuel Carvallo (1856–1945), who grasped early on the concept of linear operation (which he calls *operator*) to study systems of functional equations (see Carvallo (1891)).

2.3 The First Results of Functional Analysis

While in the twentieth century the functional became one of the mathematical objects most used, in the last decade of the nineteenth century the concept was still in its infancy. In the introduction to his article (Volterra 1889a), Volterra reported that functions of lines arise in several questions of physics and may also be related to questions of analysis. His goal in the article was to show how they could be used in the theory of functions of complex variables. In an earlier article published in 1887,[13] Volterra had already mentioned the use of functions of lines to generalise Riemann's definition of complex function. In effect, he wrote that Riemann's considerations in reference to a two-dimensional space can be extended to three-dimensional spaces provided that instead of functions defined on such a space, one begins with functions that depend on lines on this space. In Volterra (1889a), Volterra describes his notion in detail:

> In the theory of functions of a complex variable, we suppose, in a way, that the values of imaginary variables are extended on a surface, with the condition that the differential relations of the variables only depend on the points of the surface… Is it possible to

[12] The difficult relationship between Peano and Volterra forms an explosive chapter of Italian mathematics in that period. For more on this, see Guerraggio and Paoloni (2013), pp. 36–42.

[13] This is the first of a series of three notes (Volterra 1887c, d, e) on this subject published between 1887 and 1888.

> generalise this theory by referring to a three-dimensional space? This is the problem I proposed. We can solve the question, but to approach it, it is necessary to use what I have called the functions of a line. … How will the generalisation which I have just mentioned be linked to that known theory? It is easy to show that it is related to the theory of functions of several complex variables. Nearly a year ago, Mr. Poincaré, in generalising the theorem of Cauchy, proved that the integral of a uniform function of two complex variables taken on a closed surface is zero, if one can distort and reduce the surface to a point without encountering any singularities. From this it can be deduced that if the surface of integration is not closed, the integral depends on the lines that form the boundary of the surface. So we see that the integration of functions of two variables leads to functions of lines.[14] (Volterra 1889a, pp. 364–365)

Two years later, in a new article (Volterra 1891b), Volterra showed that it was possible to use functions of lines to extend to double integrals Jacobi-Hamilton's theory of the calculus of variations:

> The Jacobi-Hamilton procedure is based on the simple integral (of which the variation is to be made null) considered as a function of its limits and of the arbitrary values assigned to the unknown functions in the limits themselves… If one goes from simple integrals to the case of double integrals, instead of the two limits of the integral, we have one or more lines that form the boundary of the area of integration.[15] (Volterra 1891b, p. 464)

It is in this type of context that functions of lines come into play, permitting the construction of an element analogous to the characteristic function set forth in the Hamilton-Jacobi theory, and extending the concept of multiple integrals. In a subsequent article (Volterra 1896), Volterra explains how functions of lines make it possible to develop a general vision for the addition problem for elliptic functions

---

[14]Dans la théorie des fonctions d'une variable imaginaire, on suppose, en quelque sorte, que les valeurs des variables imaginaires sont étendues sur une surface, avec la condition que les rapports différentiels des variables ne dépendent que des points de la surface… Est-ce qu'on peut généraliser cette théorie en se rapportant à un espace à trois dimensions? Voilà le problème que je me suis proposé. On peut résoudre la question, mais pour l'aborder il faut recourir à ce que je viens d'appeler les fonctions d'une ligne… A quelle théorie connue va se rattacher la généralisation dont je viens de parler? Il est bien aisé de montrer qu'elle se rattache à la théorie des fonctions de plusieurs variables imaginaires. Il y a presque une année, M. Poincaré, en généralisant le théorème de Cauchy, à démontré que l'intégrale d'une fonction uniforme de deux variables imaginaires prise sur une surface fermée est nulle, si l'on peut déformer et réduire la surface à un point sans rencontrer de singularités. On peut déduire de là que, si la surface d'intégration n'est pas fermée, l'intégrale dépend des lignes qui forment le contour de la surface. Donc on voit que l'intégration des fonctions de deux variables conduit aux fonctions des lignes.

[15]Il procedimento Jacobi-Hamilton si fonda sull'esame dell'integrale semplice (di cui si vuole annullare la variazione) considerato come funzione dei suoi limiti e dei valori assegnati ad arbitrio alle funzioni incognite nei limiti stessi (…). Se si passa dagli integrali semplici al caso degli integrali doppi, invece dei due limiti dell'integrale, abbiamo una o più linee che formano il contorno del campo di integrazione.

and the connection with partial differential equations. Consider the following sum of multiple integrals

$$J = \iint_{\alpha_1} \frac{dydz}{\sqrt{\frac{\beta_2}{\lambda} z^2 - \frac{\beta_3}{\lambda} y^2 + \beta_1}} + \iint_{\alpha_2} \frac{dzdx}{\sqrt{\frac{\beta_3}{\mu} x^2 - \frac{\beta_1}{\mu} z^2 + \beta_2}} + \iint_{\alpha_3} \frac{dydz}{\sqrt{\frac{\beta_1}{\nu} y^2 - \frac{\beta_2}{\nu} x^2 + \beta_3}} \tag{1}$$

with the condition $\lambda + \mu + \nu = 0$. The main theorem proved by Volterra (1896) says that this sum is constant when the integration domains are limited by the projections $\alpha_1, \alpha_2, \alpha_3$ on the three coordinate planes of a curve drawn on the algebraic surface with equation

$$\lambda x \sqrt{\frac{\beta_2}{\lambda} z^2 - \frac{\beta_3}{\lambda} y^2 + \beta_1} + \mu y \sqrt{\frac{\beta_3}{\mu} x^2 - \frac{\beta_1}{\mu} z^2 + \beta_2} + \nu z \sqrt{\frac{\beta_1}{\nu} y^2 - \frac{\beta_2}{\nu} x^2 + \beta_3} = C. \tag{2}$$

Volterra then makes the following observation: the existence of a solution to the partial differential equation

$$\frac{\partial X}{\partial x} + \frac{\partial Y}{\partial y} + \frac{\partial Z}{\partial z} = 0$$

can be interpreted as the condition of existence of a first-order function of line $F[(L)]$ such that its surface derivatives $\frac{dF}{d(y,z)}, \frac{dF}{d(y,z)}, \frac{dF}{d(y,z)}$ (defined in Volterra 1887g) are respectively given by X, Y and Z. Now, such a function of line F is constant on any line L of the surface Σ defined by the equation $f =$ constant where f is solution of the partial differential equation

$$X\frac{\partial f}{\partial x} + Y\frac{\partial f}{\partial y} + Z\frac{\partial f}{\partial z} = 0.$$

Hence a possible link between the additive relation involving multiple integrals and some properties of some functions of line. Volterra remains however rather vague about the aforementioned link.

It is also interesting to note that in Volterra (1896) Volterra cites a long paragraph from a similar work by Picard and extracts from letters that show that the two mathematicians were in contact regarding this question:

> With regard to this I am pleased to present to the Academy an extract from two letters that our illustrious corresponding member Mr. Picard sent me following the communication I had sent to him of previous proposals and that he authorised me to publish: '… I believe I

recognise in the question that you show me something that appears to relate to a study that I began but which I have never gone into thoroughly and that I have never published.'[16] (Volterra 1896, p. 333)

3 Jacques Hadamard: Clairvoyant Catalyst of French-Italian Mathematical Relations

In July 1904, a few weeks before the opening of the International Congress of Mathematicians in Heidelberg, Jacques Hadamard, who knew he was working on topics similar to those of Volterra's interests and specialties, thought it prudent to ask his Italian colleague what he would speak about at the congress. This exchange marks the real beginning of their personal relationship. The themes that occupied Hadamard in those years appertained to the broad field of mathematical physics. He was specifically interested in issues surrounding the study of partial differential equations and their solutions. Such equations model the basic situations of mechanics and, more specifically, of wave problems. In fact, during the Heidelberg congress, Hamadard and Volterra co-chaired the session of 10 August 1904 devoted to these problems, during which they each also gave a talk, Volterra on wave theory and Hadamard on limit conditions in partial differential equations in physics; Arnold Sommerfeld and Robert William Genesis were the two other speakers in that session. Moreover, 2 days later, on 12 August, Hadamard gave a new talk in another session that this time he co-chaired with Tullio Levi-Civita, in which he presented a paper on the fundamental solutions of linear partial differential equations, followed by a discussion in which Volterra was the main speaker.

How did the two mathematicians come to find themselves together in Heidelberg as the two indisputed specialists in mathematical physics and PDEs? Although almost exactly the same age (Volterra was the elder by 5 years), the mathematical careers of Hadamard and Volterra had followed quite distinct trajectories. From the beginning of his professional life Volterra was fascinated, as we have seen, by questions of mathematical physics. Starting with his first articles published in 1881, as we discussed in the first section, Volterra's work included studies on the distribution of heat or electrical energy in materials, issues related to potential theory, which he made the subject of his habilitation thesis at the Scuola Normale of Pisa in 1883. On the other hand, we have amply shown that Volterra also dealt with more specifically mathematical considerations regarding questions of integration, or extensions of the concept of function (with the concept of *functions of lines*) and the development of an appropriate calculus that can permit modelling of new physical situations such as phenomena of heredity. For his French colleague, as we shall see,

[16]A questo riguardo sono ben lieto di poter presentare all'Accademia l'estratto di due lettere che il nostro illustre corrispondente signor PICARD mi ha dirette in seguito alla comunicazione fattagli delle precedenti proposizioni e che egli mi autorizza a pubblicare:' …je crois reconnaître dans la question que vous m'indiquez quelque chose qui doit avoir un rapport avec une étude que j'avais commencée mais que je n'ai pas approfondie et que je n'ai pas non plus publiée.'

the situation was quite different. In this present section we will examine how Hadamard began to take an interest in this type of problem. Such research is an outgrowth of the long and rich history of the development of the study of PDEs in the nineteenth century, which we will briefly review here.

Hadamard's work on PDEs is abundant and fundamental. One of its original aspects, very valuable for the historian, is that Hadamard consistently provided a very honest historical, or at least chronological, account to explain how his work is inscribed in the process of construction the theory, as for example at the conference he gave to 1928 International Congress of Mathematicians in Bologna (Hadamard 1928). A complete history of Hadamard's contributions to the theory of PDEs is beyond the scope of this present article; as already mentioned in the introduction, the reader would do well to consult the extensive biographical work of Vladimir Maz'ya and Tatyana Shaposhnikova devoted to the mathematician (Maz'ya and Shaposhnikova 1998), especially Chaps. 14 and 15, as well as panoramic overview by Gray (2001). For our part we will focus on a specific aspect which, although present in the works just cited, may be discussed in greater detail in the light of our own study, whose focus is more restricted.

The 1890s saw the development of intense correspondence between French and Italian mathematicians regarding PDEs, exchanges in which Volterra quickly takes centre stage. Therefore, the two questions we attempt to answer in this section, focussed on our two protagonists, are as follows. How did the Italian works, including those of Volterra, come to be known to Hadamard? How was his relationship with Volterra established in the few years between their first meeting in 1897 and the congress of 1904, where they became close friends? These questions lead us to look more closely at the role played by the previous international conferences (1897 in Zürich and 1900 in Paris) and, going beyond the example of Hadamard, to shed light on how works by Italian mathematicians during the decade between 1880 and 1890 were reviewed and extended in France at the turn of the century.

3.1 Elasticity, Waves and PDEs

3.1.1 Lamé, Riemann and Du Bois-Reymond

Although it is difficult to identify an indisputable starting point for the systematic study of PDEs in the late nineteenth century, it seems legitimate to underline, as does Gray (2001), the central importance of the study of wave propagation in elastic medium. This theory, which especially involves the study of the characteristics that describe the deformation of materials, effectively forms a bridge between different fields of mechanics: those of continuous media and fluids in particular, but also with regard to thermodynamics, by addressing the question of the compression of gasses. It had found its mathematical expression in the research of Lamé, of which here we will mention just a few brief aspects (for details, see Tazzioli (1993, 1995), Gielis et al. (2010)). Lamé's interest in elasticity stemmed from a number of mechanical studies undertaken during his long stay in Russia in the company of Benot

Clapeyron (1799–1864), notably his studies on suspension bridges. However, beyond these questions of engineering, Lamé pursued a broader goal. His conception of science and the role played by mathematics was not without echoes, and we will see it adopted later by Duhem and Hadamard. Joseph Bertrand described them thus:

> In the eyes of Lamé, science was a single entity, and the relationships, even among the individual formulas, among theories still distinct, were sure indicators of a more general doctrine that must 1 day embrace them all. The distinction between pure mathematics and applied mathematics was, in his eyes, dangerous and false.[17] (Bertrand 1878, p. 19)

Lamé's aim was to show that a theory of elasticity could be used to support a unified wave theory where elasticity of the ether served to explain both the phenomena of heat and the propagation of light. Returning to France after the consolidation of diplomatic relations with Russia that followed the July Revolution of 1830,[18] and appointed professor at the école Polytechnique in 1831, Lamé settled on a program of applying modern geometry and analysis to develop a theory of elasticity. In 1833 Clapeyron and Lamé published a note (Clapeyron and Lamé 1833) presenting the equations describing the internal equilibrium of homogeneous solid bodies: they show that the equations of elastic equilibrium (involving the directional stress introduced by Cauchy) are identical to Navier's equations for molecular forces. In 1852 appeared the *Leçons sur la théorie mathématique de l'élasticité des corps* (Lamé 1852). In the ninth lesson, Lamé studied small motions of an elastic membrane in the form of linear hyperbolic PDEs. His works were reprised and extended by Riemann (see Botazzini and Tazzioli 1995).

In 1864 appeared the great treatise du Bois Reymond (1864) by Paul du Bois-Reymond (1818–1896) on PDEs, in which the German mathematician introduced the now classic classification for second-order linear equations (hyperbolic, elliptic and parabolic). This was one of the first books that sought to impose a semblance of unity on the landscape of PDEs, which was still dominated by a catalogue of specific methods for particular cases. Du Bois-Reymond also made known the works of Riemann for the case of the equation of the propagation of sound (Riemann 1860), with the first draft of a method of characteristics that consisted in finding the curves along which the solution of the equation is obtained through ordinary differential equations. Described by du Bois-Reymond and Gaston Darboux (1842–1917) in various works of the 1870s and 1880s, this method makes it possible to obtain the value of the solution for certain hyperbolic PDEs in the plane at a point situated within a quadrilateral formed by the characteristic curves of the equation in function of the values assumed on its sides. Luigi Bianchi

[17]Aux yeux de Lamé, la science était une, et les rapprochements, même dans les seules formules, entre des théories encore distinctes étaient l'indice certain d'une doctrine plus générale qui doit un jour les embrasser toutes. La distinction entre les mathématiques pures et les mathématiques appliquées était, à ses yeux, dangereuse et fausse.

[18]Clapeyron, who had been close to the movement leading to the Decembrist Revolt in 1825, had been, like Lamé, forced to flee from the repression of the Saint-Simonians, and returned at about the same time.

(1856–1928), one of the first Italians to work on these questions, extended certain of these works to the elliptic case; in 1889 in Bianchi (1889) he described the role played by the work of du Bois-Reymond and Darboux:

> In the second volume of the fine lessons on the general theory by Mr. Darboux and in the last memoir by du Bois-Reymond, published in the 104th volume of Crelle's Journal, are contained results of great importance for the theory of second-order partial differential linear equations with two independent variables x, y. In particular, for hyperbolic equations, that is, those in which the two systems of characteristic lines are real and distinct, is proven the fundamental theorem, according to which in every quadrilateral bounded on the plane xy by four characteristics, when the values assumed by the integral along two adjacent sides of the quadrilateral are determined, the values of the integral itself in the entire interior region of the quadrilateral are identified (Darboux §364; Du Bois Reymond §15). This theorem is correlated, for the equations of the elliptic type (with imaginary characteristics), to another by which the values assumed by the integral within a connected area are generally identified by the values that the integral receives on the boundaries of the area. Several simple observations contained in the present note make it possible to establish this result with great generality. The process itself is readily extensible, as will be seen, to the case of any number of independent variables. But there is no treatment at all of the much more difficult question of whether these boundary values can effectively be arbitrarily given, a question that, with the exception of a few special cases, does not appear possible to solve at present.[19] (Bianchi 1889, pp. 35–36)

3.1.2 Kirchhoff

In 1882, the physicist Gustav Kirchhoff (1824–1887) intervened unexpectedly in the theory of PDEs. In Kirchhoff (1883), he proposed in the spherical case of the wave equation (for a space of dimension 3) a new approach with the aid of Green's formulas, in expressing solutions in integral form, extending Poisson's formula of the cylindrical case (for a space of dimension 2). Kirchhoff had come across the idea of using Green's theorem in von Helmholtz (1860), who used it to obtain the equation of vibrations of air in a tube: on these pioneering application of Green's formula, one may consult (Archibald 1988; 1989; Tazzioli 2001). The details regarding the expressions obtained by Kirchhoff can be found in Egorov and

[19]Nel tomo 2° delle belle lezioni sulla teoria generale del sig. Darboux e nell'ultima Memoria del Du Bois Reymond, inserita nel 104o volume del Giornale di Crelle, sono contenuti risultati di grande importanza per la teoria delle equazioni lineari a derivate parziali del 2o ordine con due variabili indipendenti x, y. In particolare per le equazioni del tipo iperbolico, nelle quali cioè i due sistemi di linee caratteristiche sono reali e distinti, viene dimostrato il teorema fondamentale, secondo il quale in ogni quadrilatero racchiuso sul piano xy da quattro caratteristiche, fissati i valori che l'integrale assume lungo due lati adiacenti del quadrilatero, risultano individuati i valori dell'integrale stesso in tutta la regione interna del quadrilatero [Darboux, § 364; Du Bois Reymond § 15]. A questo teorema fa riscontro, per le equazioni del tipo ellittico (a caretteristiche immaginarie), l'altro che i valori assunti dall'integrale nell'interno di un campo connesso sono generalmente individuati dai valori che l'integrale riceve sul contorno del campo. Alcune semplici osservazioni contenute nella presente Nota permettono appunto di stabilire con molta generalità questo risultato. Il processo stesso è immediatamente estendibile, come si vedrà, al caso di un numero qualunque di variabili independenti. Però non viene qui affatto trattata la questione molto più difficile se tali valori al contorno possano darsi effettivamente ad arbitrio, questione che, salvo pochi casi particolari, non sembra per ora prossima a risolversi.

Shubin (1998, p. 140 ff), and here we will only mention that, unlike Poisson's formula, for which the integral acts over the whole disk of radius *at* centered on the point under consideration—*a* being the speed of wave, *t* the time—Kirchhoff's formula in dimension 3 is an integral acting over the surface of the sphere of radius *at*. Kirchhoff then interpreted this result as the analytical expression of Huygens's principle explaining the propagation of a light wave by considering the points step by step as small localized sources of emission of spherical waves, a principle which had never in fact received a satisfactory mathematical formulation and had been considered a simple explanatory artifice; Kirchhoff's discovery finally gave it its rightful place in the arsenal of mathematical physics. The 1882 article was immediately widely circulated, despite some inaccuracies in the mathematical treatment identified by Gian Antonio Maggi (1856–1937), an Italian mathematician from Messina who had been a student of Kirchhoff in Berlin; in 1888 Maggi proposed a slightly different proof (Maggi 1888). Eugenio Beltrami (1835–1900), thanks to a slightly more sophisticated formulation of Green's theorem, arrived in 1889 in Beltrami (1889) at a correction of Kirchhoff's original proof, while retaining his original idea of seeking the solutions dependent only on the norm $r = \sqrt{x^2 + y^2 + z^2}$.

3.1.3 Pierre Duhem

There then appeared a central figure in our history, the physicist Pierre Duhem. A complex and brilliant personality, Duhem taught in Lille from 1887 to 1891, and a year in Reims, before settling down as a professor of theoretical physics at Bordeaux in 1894, where he remained until the end of his life. His thesis on thermodynamic potential had strong repercussions: it was rejected by the jury because of the ferocious opposition of Marcelin Berthelot (1827–1907), because Duhem had questioned—rightly—the principle of maximum work.[20] Berthelot became an implacable enemy, and did all that was humanly possible to hinder the career of his young colleague, blocking his access to all chairs in Paris. It may be added that this head-on opposition was also fuelled by Duhem's position in the political spectrum, there being no mystery about his strong sympathies for the far right political movement *Action Française*, his hostility to the Republican regime and his uncompromising commitment with the rigid, Gallican fringe of the Roman Catholic Church.[21]

Duhem's monumental output comprises both technical studies and texts on philosophy of science, as well as many educational treatises of the highest calibre.

In his course at the Faculty of Science at Lille, published in 1891 (Duhem 1891), Duhem expounded many aspects of physics in which PDEs played a central role,

---

[20]The principle of maximum work had been stated by Berthelot in 1879 in his *Essai de Mécanique chimique fondé sur la thermochimie* Berthelot (1879): 'Tout changement chimique accompli sans l'intervention d'une énergie étrangère tend vers la production du corps ou du système de corps qui dégage le plus de chaleur.'

[21]In particular, see Jacki (1894) which gives a very vivid picture of Duhem's philosophical-political position.

particularly the theory of elasticity. Taking stock of the results that had been achieved up to that point for hyperbolic PDEs, he showed (Duhem 1891, Tome II, Livre III, Chapitre VIII) that Kirchhoff's method discussed above can be established for dimensions 1 or 3.

Duhem was an effective promoter of his own course, sending copies out himself to many people, including Eugenio Beltrami, whom he asked to transmit it to Italian colleagues involved in mathematical physics. Beltrami did so, sending it to Ernesto Padova (1845–1896), then professor of theoretical mechanics at the University of Padua, perhaps because Padova had just published a paper in the journal *Nuovo Cimento* (Padova 1891b)[22] proposing a unitary mechanical theory of electrical, magnetic and luminous phenomena.

In a letter that he sent to Duhem in January 1892[23] to thank him for his book, Padova also insisted that Duhem express his opinion on these problems, but noted that *his friend Volterra* (who was then still in Pisa for some time to come, where he was in particular assistant to Dini) had also presented, during the same session of the Lincei, a note on electrodynamics, in which he arrived in part to the same results by another route.

The case of Ernesto Padova is a good illustration of the difficulty experienced by mathematicians nourished in the mid-nineteenth century by the milk of Lagrange and Laplace regarding the question of the universality of the mechanistic approach to physical phenomena. In March 1892, after reading Duhem's book in detail, Padova wrote him a long letter containing an impassioned appeal in support of mechanical theories, regarding which, in his opinion, Duhem was *too severe* and Poincaré too reckless:

> We cannot, because a mechanical theory adequately explains certain facts, say: things are going thus! this is how matter is constituted! that is perfectly right and you are right to fight this human vanity or presumption, but that does not prevent mechanical theories from occupying a much more important place than purely physical theories in the discovery of natural phenomena ... There is in many analysts today a tendency to reject these theories as a whole and I fear it will eventually create a gulf between them and physicists. ... In saying: if a phenomenon has a complete mechanical explanation, it will include an infinity of others that render an equally good account of all the particulars revealed by experience, doesn't Poincaré, while stating a true fact, cast discredit on all mechanical interpretations?[24] [E. Padova à P. Duhem, mars 1892]

[22]Developed further in Padova (1891).

[23]Duhem's scientific correspondence is found in the Archives of the Paris Académie des Sciences.

[24]Qu'on ne puisse, parce qu'une théorie mécanique explique suffisamment bien certains faits, dire: les choses se passent ainsi ! voilà comment est constituée la matière ! c'est parfaitement juste et vous avez bien raison de combattre cette vanité ou présomption humaines, mais cela n'empêche pas qu'aux théories mécaniques n'appartienne une place bien plus importante qu'aux théories purement physiques dans la découverte des phénomènes naturels ...Il y a dans beaucoup d'analystes aujourd'hui une tendance à rejeter en bloc ces théories et je crains qu'elle ne finisse par creuser un ab me entre eux et les physiciens. ...En disant: si un phénomène comporte une explication mécanique complète, il en comportera une infinité d'autres qui rendront également bien compte de toutes les particularités révélées par l'expérience, M. Poincaré, tout en énonçant un fait vrai ne jette-t-il pas de discrédit sur toutes les interprétations mécaniques?

In the same letter, Padova remarks that Volterra, to whom he suggested that Duhem send his works, is himself in quite the same vein as Poincaré with regard to mechanical theories.

In April 1892 Duhem sent his course to Volterra. In thanks,[25] Volterra sent his own article (Volterra 1891a) on Hertz's equations, and told him about the forthcoming publication of his article on light vibrations in birefringent media in *Acta Mathematica* (Volterra 1892b).

In July 1892, Volterra, who had read Duhem's course, took pen in hand to send his observations to the author. Having developed Duhem's fine discussion of Huygens's principle in light of Kirchhoff, he mentions having found what appears to him to be an extension of those results:

> You have shown that only two cases in the general equation of elastic vibrations
>
> $$\frac{\partial^2 V}{\partial t^2}\partial t^2 = A^2 \sum_1^m \frac{\partial^2 V}{\partial x_i^2} \quad (3)$$
>
> have int
>
> $$V = \psi F(r - At) \quad (4)$$
>
> egrals of the formwhere F is an arbitrary function and ψ is a function of r only. These are the cases where $m = 1, m = 3$. The first corresponds to the problem of vibrating strings. The second, to the question of spherical waves in 3-dimensional space. Since the Kirchhoff formula is based on the existence of the integral (4) in the case $m = 3$ we must conclude that we cannot proceed in the same way to find an analogous formula in the case of cylindrical waves or elastic membranes, and for generalising the same formulas for the vibration in an m-dimensional space. For some time I have tried to obtain formulas that, not having the same form as that of Kirchhoff, could substitute it in the case of cylindrical waves and have the same meaning and extend the results to the general case. In pursuing this objective, I have seen that if one does not pose the condition that ψ is a function of r only, we can find the integrals of Eq. (3) having the form (4), but I have shown that (apart from the cases $m = 1, m = 3$) the function ψ must have singularities (polydromes, etc.) such that starting from these integrals and using Kirchhoff's method, we find results that differ substantially from those of Kirchhoff. Due to this one cannot attain the goal. That is why I tried another route.[26] [V. Volterra à P. Duhem, 24 juillet 1892]

[25]Volterra and Duhem's correspondence is found in the archives of the Académie des Sciences de Paris and in the archives of the Accademia dei Lincei.

[26]Vous avez montré qu'il n'y a que deux cas à l'équation générale des vibrations élastiques

$$\frac{\partial^2 V}{\partial t^2} = A^2 \sum_1^m \frac{\partial^2 V}{\partial x_i^2} \quad (1)$$

possède des intégrales de la forme

$$V = \psi F(r - At) \quad (2)$$

où F est une fonction arbitraire et ψ est une fonction de r seulement. Ce sont les cas où $m = 1, m = 3$. Le premier correspond au problème des cordes vibrantes. Le second à la question des ondes sphériques dans l'espace à 3 dimensions. Puisque la formule de Kirchhoff est fondée sur

Volterra then set forth in this long letter the essence of what would constitute his note to the Accademia dei Lincei (Volterra 1892c). For a hyperbolic system in the cylindrical case (spatial dimension $n = 2$), Volterra showed that it is possible to obtain the formulas of integral representation extending the formulas of Poisson and Kirchhoff, by means of an extension of the method of characteristics (formulas A, B, D, E in Volterra 1892c, pp. 166–168).[27] For Volterra, this was a satisfactory expression of Huygens's principle, describing the wave in the elastic medium beginning with localised waves.

Duhem replied immediately and enthusiastically:

> I know of nothing analogous to the work of which you speak; I would add that it seems to me extremely interesting; the exceptional character that I recognised in the two cases $n = 1, n = 3$ struck me vividly, and I much desired greater knowledge of the general case; but as I am not a mathematician, I could not conceive of clarifying this difficult issue myself. I had recently urged my friend Paul Painlevé, whose name is certainly not unknown to you, to deal with this question. He had not yet had time to think about it. All of this will

---

(Footnote 26 continued)
l'existence de l'intégrale (2) dans le cas $m = 3$ on doit conclure qu'on ne peut pas procéder de la même façon pour trouver une formule analogue dans le cas des ondes cylindriques ou des membranes élastiques, et pour généraliser les mêmes formules pour les vibrations dans un espace à m dimensions. Depuis quelque temps, j'ai tâché d'obtenir des formules qui, n'ayant pas la même forme que celle de Kirchhoff pouvaient la substituer dans le cas des ondes cylindriques et en avoir la même signification et étendre le résultat au cas général. En poursuivant ce but, j'ai vu que si l'on ne pose pas la condition que ψ soit une fonction de r seulement, on peut trouver des intégrales de l'équation (1) ayant la forme (2), mais j'ai démontré que (en dehors des cas $m = 1, m = 3$) la fonction ψ doit avoir des singularités (polydromie etc.) telles qu'en partant de ces intégrales et en employant la méthode de Kirchhoff, on trouve des résultats qui diffèrent substantiellement de ceux de Kirchhoff. Par là on ne peut donc atteindre le but. C'est pourquoi j'ai essayé un autre chemin.

[27]Volterra's results on this type of PDE were later extended to the case of any dimension by Orazio Tedone (1870–1922) in several publications between 1893 and 1898 (see in particular Tedone 1898). Volterra's approach was taken in the case of non-constant coefficients by Jean-Marie Le Roux (1863–1949) in his thesis of 1895 (LeRoux 1895) and a newcomer from Bordeaux, Joseph Coulon (dates unknown), whose article (Coulon 1898) was published in 1898, generalizing Tedone's results for the heat equation to equation

$$\frac{\partial^2 U}{\partial x_1^2} + \cdots + \frac{\partial^2 U}{\partial x_p^2} - \frac{\partial^2 U}{\partial y_1^2} - \cdots - \frac{\partial^2 U}{\partial y_q^2} = 0.$$

These works were in turn later extended by the research of Robert d'Adhémar (1874–1941) in Adhémar (1901), d'Adhémar (1902) et d'Adhémar (1902). On these subjects, see Freda (1937). As for Coulon, an ecclesiastic, he discussed his thesis (on the integration of second-order partial differential equations by the method of characteristics) in Paris in 1902, then went to Fribourg in Switzerland to direct the French section of the Collège Saint-Michel.

> combine to show you how much pleasure and interest your memoir will bring me when you have published it.[28] [P. Duhem à V. Volterra, 30 juillet 1892]

A few months later, after reading Volterra's article, Duhem once again wrote him an enthusiastic letter:

> Kirchhoff's formula with the transformation that it must be subjected to when one wishes to extend it either to the equation $\frac{\partial^2 V}{\partial x^2} + \frac{\partial^2 V}{\partial y^2} = a^2 \frac{\partial^2 V}{\partial t^2}$ or to Lamé's equations, appears to me to be one of the finest achievements that has been made in a long time in the field of second-order partial differential equations. While Kirchhoff was a great innovator, you can, in my opinion, claim a great part of these new conquests. So accept my very sincere congratulations.[29]

3.1.4 A Fine Bordeaux Vintage

In 1893 Jacques Hadamard was appointed to the University of Bordeaux. This was his first academic position, after 3 years teaching at the Lycée Buffon in Paris, years which, if he spent them without pleasure—Hadamard's talents for secondary education do not seem to have been highly developed—had nevertheless allowed him to finish his thesis in peace, discussed in 1892 at the Sorbonne (Hadamard 1892a). They also provided him with the opportunity to find a rare pearl in the person of his young pupil Maurice Fréchet (1878–1973), whose mathematics education and future career he followed from that moment on with tireless zeal.[30] In Bordeaux, Hadamard was first assigned temporarily to teach the course of mechanics and astronomy before being named, in 1896, to the chair of astronomy and theoretical mechanics.

The scientific conditions in Bordeaux proved particularly suitable for the young Hadamard, due to the presence in town of a learned society, the *Société des Sciences Physiques et Naturelles* (SSPN), founded in 1855. The mathematician Jules Houël, who became librarian of the society in 1865, put his polyglot talents at the service of the SSPN. This made it possible for him to interact with mathematicians from around the world, to translate many works, to make the library one of the

[28]Je ne connais rien d'analogue au travail dont vous me parlez; j'ajouterai qu'il me para t extrêmement intéressant; le caractère exceptionnel que j'avais reconnu aux deux cas $n = 1, n = 3$ m'avait vivement frappé, et je désirais beaucoup une connaissance plus approfondie du cas général; mais comme je ne suis nullement mathématicien, je ne pouvais songer à élucider moi-même cette difficile question. J'avais récemment poussé mon ami Paul Painlevé, dont le nom ne vous est certainement pas inconnu, à s'occuper de cette question. Il n'avait pas encore eu le temps d'y songer. Tout cela vous marque suffisamment combien de plaisir et d'intérêt me causera votre mémoire lorsque vous l'aurez publié.

[29]La formule de Kirchhoff avec les transformations qu'il faut lui faire subir lorsqu'on veut l'étendre soit à l'équation $\frac{\partial^2 V}{\partial x^2} + \frac{\partial^2 V}{\partial y^2} = a^2 \frac{\partial^2 V}{\partial t^2}$ soit aux équations de Lamé, me semblent être une des plus belles conquêtes qui aient été faites depuis longtemps dans le domaine des équations aux dérivées partielles du second ordre. Si Kirchhoff a été un grand novateur, vous pouvez cependant, ce me semble, réclamer une belle part de ces nouvelles conquêtes. Recevez donc mes bien sincères félicitations. [P. Duhem à V. Volterra, 28 novembre 1892].

[30]See the next section.

richest scientific libraries in France, and to raise the Société's publications to a prestigious level on the international mathematical scene.[31]

In 1894, Pierre Duhem, whom Hadamard had known personally for a short time when he began his education at the école Normale Supérieure just as that of Duhem ended, arrived in Bordeaux to take over the chair of theoretical physics.

In his 1927 article on Duhem Hadamard (1927), Hadamard never misses a chance to pay tribute not only to the extraordinary scientific curiosity of the elder scholar, but also to his conception of a mathematical physicist, which had seduced him. In all likelihood it was this shared passion for scientific activity in all its forms that allowed the two colleagues in Bordeaux to spend an enormous amount of time together without their strong political differences disturbing their relationship. Since his days at the école Normale, Hadamard had perceived Duhem's enthusiasm for new theories in physics, such as Hugoniot's notion of the propagation of shock waves in fluids, which Duhem had learned of through his course at the University of Lille. That enthusiasm was visibly contagious.

Hadamard himself wrote:

> For my part our reunion at the Faculty of Sciences of Bordeaux brought me the rare good fortune of completing the reading [of Duhem's course] with a valuable and continuous exchange of views. To this reading, I owe the greater part of all my subsequent work devoted to the calculus of variations, Hugoniot's theory, to hyperbolic partial derivative equations, to Huygens's principle.
>
> Duhem himself returned to almost all these questions in the course of his immense work, and most of the theories he had so happily and so luminously expounded were suggested to him here by a particular remark, and there by additions of fundamental importance.[32] (Hadamard 1927, pp. 644–645)

3.2 From Congress to Congress

In 1897, Hadamard, appointed non-tenured professor of analytical and celestial mechanics at the Collège de France, returned to Paris. It was also during this year that he became aware of the Italian work on PDEs, and in particular during the first International Congress of Mathematicians in Zürich. Only a small number of French mathematicians attended, perhaps because they saw little use of going abroad when, in their opinion, Paris was the cardinal point of the mathematical world, the place where all that is important ended up happening; maybe they were

---

[31]On Jules Houël in Bordeaux, see Vezes (1906) and Plantade (2014).

[32]Pour ma part notre réunion à la Faculté des Sciences de Bordeaux me procura la rare fortune [de] compléter la lecture [du cours de Duhem] par de précieux et continuels échanges de vues. A cette lecture, je dois la plus grande partie de mes travaux ultérieurs tous consacrés au Calcul des variations, à la théorie d'Hugoniot, aux équations aux dérivées partielles hyperboliques, au principe de Huygens.

Duhem lui-même revenait sur presque toutes ces questions, dans la suite de son immense labeur, et la plupart des théories qu'il avait si heureusement et si lumineusement exposées lui suggérèrent ici des remarques de détail, là des compléments d'une importance fondamentale.

also reluctant to go to a place that was considered to be too much under German influence. This point probably deserves to be studied more thoroughly. Borel, who was part of the French delegation, seized every opportunity to criticize this attitude in the long and savoury account that he wrote (Borel 1897). He also emphasized the importance, in his opinion, of face-to-face encounters with his colleagues, allowing the *living word* to take precedence over *cold print.* The Italian delegation was itself an important one, and Volterra appeared among the most enthusiastic supporters of the formula. It is thus remarkable that on this occasion the two men became acquainted and struck up a friendship that lasted until Volterra's death in 1940. The fiery beginnings of their relationship are discussed elsewhere (Mazliak 2015b), and here we will cite only the first letter that Borel sent to Volterra, which illustrates the optimistic climate that followed the Congress:

> If personal relationships that have been forged between us in Zürich were to end after three years or more, the greatest and most enjoyable benefits of the Congress would be lost. But I quite intend for that not to happen for us.[33]

Hadamard was also a member of the French delegation, and he also met Volterra for the first time on this occasion, but unlike Borel, their meeting seems to have remained without any immediate effects on a personal level. Having begun various works on mathematical physics during his stay in Bordeaux, Hadamard came to Zürich to speak about PDEs, but from a prospective that was somewhat particular. In a short talk intriguingly entitled *Sur certaines applications possibles de la théorie des ensembles* (Hadamard 1897), Hadamard explained the interest that led him to study certain sets of functions in order to solve problems of extrema, and to study the properties of these sets. Inspired by the calculus of variations, it was especially in view of applications to PDEs whose solution could be solution to such a problem of extremum, namely, a function that maximizes a certain functional (Hadamard did not use the term *functional*, which would appear in his vocabulary a few years later). Hadamard especially evoked the study of the set of continuous real functions on the interval [0,1] endowed with the uniform norm and the cardinality of a covering by balls with a given radius.[34] Hadamard's talk provoked reactions. Borel pointed out (Hadamarad 1928) that he himself had considered some sets of functions in his studies of series, as the set of the functions which are the coefficients of the decomposition in power series of the solution of a PDE. But it was the short contribution by Salvatore Pincherle (1853–1936), who was also a member of the Italian delegation, which is much more significant for our purposes. Pincherle reported that several Italian mathematicians, such as Ascoli, Volterra Arzelà? and he himself, had for some years been examining these questions regarding sets of functions, envisioning functions as points of a set and even as a continuum.

[33]Si les relations personnelles qui se sont nouées à Zürich devaient s'éteindre pendant trois ans ou plus, le plus grand et le plus agréable des avantages des Congrès serait perdu. Mais je compte bien qu'il n'en sera pas ainsi pour nous. (Borel à Volterra, 14 novembre 1897).

[34]We obviously use modern terminology here. One can recognize an origin of the questions of precompacity which Fréchet will consider some years later, as we shall see in the next part.

We have no knowledge of what Hadamard's reaction might have been; Pincherle's remark seems not to have had any effect at first. It must be said that Hadamard was then absorbed in work of another kind, although this also involved PDEs, which were by that time central to his research, and in particular in the preparation of his course at the Collège de France of the years 1898–1900, which was published shortly after with title *Leçons sur la propagation des ondes et les équations de l'hydrodynamique* (Hadamard 1903a)

There, he repeated and extended Duhem's course of 1891, assembling for the first time the results obtained through solutions to hyperbolic PDEs, notably a general theory of characteristics inspired by the thesis of Jules Beudon (1869–1900) see (Beudon 1896): Fig. 1 below shows the descendence between French and Italian works on PDEs as sources for Hadamard's lectures.

On the theory of systems of PDEs. Hadamard defined the characteristic surfaces associated with an equation of arbitrary order as the location of discontinuity of

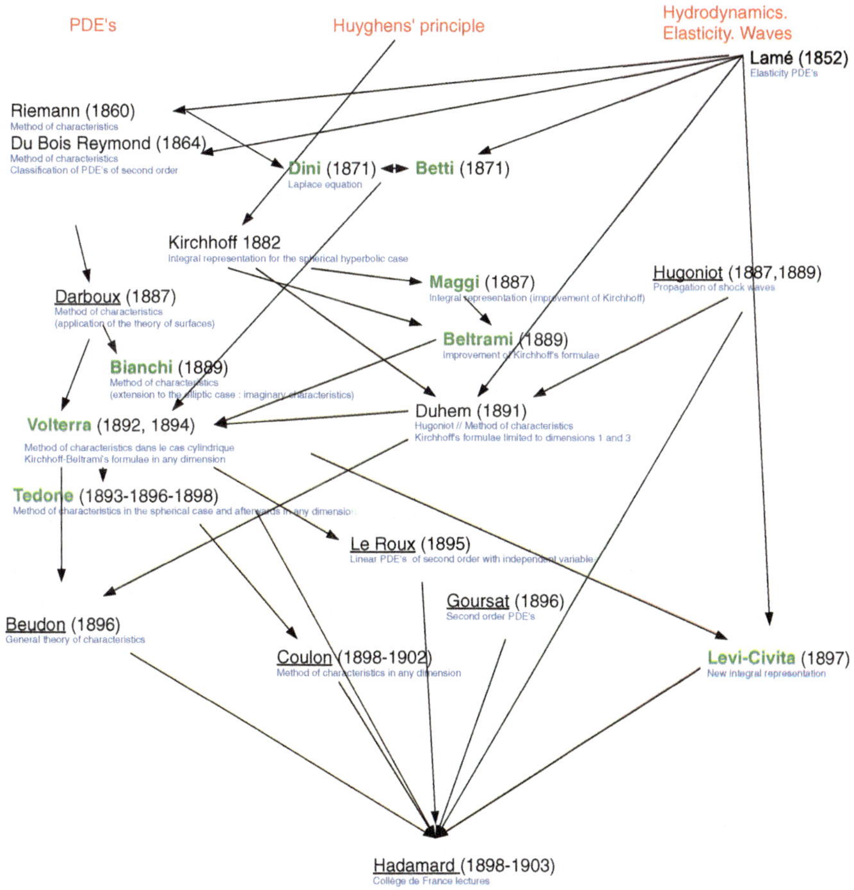

Fig. 1 The sources for the course on the PDEs that Hadamard taught at the Collège de France (1898–1900). Names of the Italians are in *bold*, those of the French are *underlined*; the dates are those of the principal publications on PDEs

certain derivatives and showed that this approach permitted a good mathematical translation of the analytical expression of the Hugoniot's considerations on shock waves (see Maz'ya and Shaposhnikova 1998, 14.2).

In 1900, Hadamard relied on Huygens's principle. In his article of that year (Hadamard 1900), he arrived at the conclusion that Volterra's interpretation was questionable from the point of view of physics.[35]

That same year the International Congress of Mathematics took place, this time in Paris. Volterra and Hadamard met once again, and gave each a talk on PDEs in the same session of 10 August 1900. Volterra's talk, soberly entitled *Sur les équations aux dérivées partielles* (Volterra 1900) reprised the integral formulas that he set out in his article of 1894. That of Hadamard, entitled *Sur les équations aux dérivées partielles à caractéristiques réelles* (Hadamard 1900a), showed that the clear separation between the elliptic case (imaginary characteristics) for which Cauchy's problem generally has no solutions, and the hyperbolic case where it in fact has some, is in fact only valid in the simple case where the equation is given in the space as a whole. In cases where the boundary conditions affect a portion of the space—such as Dirichlet or Neuman boundary conditions—the situation is more complicated, and the two types of equations have common properties.

There is no trace of discussions between the two mathematicians. We know that a few weeks after the Congress Hadamard sent his first letter to Volterra, who had just been appointed to the University of Rome, but its nature was of the most practical kind[36]:

> I have written to you in Turin, not knowing whether you have already moved to Rome, and I hope that my letter reaches you just the same. Its purpose is to beg you to send me, or better yet, send to Mr. Duporcq, Secretary of the Congress, 162 Boulevard Péreire in Paris, the text of the talk that you gave on equations with real characteristics, which we would like to have for the proceedings of the congress.
>
> I am pleased to have this opportunity to remember myself to you, and ask you accept, with all my best wishes to Mrs. Volterra, the assurance of my kindest regards.[37] [J. Hadamard à V. Volterra, août ou septembre 1900]

[35]Hadamard's criticism was based on the fact that in its original version, the principle enunciated by Huygens in fact implies that every point is at rest after the passage of the wave, which is a specific property of the equations in an odd-numbered dimension of space (such as the spherical case considered by Kirchhoff), since the integral expression of the solution is related to a sphere of radius *at*. In the case of even-numbered dimensions, to the contrary, the integral formula that extends to the entire interior of the ball prevents the point from ever returning to a state of rest and therefore makes it subject to a residual motion which gives the article its title.

[36]The correspondence between Hadamard and Volterra is conserved in the Archives of the Accademia dei Lincei, Rome.

[37]Je vous ai écrit à Turin, ne sachant si vous êtes déjà installé à Rome et j'espère que ma lettre vous joindra quand même. Elle a pour but de vous prier de vouloir bien m'envoyer ou, mieux encore, envoyer à M. Duporcq, secrétaire du Congrès, 162 boulevard Péreire à Paris, le texte de la communication que vous avez faite sur les équations à caractéristiques réelles, et que nous voudrions bien avoir pour les comptes-rendus du congrès.

Je suis heureux d'avoir cette occasion de me rappeler à votre souvenir et vous prie de recevoir, avec tous mes respects pour Madame Volterra, l'assurance de mes sympathiques sentiments.

Thus we see that Volterra and Hadamard, while being on cordial terms, do not appear to have been particularly intimate. In light of what we have shown, we can conclude that while Hadamard esteemed his Italian colleague, attributing to him a place of merit among those investigating the methods for solving hyperbolic PDEs, he expected nothing especially grandiose of Volterra's work. That situation would change in the years between then and the Heidelberg Congress of 1904.

3.3 The Discovery of Functions of Lines

In 1902, Hadamard wrote the article (Hadamard 1902a), his first publication using a formalism where functions are taken as variable elements, 5 years after his talk at the Zürich conference and Pincherle's observations, mentioned above. In his 1945 book about the psychology of a mathematician during the process of research (Hadamard 1945), he also testifies to his own surprise over how that concept intervened in his work:

> Much more surprising is the fate of the extension given to that initial conception [of calculus of variations] in the last part of the nineteenth century, chiefly under the powerful impulse of Volterra. Why was the great Italian geometer led to operate on functions as infinitesimal calculus had operated on numbers, that is, to consider a function as a continuously variable element? Only because he realized that this was a harmonious way of completing the architecture of the mathematical building, just as the architect sees that the building will be better poised by the addition of a new wing. One could already imagine that such a harmonious creation could be of help for solving problems concerning functions considered in the previous fashion; but that "functionals", as we called the new conception, might be directly related to reality could not be thought as otherwise than mere absurdity. Functionals seemed to be an essentially and completely abstract creation of mathematicians. Now, precisely the absurd has occurred. Hardly intelligible and conceivable as it seems, in the ideas of contemporary physicists (in the recent theory of "wave mechanics"), the new notion, the treatment of which is accessible only to students already familiar with very advanced calculus, is absolutely necessary for the mathematical representation of any physical phenomenon. Any observable element, such as a pressure, a speed, etc., which one used to define a number, can no longer be considered as such but is mathematically represented by a functional! (Hadamard 1945, pp. 129–130)

Hadamard embellished the situation somewhat by attributing an ethereal aesthetic vision to Volterra's invention of functions of lines. As we have seen, in his 1887 article (Volterra 1887i), the Italian mathematician had underlined how in numerous problems of mechanics and physics there are many naturally occuring quantities that depend on all the values given by one function or another, a situation that he illustrated again and again with the example of the temperature at a point of a conductive blade, which depends on all the temperatures on the edge. In his article Hadamard (1902a), Hadamard dealt with the notion of derivative of a line function such as Volterra had formulated. Under a certain number of assumptions of regularity, he had given an integral expression to it. However, Hadamard remarked that the derivative in question satisfies the properties of a *fonction linéaire* (linear function), a notion that he himself had introduced in 1901 in his slender, surprisingly little known book on Taylor series (Hadamard 1901). In Chapter "The French 'Analysts' and Peano's Mathematical Logic:Couturat's Remarques to Borel, Baire and Lebesgue", Hadamard observes that in a number of situations of extension of analytic functions, it is necessary to pay attention to functional transformations, such as Borel's method, who to $f(x) = \sum a_m x^m$ made correspond the *fonction associée* (associated function) $F(x) = \sum a_m x^m / m!$ Setting $\tilde{f}(x) = \int_0^\infty e^{-t} F(tx) dt$, one thus defines a function that in good cases extends f beyond the circle of convergence. In Chapter "Louis Rougier's reception of the Peano School", Hadamard then systematically introduces, following Bourlet (and Pincherle), functions that associate a function to another function, which he names *transmutations*, an example of which is the operator derivative. A transmutation is defined on a class of functions that Hadamard called a *functional field.* He then introduced a particular case of a linear transmutation A verifying $A(f_1 + f_2) = A(f_1) + A(f_2)$. The same situation was described as *distributive* by Pincherle and *additive* by Bourlet, and these are the only two authors who appear in the chapter's bibliography.

The notion of linear transformation thus makes it possible to envision quite general forms for the derivatives of a line function. Focusing specifically on how it is possible to define the second derivative of a functional, Hadamard showed that it requires a more subtle treatment and that it cannot be expressed in general in the form of an integral by means of the second partial derivative as a naive extension of the form given by Volterra to the first derivative might suggest.[38]

However, it is especially in a paper published in Hadamard (1903c) that Hadamard appears to have begun to see the advantages deriving from the theory of functions of lines, since he himself said that it seemed most practical for studying certain situations by directly manipulating a function as a variable, without having to be limited to analytic functions to go back to sequences of scalar coefficients. Extending by analogy Volterra's formalism to surface functions, Hadamard obtains different representations for linear functionals in the sense that he had introduced previously. A key aspect was the presentation of a fundamental example in which this type of formalism could be used: the first variation of Green's function g_A^B relative to two points A and B within a surface S that is deformed.[39]

---

[38] It is important to place this discussion in its proper context, to recall that here we are still several years before Fréchet established the notion of differential precisely to deal with this type of situation.

[39] The Green's function g associated with the Laplacian Δ in a domain D is defined by means of the impulse responses defined on the domain. Taking as an argument a pair of points (A, B) of the domain D, $g(A, B)$ (in Hadamard's (notation g_A^B) denotes the value in B of the harmonic function in the domain deprived of point A, null on the boundary and infinite in A. It was introduced in 1830 by George Green to obtain by convolution with a function f the solution of the equation $\Delta u = f$. In the modern approach to linear PDEs in the framework of the theory of distributions, it is generally preferred to use the concept of fundamental solution, which is closely connected to it (see Taylor 2011, p. 241ff).

Hadamard now became aware of having found terrain on which to stand with Volterra, or even compete with him. There is a touch of anxiety in the letter he sent him a few weeks before the conference in Heidelberg:

> I have always forgotten to ask you what you intend to speak about at the Congress of Heidelberg (in the section of applied mathematics). May I ask you to tell me about this, so I will not go over ground you have already broken?
>
> Please accept my apologies for the bother, and the assurance of my devoted very sentiments.[40] [J. Hadamard to V. Volterra, Spring 1904]

Volterra's generous reply on the eve of the congress, of which a note is conserved, probably reassured Hadamard.

> Thank you very much for your kind letter. I intend to say a few words at the Heidelberg Congress on wave theory. The subject is so vast that I am sure that if you also wish to talk about it there will be no interference between the two talks. Here are more details about my program. I begin by looking at whether the limits of what is called wave theory are well marked and I will attempt to treat this aspect by considering a few questions. I will then attempt to show some lemmas that I believe I will find in the analytical theory of dual distribution and I will try to compare Kirchhoff's method with Mrs. Kovalevsky's method. In relation to the vibrations of membranes, which I taught this year in my course, one can use the method of images when the membrane is rectangular and in other cases where it is bounded by straight lines (you have already touched on this subject in your paper). I wish to note that this method gives much simpler results in analogous problems of heat and electricity because it does not reduce to series. This usually occurs for hyperbolic equations. I will not fail in this regard to cite your paper[41] of the French Mathematical Society of 1903, in which when you use the wave method. If I have time I would like to touch on the relationship between vibration of membranes and wave theory. It might be that my plan is too vast. If you would like to make some remarks about it for me, I will be much obliged. I look forward to seeing you in Heidelberg. Mrs. Volterra will accompany me and she will be pleased to meet Mrs. Hadamard.[42] [Volterra to Hadamard, 27 July 1904]

---

[40]J'ai toujours oublié de vous demander ce que vous avez l'intention de traiter au Congrès de Heidelberg (Section des mathématiques appliquées). Puis-je vous demander de me renseigner sur ce point, afin que je n'aille point sur vos brisées?

Recevez avec mes excuses pour le dérangement que je vous cause l'assurance de mes sentiments bien dévoués.

[41]Volterra refers to Hadamard (1903).

[42]Je vous remercie beaucoup de votre aimable lettre. J'ai l'intention de dire quelques mots au congrès de Heidelberg sur la théorie des ondes. Le sujet est si vaste que je suis sûr que si vous voulez parler aussi du même sujet il n'y aura pas que des interférences entre les deux communications.

Voilà à peu près quel est mon programme. Je commence par chercher si les limites de ce qu'on appelle théorie des ondes sont bien marquées et je pense le [traiter en?] quelques questions. Je tâche après de montrer quelques lemmes que je crois trouver dans la théorie analytique de la double distribution et je tâche de comparer la méthode de Kirchhoff avec la méthode de Mme Kovalevski.

Par rapport aux vibrations des membranes j'ai enseigné cette année dans mon cours qu'on peut employer la méthode des images lorsque la membrane est rectangulaire et que d'autres cas où elle est limitée par des lignes droites (vous avez déjà touché ce sujet dans votre note). Je désire faire la remarque que cette méthode donne des résultats beaucoup plus simples dans les problèmes analogues de la chaleur et de l'électricité car on ne tombe pas sur des séries. Cela arrive en général pour les équations de type hyperbolique.

This new encounter in Heidelberg (their third Congress together!) between Volterra and Hadamard was a good one, and their relationship, like that between Borel and Volterra, was now one of esteem and mutual affection.[43]

In several subsequent articles, Hadamard pursued his variational approach to the physics of vibrations. He sought, for example, the law of variation of functions which are solutions to equations of the type $\Delta\Delta V = kV$ when one varies the shape of the boundary of the domain, a type of formulation that occurs naturally in the theory of elasticity when one considers the equilibrium conditions of fixed elastic plates. Consequently, Hadamard's functional equations shed light on the role of evolution equations that one might attempt to study to obtain the form of the various physical parameters. Taking stock of all these questions, Hadamard finally composed the lengthy paper (Hadamard 1907), which was awarded the Prix Vaillant. We find here the origin of the works by Paul Lévy (1886–1971), who was present at Hadamard's lectures at the Collège de France in 1909, to whom he suggested the idea of a thesis on the systematic study of functional equations.[44] Hadamard always maintained the highest regard for the work carried out by Lévy in his thesis.[45]

4 Maurice Fréchet: In Search of a General Functional Analysis

In the previous section we mentioned the fateful encounter at the Lycée Buffon in Paris between Hadamard and his young pupil Fréchet, whose progress he followed untiringly. When Fréchet, some 16 years after the older mathematician, returned to the école Normale Supérieure, he benefited from all the relationships that had been established between Hadamard and Volterra, and the new interest of French mathematicians in the work of Italian analysts, especially Pincherle and Volterra.

In 1902 Hadamard published his article (Hadamard 1902a) and Fréchet rapidly worked his way through Volterra's articles of 1887 on line functions. By 1904, Fréchet had published in his turn an article of his own on the subject (Fréchet 1904a).

(Footnote 42 continued)

Je ne manquerai pas de noter à ce propos votre note de la Société math. de France de 1903 où vous employez la méthode des ondes. Si j'aurai le temps je voudrais toucher à une relation entre les vibrations des membranes et la théorie des ondes. Peut être le programme est trop vaste. Si vous voulez bien me faire quelques remarques je vous en serai fort obligé.

Je serai heureux de vous voir à Heidelberg. Mme Volterra m'accompagne et elle sera heureuse d'y rencontrer Mme Hadamard.

[43]An overwhelming expression of friendship is shown during the tragic moments that Borel and Hadamard went through during World War II (see Mazliak and Tazzioli 2009). Further, when Volterra was in the grips of a struggle with the Fascist regime in 1930, Borel and Hadamard did everything they could to aid him.

[44]'Puisque vous avez mis la main sur le sujet, je vous l'abandonne' (since you have put your hand to the subject, I leave it to you); this is how Lévy described Hadamards's reaction in his autobiography Lévy (1970) (p. 42).

[45]See Mazliak (2015b) and also what was said by Hadamard in his talk at the Bologna congress in 1928 (Hadamard 1928a, p. 152). For more details on Lévy's work, see also Barbut (2013), Sect. 6.

The ideas that influenced and nourished Fréchet's work on functional analysis have been the subject of several historical studies that have focused specifically on the essential concept underlying his vision: the notion of 'abstract space' in analysis. In particular, the interested reader may consult the pioneering thesis by Reinhard Siegmund-Schultze (1990) and the very lengthy, three-part work by Taylor (1982), (1985) and (1987).

In his study, Taylor describes as 'indirect' Volterra's influence on the concept of abstract spaces:

> Fréchet's short paper of 1904 [on Weierstrass's theorem] broke absolutely fresh ground. Although Vito Volterra's work certainly had some influence on Fréchet's work taken as a whole, I think a good deal of it was exerted indirectly, through Hadamard. I see little or no reason for thinking that Volterra contributed directly to the shaping of Fréchet's ideas on L-classes, V-classes, or E-classes (Taylor 1982, pp. 286–287).

The paper of 1904 to which Taylor refers to here (Fréchet 1904b) introduced for the first time in Fréchet's work a very general vision that allows him to conceive 'functional operations' on 'sets' constituted of certain 'categories of arbitrary elements' ('numbers, surfaces, etc.') in bringing into play elements of topology. Naturally Fréchet did not use the word 'topology', but what matters, for the quite special purposes that we describe in what follows, are the ideas that we now would classify in the field of 'topology', such as compactness, or the properties of one or another particular type of convergence of series. According to Taylor, that paper marks the starting point of Fréchet's progression to abstract spaces, and he in some way relegates the studies carried out at the same time on line functions to a place among his youthful works. However, we believe there is legitimate reason to reconsider the importance of Fréchet's reading of Volterra's work for his notion of abstract spaces and, more generally, for his work in functional analysis. In particular, we will show that it is essentially the search for a general means of understanding the 'new' functions set forth by Volterra that will, starting with the publications of 1904, give rise to two types of developments which would only come together little by little into a general framework.

On the one hand, by formulating the idea of function that Volterra put forward in a simple manner, Fréchet developed a theory in line with the definition of Dirichlet (cf. Dirichlet 1837) and Weierstrass, who would take a very general form obtained by process of abstraction characteristic of the approach of the young French mathematician. The notion of abstract space appears here as a true preliminary to a general theory of functions or, better, 'operations', a term that Fréchet used in the form of 'operation' or 'functional operation' to designate a function in a wider sense. At the beginning of his thesis he specifies the terms: if E is a set formed of arbitrary elements (numbers, points, features, lines, surfaces, etc.), a 'functional operation' in E is a mathematical object that to any element A of this set makes correspond a specific number, $U(A)$. The study of these operations is the subject of calcul fonctionnel (cf. Fréchet 1906). In his dissertation Fréchet uses the term 'function' to describe what becomes a special case of 'operation', namely, a classic function of one or more real variables.

On the other hand, following the path opened by Volterra, Fréchet will elaborate tools suitable for dealing with problems of variations arising in mathematical physics. We will analyze below the process used by Fréchet in seeking to bring the problems in this area together within a single conceptual framework, particularly by means of the notion of 'linear operation', and the search for the representation theorems of these functions. Our aim is to show that these two channels were developed in parallel, and took root in a rereading of Volterra's ideas, before they gradually came together again in a general theory of linear operations defined on topological vector spaces. Abstract spaces will be at the centre of much of the subsequent research and are still today of paramount importance in functional analysis.

4.1 The First Contact with Line Functions

Shortly after finishing his training at the école Normale in autumn 1903, Fréchet wrote a long letter to Volterra, on the advice of both Borel and Hadamard, who had suggested that he would probably find material in the Italian geometer's work that could supply a subject for a thesis. In his letter Fréchet stated that, starting from Volterra's articles, he had already begun working on line functions, which he had learned about through Hadamard's course, mentioned in the previous section. He then goes into a technical discussion of these line functions in order that Volterra might help him decide upon an interesting subject for his thesis. This might lead us to think that, although Fréchet makes only a passing reference in the first letter, he also knew of Pincherle's approach through having read his articles, as well as through the 1901 treatise that Pincherle wrote with Amaldi and Pincherle (1901).

Although the letter was not in the form of an article, it offers a very detailed exposition and shows that Fréchet had already given much thought to Volterra's conceptions. In fact, in essence it contains the results that would be published soon after in Fréchet (1904a). Fréchet defines the notion of extended function in terms similar to those of Volterra:

> Mathematical physics has led to the study of functions that is much more general than the functions depending on the value of one or more variables (functions that I will call *ordinary*). I wish to speak of expressions that are determined only by the knowledge of all the values of one or more *ordinary functions*.[46] (Fréchet 1904a, p. 557)

However, as soon as this general approach is outlined, Fréchet limits its exploration to the most 'simple' cases, and specifies the framework that his study covers:

[46]La physique mathématique conduit à l'étude de fonctions beaucoup plus générales que les fonctions dépendant de la valeur d'une ou de plusieurs variables (fonctions que j'appellerai *ordinaires*). Je veux parler des expressions qui ne sont déterminées que par la connaissance de toutes les valeurs d'une ou de plusieurs *fonctions ordinaires*.

> We restrict ourselves to the case of the function U_L whose value varies only with the *shape* of a line L planar or curved, continuous, closed, and thus whose tangent varies continuously, except at a finite number of isolated points.[47] (Fréchet 1904a p. 557)

Although he does not say so explicitly, Fréchet envisions only U_L functions of real or complex values. The point of view adopted for the variable is that of the parameterized family of curves that make it possible to define the (uniform) limits and the variation of the function U_L:

> Consider a family G of these lines, dependent on a parameter α such that if α tends to α_0, L tends uniformly at all points to the corresponding points of L_0. For these lines L, U_L Footnotes have been renumbered. Kindly check and confirm.will be a function of α and we add the assumption that it is a function that is continuous and differentiable in α. Under these conditions, we can speak, as in the calculus of variations, of the first variation of U_L: δU_L,[48] which depends, of course, on the family G under consideration.[49] (Fréchet 1904a, p. 557)

In the study Fréchet makes direct reference to Volterra's works and in particular (Volterra 1889b), where the notion of function lines made it possible to formulate a framework for generalizing the study of functions, extending the concept of variable that varies from point to the curve, and then to the surface etc. Yet the approach of Fréchet does not resume directly to their account this view. He questions the best overall design as possible to deal with problems that have been identified as functional problems, that is to say, whose questioning relates to functions.[50] The presentation of Fréchet suggests that he is seeking from that article in a way to see

---

[47]Nous nous bornerons au cas des fonctions U_L dont la valeur varie seulement avec la *forme* d'une ligne L plane ou gauche, continue, *fermée* et dont la tangente varie d'une manière continue, sauf en des points isolés en nombre fini.

[48]The reader can consult the course of Joseph Alfred Serret (1819–1885) (Serret 1868), published the first time in 1868 and reprinted many times thereafter, for an overview of notations and definitions of the terms 'derived', 'augmented,' 'differential', etc. in use at the time for ordinary functions. The texts of Volterra, Hadamard or Fréchet never redefine these concepts, which they use in new, different contexts. We must precisely distinguish and analyse them. In particular here δU_L denotes the differential relative to the parameter α (see Chap. XII of Serret 1868).

[49]Considérons une famille G de ces lignes, dépendant d'un paramètre α de façon que, si α tend vers α_0, L tende *uniformément* en tous ses points vers les points correspondants de L_0. Pour ces lignes L, U_L sera une fonction de α et nous ajoutons l'hypothèse que ce soit une fonction continue et dérivable en α. Dans ces conditions, nous pourrons parler, comme dans le calcul des variations, de la variation première de U_L : δU_L (On peut consulter le cours de Joseph-Alfred Serret (1819–1885) (Serret 1868) publié pour la première fois en 1868 et réédité à de nombreuses reprises par la suite pour avoir un aperçu des notations et des définitions de *dérivée*, *accroissement*, *différentielle* etc. en usage à l'époque pour les fonctions ordinaires. Les textes de Volterra, Hadamard ou Fréchet ne redéfinissent pas ces notions qu'il nous faut précisément distinguer et analyser ici. En particulier ici δU_L désigne la différentielle relative au paramètre α (cf. Ch XII de Serret (1868))), laquelle dépend, bien entendu, de la famille G considérée.

[50]For an overview of the issues at the core of the research discussed here, one can see Hadamard's 1898–1899 course at the Collège de France, published in 1903 (Hadamard 1901). It may be interesting to read in parallel the Leçons sur le calcul des variation (Hadamard 1910) published later, in 1910, which gives an broader panorama of the questions embraced by this term at the beginning of the twentieth century.

the functions of general lines and as a tool to read through a single abstract concept all the situations encountered.

As we have seen, in 1887 Volterra had attempted to adapt the classic strategy of studying the functions of the real variable: a dependence, the 'function', which he viewed in a broad sense, hence a variation of the variable that engenders a variation the value of the function that in turn provides the notion of derivative. Finally, in this sequence one obtains an expression of the variation of which the first-order terms give the differential. The analysis in the case of lines is done in the same way by Volterra, starting with the deformation of a line L and observing the variation produced for the function U_L that will give the form of the differential.

Fréchet's work proposes an approach that reveals a genuine difference in perspective that he justifies by noting that Volterra's presentation does not make it possible to treat the classic functionals arising from the calculus of variations. This different point of view appears even in the choice of the notations used for line function: while Volterra used the notation $\varphi|[L]|$ to denote the line function, Fréchet goes back to a notation that had been used earlier by Hadamard, U_L, where the connection with the concept of (line) function is not completely transparent. Here the symbol U_L is used instead by Fréchet to denote certain elements of the calculus of variations, elements he eventually identifies as line functions and whose properties he seeks to make evident.

The strategy for generalisation adopted by Fréchet is that of considering in the first place the simplest functionals encountered in the calculus of variation. Thus Fréchet begins with an integral of the type $I_L = \int_L P(x,y,z)dx + Q(x,y,z)dy + R(x,y,z)dz$ the variation of which can be written $\delta I_L = \int_L (I'_x \delta x + I_{y'} \delta y + I_{z'} \delta z) ds$. Hence, one way to generalise consists in taking that property of a particular functional as a general definition. It is nevertheless necessary to ensure that the new definition covers new cases, which Fréchet does by providing specific examples. Line functions are thus defined beginning with the variation that Fréchet names after Volterra:

> We will thus give the name 'Volterra's function', or function $(\mathcal{V})$, to all closed line functions U_L, satisfying the condistion that we posed in no. 1, and such that one has
>
> $$\delta U_L = \int_L (U'_x \delta x + U'_y \delta y + U'_z \delta z) ds,$$
>
> $U_{x'}$, $U_{y'}$, $U_{z'}$ having quantities determined in each point M on the entire closed line L.[51] (Fréchet 1904a, p. 560)

[51]Nous appellerons donc *fonction de Volterra*, ou *fonction* $(\mathcal{V})$, toute fonction de ligne fermée, U_L, satisfaisant aux conditions que nous avons posées au n. 1 et telle que l'on ait

$$\delta U_L = \int_L (U'_x \delta x + U'_y \delta y + U'_z \delta z) ds,$$

$U_{x'}$, $U_{y'}$, $U_{z'}$ étant des quantités déterminées en chaque point M de toute ligne fermée L.

Further, this definition permits defining several other classes of functions: functions $(\mathcal{V})$ of the first degree, that is, those of the form I_L, which were used to forge the extended definition, and 'simple functions $(\mathcal{V})$', which are presented as ordinary functions of a single function $(\mathcal{V})$ of the first degree.

This approach leads Fréchet to find the characteristic properties of different classes of functions $(\mathcal{V})$. In particular, here we find, in a second moment, the notion of linearity that Volterra had also put forward:

> Let us recall, in order to generalise it, a theorem of Mr. Volterra's. We denote, as he did, by $L+L'$ a closed contour (which may include several closed curves) and constituted by the contours L and L', where the common parts (if any exist) have been removed, travelled in the opposite direction (for U_L depends in general on the direction of travel of L).
>
> Mr. Volterra has shown that the functions of first degree are the only functions $(\mathcal{V})$ that verify the functional equation
>
> $$U_{L+L'} = U_L + U_{L'}.$$
>
> More generally, we will now show that the simple functions are the only functions $(\mathcal{V})$ that verify the functional equation
>
> $$U_{L+L'} = \varphi(U_L, U_{L'}),$$
>
> φ being an ordinary function (continuous and derivable) of U_L and $U_{L'}$.[52] (Fréchet 1904a, p. 562)

These objects will become the focus of a series of studies carried out by Fréchet in the years to follow, and we will show in a later section that this development is not without a connection to the ideas that made it possible to forge his abstract spaces.

Finally, we believe it is important to emphasise how Fréchet positioned himself and the strategy he adopted in the search for a general framework of study, and to compare these elements with the ideas chosen by Volterra before him.

Volterra conceived the notion of function that depends on other functions in a very general way in order to conceive all the situations that arise in problems of variations. Let us recall, as we have seen in the first section, that it was only in a later moment that he restricted his point of view to particular cases for the study of

---

[52]Rappelons, pour le généraliser, un théorème de M. Volterra. Désignons avec lui par $L+L'$ un contour fermé (pouvant comprendre plusieurs courbes fermées) et constitué par les contours L et L', où l'on a supprimé les parties communes (s'il en existe) parcourues en sens contraire (car U_L dépend en général du sens de parcours de L). M. Volterra a démontré que les fonctions du premier degré sont les seules fonctions $(\mathcal{V})$ qui vérifient l'équation fonctionnelle

$$U_{L+L'} = U_L + U_{L'}.$$

Plus généralement, nous allons montrer que les fonctions simples sont les seules fonctions $(\mathcal{V})$ qui vérifient l'équation fonctionnelle

$$U_{L+L'} = \varphi(U_L, U_{L'}),$$

φ étant une fonction *ordinaire* (continue et dérivable) de U_L et $U_{L'}$.

problems of variations, by means of the notion of line functions and the property of linearity. Moreover, in order to transpose the usual properties of ordinary functions, such as continuity or differentiability, he had conceived a notion of neighbourhood or of perturbation, which he declined in function of the situations, from the most general to the most particular. For functions that depend on other functions, one cannot properly speak of neighbourhood and a perturbation is a function θ whose values are everywhere small. In the case of line functions, Volterra develops a much more geometric version of neighbourhood, which he calls the domain of line L (Volterra 1889b, pp. 237–238), consisting in a tube around the line L.

Fréchet's approach is significantly different. He uses a strategy of abstraction that permits him to take the classic situations into account in order to extract an effective general point of view. This strategy will be applied to the reading of Volterra's papers and will result in two types of developments.

The first arises from the notion of function which is based on Volterra, and appears to be a major factor in what might be called the process of generalisation in analysis (from finite to infinite, as the protagonists themselves say). Fréchet will therefore rethink the notion of function and attempt to extract the essential elements that make it possible to develop a general theory, doing away with the necessity of seeing the variable as a real number, hence embarking all the topological tools then known on $\mathbb{R}$. This first approach will be developed in Fréchet (1904a) and will constitute the heart of Fréchet's doctoral thesis.

On the other hand, in discussing Volterra's ideas, Fréchet proposes a second point of view: he decides this time to begin with the notion of line function, which is also the idea Hadamard had had (see Hadamard 1902a). In his case he begins with a simple case where the parallel with the results already established in some particular case of the theory of variations is clear and from which he extracts the elements that could make it possible to conceive a general framework. Thus, as we have shown earlier, he decided to define first of all a type of line function according to the form of variation. It is thus by beginning with the analysis of this simple case that Fréchet, by a process of abstraction, puts forward the notion of linear functional. He thus deals with two related problems: on the one hand, that of the representation of linear operations, and on the other hand, integration, that is to say, how to find the function when its differential is known, which is none other than a line function. These problems will essentially be developed in three papers entitled 'Sur les Opérations linéaires' published between 1904 and 1907 (Fréchet 1904c, 1905, 1907).

4.2 The Extension of the Study of Functions

4.2.1 The Generalisation of a Theorem by Weierstrass

In Fréchet (1904b), Fréchet proposes a first direction for the general study of functions adapted to variational problems. The introduction positions the article along the lines of Hadamard's work, not only for the notation U of a functional, but above all for the questions of minima that he raised in 1897 (see Hadamard 1897), as we saw earlier in our second section:

> One knows how important it would be, in a large number of problems, to know whether a quantity U depending on certain elements (points, functions, etc.) had effectively reached a minimum in the field under consideration. Dirichlet's principle offers one of the most striking justifications of this remark ...[53] (Fréchet 1904b, p. 848)

Fréchet immediately recalls the result he has in mind and would like to retain in the more general framework:

> The problem is solved in the particular case where U is a simple function of x (or of several independent variables). Weierstrass has in effect shown that any continuous function in a bounded interval attains its maximum at least once.[54] (Fréchet 1904b, p. 848)

This property depends on many of the ingredients that Volterra himself had relied on in his approach, namely the central use of a notion (extended) of function and continuity. However, it appears to Fréchet that in their approaches analysts relied on the intervention of the nature and certain of the elements that serve as object or variable to obtain the modern results of the usual theory of functions. He thus conceived a general framework by taking these aspects into account, and the definition obtained appears to be an extension of that of Dirichlet:

> We assume given a certain class[55] C of arbitrary elements (numbers, surfaces, etc.), in which it is possible to discern the distinct elements. We can say that U_A is a uniform function (or functional operation) in a set E of elements of C, if to any element A of E there corresponds a well-determined number U_A.[56] (Fréchet 1904b, p. 849)

We can remark a somewhat similar approach in Fréchet (1904a), in the introduction to which Fréchet had already described the set of lines on which the functional acted, as we have mentioned earlier.

In Fréchet (1904a), as did Volterra before him, Fréchet defines continuity in a particular way for each situation. In contrast, in Fréchet (1904b) Fréchet takes a much more general approach in which a notion of sequential convergence is given a priori:

---

[53]On sait l'importance qu'il y aurait, dans un grand nombre de problèmes, à savoir si une quantité U dépendant de certains éléments (points, fonctions, etc.) atteint effectivement un minimum dans le champ considéré. Le principe de Dirichlet offre une des justifications les plus frappantes de cette remarque...

[54]La question est résolue dans le cas particulier où U est une simple fonction de x (ou de plusieurs variables indépendantes). Weierstrass a en effet démontré que toute fonction continue dans un intervalle limité y atteint au moins une fois son maximum.

[55]Care must be taken not to attribute a modern meaning to the term "class" (*catégorie*). Here 'class' is used to form a certain typology. It follows that in this statement a set is formed of elements of a certain determined type ('numbers', or 'surfaces', etc.).

[56]Nous supposons donnée une certaine catégorie C d'éléments quelconques (nombres, surfaces, etc.), dans laquelle on sache discerner les éléments distincts. Nous pourrons dire que U_A est une fonction (ou opération fonctionnelle) uniforme dans un ensemble E d'éléments de C, si à tout élément A de E correspond un nombre bien déterminé U_A.

> To arrive at the notion of continuity of such a function, we presume having arrived at a definition that gives a precise meaning to this phrase: *the infinite set* $A_1, A_2, \ldots, A_n, \ldots$ *of the elements of C has a limit B*. This definition will suffice for us, in any case, when the following two conditions are satisfied: 1st, if the sequence $A_1, A_2, \ldots, A_n, \ldots$ has a limit, then any sequence $A_{p_1}, A_{p_2}, \ldots, A_{p_n}, \ldots$ formed of elements of the first sequence with increasing indices also has the same limit; 2nd, if none of the elements $A_1, A_2, \ldots, A_n, \ldots$ of an arbitrary sequence is different from A, that sequence has a limit which is A.[57] (Fréchet 1904b, p. 849)

The abstract nature of this notion of convergence will then make it possible to define the notions of closed set, continuity and compact set in a way that is likewise independent of the nature of the elements of C. For example, the notion 'compact set' is defined by Fréchet as follows:

> We will call *compact set* any set E such that there is at least one element common to an arbitrary infinite sequence of sets $E_1, E_2, \ldots, E_n, \ldots$ contained in E, where these (possessing at least one element each) are closed and each is contained in the previous.[58] (Fréchet 1904b, p. 849)

In this framework, Weierstrass's theorem receives a formulation whose generality guarantees its applicability in many situations:

> THEOREM. Any functional operation U_A uniform and continuous in a compact and closed set E: 1st, is bounded in E; 2nd, attains at least once its upper limit.[59] (Fréchet 1904b, p. 849)

This idea will be developed and specified in Fréchet's thesis, which we will discuss below. However, it is also his strategy for seeking a general framework for treating the problems that will interest us in the rest of in our study. It illustrates in effect the spectacular way that Fréchet steered the concepts of his Italian predecessor into new territory.

4.2.2 Fréchet's Doctoral Thesis

Fréchet's doctoral thesis was published in 1906 under the title 'Sur quelques points du calcul fonctionnel' (Fréchet 1906). We will not give an extensive presentation, but will limit ourselves to highlighting a few elements that extend the ideas we have just mentioned.

---

[57]Pour arriver à la notion de continuité d'une telle fonction, nous supposerons acquise une définition qui donne un sens précis à cette phrase: *la suite infinie* $A_1, A_2, \ldots, A_n, \ldots$ *d'éléments de C a une limite B*. Il nous suffira que cette définition, d'ailleurs quelconque, satisfasse aux deux conditions suivantes: 1° si la suite $A_1, A_2, \ldots, A_n, \ldots$ a une limite, toute suite $A_{p_1}, A_{p_2}, \ldots, A_{p_n}, \ldots$ formée d'éléments d'indices croissants de la première suite a aussi une limite qui est la même : 2° si aucun des éléments $A_1, A_2, \ldots, A_n, \ldots$ d'une suite quelconque n'est distinct de A, cette suite a une limite qui est A.

[58]Nous appellerons *ensemble compact* tout ensemble E tel qu'il existe toujours au moins un élément commun à une suite infinie quelconque d'ensembles $E_1, E_2, \ldots, E_n, \ldots$ contenus dans E, lorsque ceux-ci (possédant au moins un élément chacun) sont fermés et chacun contenu dans le précédent.

[59]THÈORÈME. Toute opération fonctionnelle U_A uniforme et continue dans un ensemble compact et fermé E : 1° est bornée dans E; 2° y atteint au moins une fois sa limite supérieure.

Fréchet's ambition is to construct a general framework for modern analysis that would encompass a number of classes of functions with specific properties. After recalling that in the previous decade several mathematicians (Le Roux, Volterra, Arzelà, Hadamard) had generalized the notion of function by considering increasingly extensive cases, Fréchet engages in a strategy that consists in conceiving the broadest and most indeterminate possible for the variable.

> We say that a functional operation U is defined in a set E of arbitrary elements (numbers, curves, points, etc.) when to any element A of E there corresponds a determined numeric value of U: $U(A)$. The search for the properties of such operations constitute the object of functional calculus.[60] (Fréchet 1904a, p. 1)

From the beginning, Fréchet remarks that nothing seems to play naturally and in a uniform way for all situations the role of intervals, a notion that occupied a determinant role in analysis for the functions of a real variable. In a second step, like Volterra, he thought it necessary to extend the notion of continuity.

> We say that a functional operation V uniform in a set E of elements of a class (L) is continuous in E if, whatever element A in E is the limit of a sequence of elements $A_1, A_2, \ldots, A_n, \ldots$ of E, we always have: $V(A) = \lim_{n=\infty} V(A_n)$.[61]

(Fréchet 1906, p. 7)

In passing, Fréchet notes that this concept is too general to permit defining uniform continuity in arbitrary sets. He will then, by means of the special cases that he deems to be most significant, progressively consider new conditions that make it possible to develop a complete theory of functions. The examination of particular fields and of uniform continuity (which will play a key role in his approach) spur him to define a notion of neighborhood and to isolate the sets where this notion of neighborhood is defined, which he calls V classes.

Likewise, to use topological concepts in sequential form, Fréchet places himself in a framework of completeness. There again, the generalisation works by conserving a property obtained in the classic case as a result of a theorem, this property now characterising a general class:

> We say then that a class (Y) admits a generalisation of Cauchy's theorem if any sequence of elements of that class, which satisfies Cauchy's conditions, has a (necessarily unique) limit element.[62] (Fréchet 1906, p. 23)

---

[60] Nous dirons qu'une opération fonctionnelle U est définie dans un ensemble E d'éléments de nature quelconque (nombres, courbes, points, etc.) lorsqu'à tout élément A de E correspond une valeur numérique déterminée de $U : U(A)$. La recherche des propriétés de ces opérations constitue l'objet du Calcul Fonctionnel.

[61] Nous dirons qu'une opération fonctionnelle V uniforme dans un ensemble E d'éléments d'une classe (L) est continue dans E, si, quel que soit l'élément A de E limite d'une suite d'éléments $A_1, A_2, \ldots, A_n, \ldots$ de E, on a toujours: $V(A) = \lim_{n=\infty} V(A_n)$.

[62] Nous dirons alors qu'une classe (Y) admet une généralisation du théorème de CAUCHY si toute suite d'éléments de cette classe, qui satisfait aux conditions de CAUCHY, a un élément limite (nécessairement unique).

Finally, Fréchet introduces one final, more restrictive class denoted by (E) thanks to a particular neighborhood he calls "écart" and corresponds to our notion of distance. Nevertheless, at this point in time Fréchet still thinks that the proper level of generality is represented by his broader notion of class (V).

> In the majority of proofs of known theorems, the property (b) of the distance [*écart*] intervenes in the reasoning. However, the theory developed in this chapter shows that it is not indispensable, and that it suffices to make use of neighborhood without the need for it to significantly complicate the reasoning.[63] (Fréchet 1906, p. 30)

In the remainder of his thesis, Fréchet examines the consequences that can be drawn from his new concepts, notably by revisiting a number of classic cases. We will not dwell on this here. We will, however, complete this section by showing how Fréchet, beginning with the concept of generalised function set forth by Volterra, developed his ideas of abstract space and classes (L), (V), (E), etc. in order to outline the general framework that seemed essential to relieve analysis of unnecessary encumbrances. The result of this choice of a very general framework is that it provides the opportunity for Fréchet to give his theorems definitive postulates that prove exceptionally close to those utilised today. Hadamard did not exaggerate when he wrote in a report on Fréchet's work on the occasion on his candidacy for membership in the Académie des Sciences,[64]

> Mr. Fréchet taught us to reason about sets that are completely abstract, that is to say, composed of elements about which one makes, at least in the beginning, no hypothesis. He goes in a stroke to extreme generality, a generality which, by definition, can never be exceeded.[65]

4.3 Linear Operations

We will now examine the second development that grew out of Fréchet's article (Fréchet 1904a) ("Fonctions de lignes fermées") and thus directly from Volterra's legacy.

In his three papers on linear operations (Fréchet 1904, 1905, 1907), Fréchet does not adopt the general framework for studying Volterra functions. These articles directly concern a problem that Fréchet had mentioned in Fréchet (1904a). We recall that Fréchet had chosen, following his reading of Volterra, to elaborate

[63]Dans la plupart des démonstrations des théorèmes connus, la propriété b) de l'écart intervient dans les raisonnements. Cependant la théorie développée dans ce Chapitre montre qu'elle n'est pas indispensable et qu'il suffit de se servir du voisinage sans avoir besoin pour cela de compliquer notablement le raisonnement.

[64]The quote is taken from that report (1934), which is conserved in the Archives of the Académie des Sciences in Paris.

[65]M. Fréchet nous a appris à raisonner sur des ensembles entièrement abstraits, c'est-à-dire composés d'éléments sur lesquels on ne fait, tout au moins en commençant, aucune hypothèse. Il va d'un coup à l'extrême généralité, une généralité qui, par définition, ne pourra jamais être dépassée.

functionals defined by their variation, especially when that variation is itself a linear functional. This then translated into an equality of the type:

$$\delta U_L = \int_L (U'_x \delta x + U'_y \delta y + U'_z \delta z) ds.$$

Besides its focus on the linear aspect of the functional, this equation poses the problem of the integration of this formula. The integration consists in determining the functional U_L solely by starting with its derivatives U'_x, U'_y, U'_z and the relationship we have referred to above. Fréchet had devoted the last section of his 1904 article on line functions to this question, and arrived at a necessary and sufficient condition that a function $(\mathcal{V})$ is of the first degree (up to an additive constant), stipulating that U'_x, U'_y, U'_z depend only on the point M and the tangent to L at this point (Fréchet 1904a, p. 570). These questions then find in Fréchet a natural extension in the general problem of the representation of linear functionals, in the form of an integral, or in the form of a Taylor series, as Volterra had already suggested.

It is this sequence of ideas that Fréchet intends to pursue and develop in his three papers (Fréchet 1904, 1905, 1907), beginning with the works by Hadamard, which he first seeks to extend.[66]

At the beginning of Fréchet (1904) Fréchet explains the perspective in which he regards operations by making reference to Hadamard:

> We say that an operation is defined if one makes a determined and finite real number U_f correspond to any function $f(x)$ that is real and continuous between two fixed numbers a and b. We call with Mr. Hadamard a linear operation any operation that enjoys the following two properties:

(1) it is distributive, that is, if f_1 and f_2 are two continuous functions between a and b, one always has

$$U_{f_1+f_2} = U_{f_1} + U_{f_2}$$

(2) it is continuous, that is, that U_{f_1} tends to U_{f_2} when the function f_1 tends uniformly to the function f_2 between a and b.

[67] (Fréchet 1904, p. 493])

---

[66]These notes are of utmost importance for understanding the evolution of ideas in the birth of functional analysis. For a precise analysis of the elements involved in this perspective, see JaËck (2010).

[67]Nous dirons qu'une *opération* est définie si l'on fait correspondre un nombre réel déterminé et fini U_f à toute fonction $f(x)$ réelle et continue entre deux nombres fixes a et b. Nous appellerons avec M. Hadamard *opération linéaire* toute opération qui jouit de deux propriétés suivantes:

(1) elle est *distributive*, c'est-à-dire que si f_1 et f_2 sont deux fonctions continues entre a et b, on a toujours

The views we see expressed here differ from what was proposed in the thesis. In this present context, an operation has as an argument an ordinary continuous function and the continuity of the operation is given by the uniform convergence naturally available for these functions. These elements are consistent with the ideas exchanged between Volterra and Fréchet, the latter, as we have seen, nourished by Hadamard's subtle remarks. The ideas are those used to reread known situations in the light of functionals.

The first paper (Fréchet 1904) then reprises the result of representation of such an operation given by Hadamard in the form of limit. The essential aim of the article consists in finding a development similar to Taylor series for such operations. Here Fréchet reprises Hadamard's ideas, and particularly his approach centred on the decomposition of functions in Taylor series (Hadamard 1892b, 1901, 1902b).

Fréchet then proposes a series of generalisations essentially based on the linearity of U, its continuity for uniform convergence and the possibility to decompose the functions in a series that converges uniformly. In particular, he uses the uniform convergence of Césaro means proven by Fejer to study the convergence of the Fourier series associated with a continuous periodic function (Hawkins 2001, pp. 166–167). The idea is notably not to appeal to a hypothesis of analyticity, which we have long seen that Fréchet considered too restrictive to deal with problems of mathematical physics.

Finally, thanks to the development of functions into series that are increasingly more general, Fréchet rediscovers Hadamard's theorem in all its generality:

$$U_f = \lim_{n=\infty} \int_0^{2\pi} f(y) H_n(y) dy.$$

Fréchet does not stop there; he goes on to analyse the formula obtained. He thus asks what are the functions H_n for which this expression defines a linear operation, and what are the functions for which the limit itself admits an integral representation with the aid of a continuous function H. By showing that the 'pathological' functions H_n nevertheless make it possible to give sense of such an integral, Fréchet will again consider expressions of the type $\int_0^{\pi} f(y)H(y)dy$, an expression in which it is not even supposed that H is Lebesgue integrable.

Again we find a process dear to Fréchet: isolating a form that was itself obtained as a consequence of a traditional approach (here when one has recourse to a theorem of uniform convergence), in order to construct a generalised vision.

Fréchet's second paper (Fréchet 1904d) will focus, on the one hand, on refining these considerations on the integral representation but will also finish by setting off in a new direction and reflecting on the nature of the functions f that can be

(Footnote 67 continued)

$$U_{f_1+f_2} = U_{f_1} + U_{f_2}$$

(2) elle est *continue*, c'est-à-dire que U_{f_1} tend vers U_{f_2} lorsque la fonction f_1 tend *uniformément* vers la fonction f_2 entre a et b.

considered as the object of the operation. This is an important step to which Fréchet devotes a separate section in his article, entitled 'Importance du champ fonctionnel dans lequel on définit une opération linéaire'. At the end of the paper, Fréchet shows in particular that it is possible to define functionals for non-continuous functions.

This new direction will bring Fréchet to write a third paper (Fréchet 1907) in which the field of all the functions f become the central focus. The two developments that came out of his reading of Volterra that we have followed are both found here, each extensively modified or inclined towards the search for a general analysis. What in fact occurs here is a complete reorganization of the discourse. This article, in the image of the doctoral thesis, reprises a presentation that begins with the domain of definition of the functional—presuming by now that it is a completely arbitrary field of functions—and supposing the existence of a property of the type $U_{cf} = cU_f$ for any real c that ensues from the fact of the continuity expressed in a particular way for each functional field. In contrast to the topological aspects seen in the thesis, the emphasis here is on the structural elements of vector space. Nevertheless, this very much involves the conditions necessary for the expression of linearity that Fréchet wishes to maintain here.

> Definition of the field. Consider a field of functions of variable x defined in the interval $(0, 2\pi)$. I will assume that if two functions belong to the field, it is the same of their sum. To any function of the field, $f(x)$, we can make correspond a well determined number U_f. We thus define an operation in this field functions. We say that this operation is distributive if given any functions $f_1(x), f_2(x)$ of the field, one has:
>
> $$U_{f_1} + U_{f_2} = U_{f_1+f_2}$$
>
> One can conclude in particular that one has for any distributive operation:
>
> $$cU_f = U_{cf}$$
>
> for any rational constant c. For this relationship to occur even for an irrational c, it is sufficient that U_f satisfies a certain complementary condition. We will state below this complementary condition, but we will state it in a particular way for each field of functions that we will then examine.[68] (Fréchet 1907, p. 433)

[68]Définition du champ. Considérons un champ de fonctions de la variable x définies dans l'intervalle $(0, 2\pi)$. Je supposerai que si deux fonctions appartiennent au champ, il en est de même de leur somme. A toute fonction du champ, $f(x)$, nous pourrons faire correspondre un nombre bien déterminé U_f. Nous définirons ainsi une opération dans ce champ de fonctions. Nous dirons que cette opération est *distributive*, si quelles que soient les fonctions $f_1(x), f_2(x)$ du champ, on a:

$$U_{f_1} + U_{f_2} = U_{f_1+f_2}$$

On en conclut en particulier que l'on a pour toute opération distributive:

$$cU_f = U_{cf}$$

quelle que soit la constante rationnelle c. Pour que cette relation ait lieu même pour c irrationnel, il suffit que U_f satisfasse à une certaine condition complémentaire. Nous allons énoncer plus loin

As we have seen, Fréchet makes no reference to any notion of linear space, although this is already found in the works of Peano and Pincherle. In effect, at the end of Peano's 1888 book (Peano 1888) there is a chapter entitled 'Trasformazioni di sistemi lineari' (Transformations of linear systems). Peano first of all defines the notion 'linear system' in terms very close to those used today to define vector spaces. Finally, in a second moment, he defines the 'operations', which make an element of one system correspond to an element of another system (or possibly of the same system). Immediately the notions of linearity or distributivity (both terms are used by Peano) are set forth. In this optic, Peano evokes, by means of the structure of 'linear system', the minimum elements required to develop a theory of 'linear transformations' in which all the properties necessary for establishing the proofs are explicitly set forth in the definitions introduced at the beginning. A few years later, in 1901, Pincherle and Amaldi published a book entitled 'Le operazioni distributive e le loro applicazioni all'analisi' (Pincherle and Amaldi 1901b), which opens with a construction similar to that of Peano, explicating first of all 'general sets of linear systems' (insieme/sistema lineare generale), that is, sets that possess a number of properties that define what we call vector spaces (the wording differs only slightly from that of Peano). It is within this framework that is defined, in the second chapter, the notion of 'operation' in the form of 'correspondence' (corrispondenza) between the elements of two linear systems.

We can thus see that Fréchet positions himself very differently, and does not reprise in his account the ideas of his predecessors: although the previous quote begins with the words 'Definition of field', Fréchet does not select any particular property common to fields of functions that he will use. The collections of functions he mentions, besides being stable by addition in order to define distributive operations, are considered with all of their properties without discrimination. We do not know at this point what properties of the sets of functions will be useful in establishing the proofs of the theorems. It is the notion of function field, which is not defined abstractly here, that integrates all the properties necessary to develop a general vision, that is, a vision that embraces all the particular cases of function fields that Fréchet envisions at the end of the article. Finally, the definition of 'distributive operation' is introduced explicitly and synthetically in this context.

The end of Fréchet (1907) allows Fréchet to specify in different functional fields the particular nature of the concepts of continuity and of convergence.

The initial motivations that led Fréchet to conceive this organisation are significantly different from those that motivated the ideas expounded in his thesis. Nevertheless, in both we see a common goal underlying the mathematician's thinking: Fréchet's aim is to propose a general and abstract vision of analysis, and more particularly at this stage for functional analysis, which he conceives beginning with diverse situations identified as belonging to a single domain. Moreover—and this links him strongly to Volterra—the notion of generalized function, which little

(Footnote 68 continued)
cette condition complémentaire, mais nous l'énoncerons d'une manière particulière pour chacun des champs de fonctions que nous allons examiner.

by little becomes functional operation and shortly after operator, remains the key, determinant element in this general vision.

5 Conclusion

We have aimed at showing two outcomes of French mathematicians' readings of the work of Volterra. Hadamard was probably attracted by Volterra's progress in a mathematical vision that harmonised with his own, and a way of interpreting problems that was immediately compatible with his own work. He saw in the work on line functions a program that was ambitious and suitable to problems of PDEs arising from mathematical physics that occupied him in the early twentieth century. The acceptance of the ideas developed by his Italian colleague was total and within a few years the relationship between the two men turned into a genuinely personal as well as scientific friendship. Later Hadamard and Volterra would meet often, especially at international congresses, and their works on functional analysis often contain laudatory references to each other. In 1909 Hadamard was appointed professor at the Collège de France and in 1912 was elected to the Paris Académie des Sciences, two strategic positions that allowed him to help disseminate the ideas of his Italian colleague. The joint efforts of Hadamard and Borel culminated in 1912 with the invitation to Volterra to give a series of lectures at the Sorbonne on line functions, published shortly after in a book edited by Joseph Pérès (Volterra 1913).

We have seen how, in a sort of benevolent gesture,[69] Hadamard had encouraged Fréchet to take an interest in Volterra's work, and how Borel willingly facilitated contacts between the young mathematician and his friend Volterra. A remarkable result of these strong personal ties between Borel, Hadamard and Volterra was the establishment, in the early 1910s, when several new modes of funding became available in France, of scholarships for French students to go to Rome to study with Volterra. Pérès was the first student to make the journey in 1912, followed the next year by Gateaux.[70] This trend was quickly disrupted by the outbreak of World War I. The war and its attendant tragedies also played a large role in the evolving relationship between Volterra and Hadamard (and more generally between Volterra and all his French colleagues).[71]

We have observed that the young Fréchet was rereading the works on line functions at the very time when he was forging his own research program and inaugurating his conception of functional analysis in a general and abstract sense. It seems that Fréchet very quickly decided to construct a new mathematical edifice that he named 'general analysis', and which he could not conceive of as the culmination of the program proposed by Volterra. A veritable divergence of points of view was in fact born, a divergence that led to a certain degree of tension between

[69]On this subject, see Brechenmacher (2014).

[70]For more on this, see Mazliak (2015b).

[71]For this, we refer to Mazliak and Tazzioli (2009).

the two mathematicians. We find the traces of this friction in an exchange of letters between Volterra and Fréchet in spring 1913. On the occasion of the publication of Volterra (1913), Fréchet wrote a letter to Volterra to say that he himself was going to publish on the subject:

> My dear Sir and colleague,
> As you have published a book on line functions, I wanted to send you two articles that I have written on the subject. Unfortunately one is printed, but has not yet appeared, the other is not yet printed. The first in any case is the development two notes published in Comptes Rendus Ac. Sc. in Paris (1911, vol. 152, p. 854[72] and 1050), or rather the parts of these two papers regarding to functional calculus. It may be of interest to you for me to summarise the essential ideas. If by chance you are interested in these reflections, I could send you the manuscript of my paper [presented] to the Congress of Sociétés Savantes in 1912, which as I mentioned is in press but has not yet been issued. It is 19 pages long.[73] [M. Fréchet to V. Volterra, 26 May 1913]

What ensued was a tense discussion between the two mathematicians and an exchange of letters that show how each wished to make his point of view known. The publications that Fréchet mentioned in the previous letter (Fréchet 1911a, b) are both entitled 'Sur la notion de differentielle'. They certainly do not make any reference to Volterra's work on functional calculus; they are two very brief papers that do not go into detail, but which set forth a general way of conceiving the differential. Paper Fréchet (1912) is more detailed and focuses over fourteen pages on elements that might bring Volterra to mind. Volterra's works are cited in passing, and this probably caused a change of humour in the Italian mathematician.

> To arrive at a generalisation of the theorems of differential calculus, one must first of all generalise the notion of derivative where differential. One might base oneself, to carry out this extension, on the method used in the calculus of variations, which is but a chapter of functional calculus.
> This is the path followed by Mr. Volterra, who had the merit of developing the first coherent theory of differential functional calculus. It consists in operating with the variation of the functional in the sense of the term as used in the calculus of variations. U_f is the functional defined in the field of continuous functions within a given interval I. Mr. Volterra considers the case in which the quantity $U_{f(x)+\epsilon\varphi(x)}$ has a differential with respect to ϵ for $\epsilon = 0$: this differential will by definition the variation of U_f for the argument $f(x)$. He remarks that, under certain simple conditions, this variation is of the form

[72]One in fact finds this on p. 845, and this is probably a slip of the pen in Fréchet's letter.

[73]Monsieur et cher collègue

Comme vous publiez un livre sur les fonctions de lignes, j'aurais voulu pouvoir vous envoyer deux articles que j'ai écrit à ce sujet. Malheureusement l'un sous presse tarde à paraitre, l'autre n'est pas encore imprimé. Le premier est d'ailleurs le développement de deux notes parues dans les Comptes Rendus de l'Ac. des Sc. de Paris (1911, tome 152, pp 854 et 1050) ou plutôt des parties de ces deux notes relatives au Calcul Fonctionnel.

Peut-être vous intéresserait-il que j'en résume l'essentiel. Si par hasard ces réflexions vous intéressaient, je pourrais vous envoyer le manuscrit de ma note au Congrès des Soc. Savantes de 1912 qui déjà imprimée comme je le disais plus haut n'est pas encore parue. Elle a 19 pages.

$$\epsilon \int_a^b \varphi(y)k(y)dy$$

$k(y)$ being an independent function of $\varphi(y)$.

However, Mr. Hadamard observed that the calculus of variations already offers us the example of a very simple functional whose variation cannot be put in this form. He therefore reduces Volterra's condition to: the variation of U_f must simply be a linear functional… with respect to the increase $\epsilon\varphi(x)$ of the point $f(x)$.

This is the essential indication which will form my point of departure. Nevertheless, it appears to me necessary to make the existence of the variation derive from that of the differential and in consequence define first of all the differential of a functional.[74] (Fréchet 1912, p. 47)

This quote allows us to see the issues of this dispute, which will play out over several months without ever really being resolved. In the first place, Fréchet fails to cite all of the works of Volterra, who had himself considered singular points and forms of variations taking into account these particular cases. Volterra sent a written request to Fréchet asking him kindly to rectify this omission. In a letter dated 17 November 1913 he writes, 'I also hope that you will correct what you have said in relation to singular points. I attach great importance to this point, which I have emphasised since my first memoirs of 1887'.[75]

[74]Pour arriver à généraliser les théorèmes du calcul différentiel, il faut généraliser d'abord la notion de dérivée où de différentielle. On pourrait se baser, pour effectuer cette extension, sur la méthode employée dans le Calcul des Variations, qui n'est qu'un chapitre du Calcul Fonctionnel.

C'est la voie suivie par M. Volterra, qui a eu le mérite de développer le premier une théorie cohérente du Calcul Différentiel Fonctionnel. Elle consiste à opérer avec la variation de la fonctionnelle au sens où on entend ce mot dans le Calcul des Variations. Soit la fonctionnelle U_f définie dans le champ des fonctions continues dans un intervalle donné I. M. Volterra considère le cas où la quantité $U_{f(x)+\epsilon\varphi(x)}$ a une différentielle par rapport à ϵ pour $\epsilon = o$: cette différentielle sera par définition la *variation* de U_f pour l'argument $f(x)$. Il remarque que, sous certaines conditions simples, cette variation est de la forme

$$\epsilon \int_a^b \varphi(y)k(y)dy$$

$k(y)$ étant une fonction indépendante de $\varphi(y)$.

Mais M. Hadamard a fait observer que déjà le Calcul des Variations nous offre l'exemple de fonctionnelles très simples dont la variation ne peut se mettre sous cette forme. Il réduit donc la condition de Volterra à celle-ci : la variation de U_f doit être simplement une fonctionnelle linéaire […] par rapport à l'accroissement $\epsilon\varphi(x)$ de l'argument $f(x)$.

C'est là l'indication essentielle qui formera mon point de départ. Cependant il me parat nécessaire de faire découler l'existence de la variation de celle de la différentielle et par conséquent de définir d'abord la différentielle d'une fonctionnelle.

[75]J'espère aussi que vous aurez corrigé ce que vous avez dit par rapport aux points singuliers. Je tiens beaucoup à ce point que j'ai mis en évidence depuis mes premiers mémoires de 1887.

Fréchet had quickly, from the very first letters of Volterra and with much deference, promised to rectify his articles by mentioning the exact extent of Volterra's results. He was in fact able to do so in Fréchet (1914), which appeared in 1915. Yet even in this article citing Volterra, Fréchet introduced nuances and clearly marks how he differs:

Historical Overview

> The first attempt to apply to functionals the procedures of differential calculus appears to be due to Mr. Volterra…
>
> Mr. Volterra did not fail to remark that such a definition was not entirely satisfactory, since it leaves out many of the expressions that intervene in the calculus of variations, namely those variations of definite integrals where the limits are not fixed. Such variations have in effect, besides a definite integral of the form (1) [$\delta U_L = \int_L U_{L,x} \delta y dx$], finite terms at the limits. He therefore agreed to add to the second member of (1) the terms that depend, in a special way, as he phrased it, on certain exceptional points. By adopting the definition of Mr. Volterra, one is inspired by the first applications that have presented themselves and that Mr. Volterra has treated with a success that practically justifies his definition. However, it was desirable from the point of view of logic and to ensure the future development of the theory to deduce the definition of a unique and general principle. Mr. Hadamard thus proposed to 'consider as a functional to which one can extend the methods of infinitesimal calculus, all functionals U_y whose variation is a linear function of the variation of y'.[76] (Fréchet 1914, p. 136)

The expression 'appears to be due to Mr. Volterra' enabled Fréchet to imply that the subject did not belong to anyone, and shows that by now he felt prepared to take it on in a way that he believed to be innovative with respect to what Volterra and Hadamard had proposed. The dispute between the two mathematicians was actually of short duration, even though Fréchet adopted a somewhat ambiguous line of conduct. In his publications, he sometimes failed to cite Volterra as the initiator of functional analysis, while writing him letters filled with of tributes and acknowledgments:

> Dear Sir, I should be sorry if you might believe that I do not appreciate the true value of your essential contribution to functional calculus. First of all, I see as much as you do no

[76]Aperçu historique

Le premier essai pour appliquer aux fonctionnelles les procédés du Calcul Différentiel semble être dû à M. Volterra. […]

M. Volterra ne manqua pas de remarquer qu'une telle définition n'était pas entièrement satisfaisante, puisqu'elle laisse de côté une grande partie des expressions qui interviennent dans le Calcul des Variations, à savoir celles des variations des intégrales définies où les limites ne sont pas fixes. De telles variations comportent en effet, outre une intégrale définie de la forme (1) [$\delta U_L = \int_L U_{L,x} \delta y dx$], des termes finis aux limites. Il convint donc d'ajouter au second membre de (1) des termes qui dépendent, d'une manière spéciale, selon son expression, de certains points exceptionnels. En adoptant la définition de M. Volterra, on s'inspire des premières applications qui se sont présentées et que M. Volterra a traitées avec un succès qui justifie pratiquement sa définition. Mais il était souhaitable au point de vue logique et pour assurer le développement futur de la théorie de déduire la définition d'un principe unique et général. M. Hadamard proposa donc de "considérer comme fonctionnelles auxquelles on peut étendre les méthodes du Calcul Infinitésimal, toutes les fonctionnelles U_y dont la variation est une fonctionnelle linéaire de la variation de y."

'difference between your definition of function that depends on all the values of another function, and functional, as well as between their calculus'....

If my current ideas may be of some interest, in any case, I myself recognise it as a secondary refinement of the beautiful theory that you have constructed.

It is precisely because I consider your theory to have remained well enough known that I have not believed it necessary to insist on its importance and have directly proposed some refinements of the details.[77] [M. Fréchet à V. Volterra, 21 novembre 1913]

The short-lived dispute aside, the exchanges between the two mathematicians reveal the significant difference in vision that we have highlighted in our study, which marks, we might say, a change of generation. While the idea of function allowed the two mathematicians to develop in each case a general vision and conceive a rereading of problems of functionals in an abstract setting, Fréchet had an early desire to build a genuinely general theory which naturally embraced the problems of mathematical physics dear to his masters. It is in this quest for a general analysis that he felt the need to break away from Volterra's vision. This shows no lack of respect or recognition, but a bifurcation that was necessary to carry these ideas forward. An inevitable consequence was that the centre of gravity for subsequent studies in functional analysis moved far away from the sunny skies of Rome.

Translated from the French by Kim Williams.

References

d'Adhémar R. Sur une intégration par approximations successives. Bulletin de la Société Mathématique de France, 1901, 29: 190–199.

d'Adhémar R. Sur l'integrations d'une équation aux dérivées partielles du second ordre, du type hyperbolique, à plus de deux variables indépendantes. C. R. 135, 1902a: 1100–1102.

d'Adhémar R. Sur une classe d'équation aux dérivées partielles, intégrables par approximations successives. C. R. 134, 1902b: 407–409.

Amaldi U. and Pincherle S. Le operazioni distributive et le loro applicazioni all'analisi. Bologne, 1901.

Archibald T. Tension and potential from Ohm to Kirchhoff. Centaurus, 1988, 31 (2): 141–163.

Archibald T. Connectivity and Smoke-Rings: Green's second identity in its first fifty years. Mathematics Magazine, 1989, 62, 4: 219–232.

Archibald T. and Tazzioli R. Integral equations between theory and practice: the cases of Italy and France to 1920. Archive for history of exact science, 2014, 68: 547–597.

Barbut M., Locker B. and Mazliak L. Paul Lévy - Maurice Fréchet, 50 years of mathematics, Springer, 2013.

[77]Cher Monsieur, Je serais désolé si vous pouviez croire que je n'apprécie pas à sa juste valeur votre contribution essentielle au Calcul Fonctionnel. Tout d'abord pas plus que vous je ne vois "de différence entre votre définition de fonction qui dépend de toutes les valeurs d'une autre fonction et les fonctionnelles ainsi que dans leurs calculs" [...]

Si mes idées actuelles peuvent avoir quelque intérêt, ce n'est en tout cas, je le reconnais moi même, que comme perfectionnement secondaire de la belle théorie que vous avez édifiée.

C'est précisément parce que je considère que votre théorie est maintenant suffisamment connue que je n'avais pas cru inutile d'insister sur son importance et que j'avais proposé directement quelques perfectionnement de détails.

Beltrami E. Sul Principio di Huygens, Rend. Istituto Lombardo, II, XXII, fasc. X, 1889.
Bertrand J. Eloge de Lamé. Séance publique du 28 janvier 1878, Académie des Sciences de Paris.
Berthelot M. Essai de Mécanique chimique fondée sur la thermochimie. Tome premier : Calorimétrie. Dunod éditeur, Paris, 1879.
Beudon, J. Sur les systèmes d'équations aux dérivées partielles dont les caractéristiques dépendent d'un nombre fini de paramètres. Ann. de l'éc. Norm. (3) 13, Suppl., 3–51 (1896).
Bianchi L. Sulle equazioni lineari a derivate parziali del 2o ordine, Rendiconti Accademia dei Lincei, Serie IV, Vol. V, fasc.2, 35–44, 1889.
du Bois Reymond P. Beiträge zur Interpretation der partiellen Differentialgleichungen mit drei Variabeln. Leipzig, 1864.
Borel E. Verhandlungen des ersten Internationalen Mathematiker- Kongresses in Zürich, 204–205, 1897.
Bottazini U. and Tazzioli R. Naturphilosophie and its role in Riemann's mathematics. Revue d'histoire des mathématiques, 1, 3–38, 1995.
Brechenmacher F. Jacques Hadamard, passeur. http://images.math.cnrs.fr/Jacques-Hadamard-passeur.html; mai 2014.
Carvallo E. Sur les systèmes linéaires, le calcul des symboles différentiels et leur application à la physique mathématique. Monatshefte für Mathematik und Physik Volume 2, Issue 1, December 1891, Pages 177–216.
Clapeyron B. and Lamé G. Mémoire sur l'équilibre intérieur des corps solides homogènes, Memoire des Savants Etrangers, IV, 463–562, 1833.
Cooke R. The Mathematics of Sonya Kovalevskaya. Springer, New-York, 1984.
Coulon J. Sur les équations aux dérivées partielles du second ordre à caractéristiques réelles. Procès Verbaux de la Société des Sciences Physiques et Naturelles de Bordeaux, 85–88, 1898–1899.
Dini U. Fondamenti per la teorica delle funzioni di variabili reali. T. Nistri, Pisa, 1878.
Duhem P. Cours de physique mathématique et de cristallographie de la Faculté des sciences de Lille. Hydrodynamique, élasticité, acoustique : I. Théorémes généraux, corps fluides; II. Les Fils et les membranes, les corps élastiques, l'acoustique. A. Hermann,1891.
Egorov Y.V. and Shubin M.A. Partial Differential Equations I. Foundations of the Classical Theory. Springer, 1998.
Fréchet M. Fonctions de Lignes Fermées. *Annales de l'ÉNS*, 1904.
Fréchet M. Généralisation d'un Théorème de Weierstrass. *C.R. Ac. Sc., t.139*, Paris, 21 Nov. 1904.
Fréchet M. Sur les Opérations linéaires (Note 1) Transactions of the American Mathematical Society, Vol. 5, No. 4 (Oct., 1904), 493–499.
Fréchet M. Sur les Opérations linéaires (Note 2) Transactions of the American Mathematical Society, Vol. 6, No. 2 (Apr., 1905), 134–140.
Fréchet M. Sur quelques Points du Calcul fonctionnel *Thèse*, Université des Sciences, Paris, 1906.
Fréchet M. Sur les Opérations linéaires (Note 3) Transactions of the American Mathematical Society, Vol. 8, No. 4 (Oct., 1907), 433–446.
Fréchet M. Sur la notion de différentielle, Comptes rendus de l'Académie des Sciences, Paris, 152, 845–847 (1911).
Fréchet M. Sur la notion de différentielle, Comptes rendus de l'Académie des Sciences, Paris, 152, 1050–1051 (1911).
Fréchet M. Sur la notion de différentielle dans le calcul fonctionnel. Congrès Soc. Sav. Paris, 1912.
Fréchet M. Sur la notion de différentielle d'une fonction de ligne. Transactions of the American Mathématical Society, 15, 1914, 135–161.
Freda H. Méthode des caractéristiques pour l'intégration des équations aux dérivées partielles linéaires hyperboliques. Mémorial des sciences mathématiques. Fascicule 84. Gauthier-Villars, 1937.
Gielis J., Caratelli D., Haesen S. and Ricci P.E. Rational mechanics and science rationnelle unique p. 29–43 *in* Stephanos A. Paipetis and Marco Ceccarelli (Editors). The Genius of Archimedes. 23 Centuries of Influence on Mathematics, Science and Engineering. Springer, 2010.

Gispert H. La correspondance de G Darboux avec J Houël: chronique d'un rédacteur (déc. 1869-nov. 1871). Cahiers du séminaire d'histoire des mathématiques 8, Inst. Henri Poincaré, 1987, 67–202.

Goodstein J.R. The Volterra Chronicles. The Life and Times of an Extraordinary Mathematician, 1860–1940, AMS/LMS, 2007.

Gray J. Mathematics and natural science in the nineteenth century: the classical approaches of Poincaré, Volterra and Hadamard. Chapter 6 (p. 113–135) in U.Botazzini and Amy Dahan-Dalmedico. Changing images in mathematics. From the French Revolution to the New Millenium. Studies in the History of Science, Technology and Medicine. Routledge, 2001.

Guerraggio A. and Paoloni G. Vito Volterra. Springer Verlag, 2013.

Hadamard J. Essai sur l'étude des fonctions données par leur développement de Taylor, Thèse de l'Université de Paris, 1892.

Hadamard J. Essai sur l'étude des fonctions données par leur développement de Taylor. 1892.

Hadamard J. Essai sur l'étude des fonctions données par leur développements de Taylor. Étude sur les propriétés des fonctions entières, et en particulier d'une fonction considérée par Riemann. (Mémoire couronné par l'Académie des Sciences). Paris : Hermann. 132 S. 4e, 1902.

Hadamard J. Sur certaines applications possibles de la théorie des ensembles" (Verhandlungen des ersten Internationalen Mathematiker- Kongresses in Zürich, 201–202, 1897.

Hadamard J. Sur l'intégrale résiduelle. Bulletin de la Société Mathématique de France, 28, 69–90, 1900.

Hadamard J. Série de Taylor et prolongement. Scientia: Physique-Mathématique n°12. Mai 1901.

Hadamard J. Sur les équations aux dérivées partielles à caractéristiques réelles. Compte Rendu du Deuxième Congrès International des Mathématiciens, tenu à Paris du 6 au 12 août 1900. Gauthier-Villars, 1902, 373–376.

Hadamard J. Sur les dérivées des fonctions de lignes. Bulletin de la Société Mathématique de France. 30, 40–43, 1902.

Hadamard J. Leçons sur la propagation des ondes et les équations de l'hydrodynamique. Hermann, 1903.

Hadamard J. Sur les opérations fonctionnelles, 136, 351–354, 1903.

Hadamard J. Mémoire sur le problème d'analyse relatif à l'équilibre des plaques élastiques encastrées. Mémoires présentés par différents savants à l'Académie des Sciences, XXXIII, 4, 1907.

Hadamard J. Leçons sur le calcul des variations. Tome premier : La variation première et les conditions du premier ordre; les conditions de l'extremum libre. Paris: A. Hermann et Fils. VIII + 520 S. 4e, 1910.

Hadamard J. L'œuvre de Duhem dans son aspect mathématique. Mémoire de la Société des sciences physiques et naturelles de Bordeaux, 1, 637–665, 1927.

Hadamard J. Le développement et le rôle scientifique du calcul fonctionnel, Atti del Congresso dei Matematici di Bologna, Volume 1, 143–161, 1928.

Hadamard J. Le développement et le rôle scientifique du calcul fonctionnel. Atti del Congresso Internazionale dei Matematici. Bologna 3–10 Settembre 1928, Tomo I, Nicola Zanichelli, Bologna, 1928.

Hadamard J. An essay on the Psychology of Invention in the Mathematical Field. Princeton University Press, 1945.

Hawkins T. Lebesgue's theory of integration,its origin and development, AMS, 2001.

Helmoltz H.V. Theorie der Luftschwingungen in Röhren mit offenen Enden. Journal für die reine und angewandte Mathematik, 57, 1., 1–72. 1860

Jacki S.L. Lettres de Pierre Duhem à sa fille Héléne, Beauchesne, 1894.

Jaëck F. Éléments structurels en analyse fonctionnelle : trois notes de Fréchet sur les opérations linéaires. Archive for History of Exact Sciences, 2010, 64, 461–483.

Kirchhoff G. Zur Theorie des Lichtstrahlen. Annalen der Physik, Volume 254, Issue 4, 663–695, 1883.

Kowalevski S. Über die Brechung des Lichtes In Cristallinischen Mitteln. Acta Mathematica 6, 1885, 249–304.

Lamé G. Leçons sur la théorie mathématique de l'élasticité des corps. Bachelier, 1852. (2ème édition. Gauthier-Villars, Paris, 1866).

Le Jeune Dirichlet G. Über die Darstellung ganz willkürlicher Functionen durch Sinus- und Cosinusreihen. Repertorium der Physik, unter Mitwirkung der Herren LEJEUNE DIRICHLET, JACOBI, NEUMANN, RIESS, STREHLKE, herausgegeben von HEINRICH WILHELM DOVE und LUDWIG MOSER. Bd. I, 1837, pp. 152–174.

Le Roux J. Sur les intégrales des EDP linéaires du second ordre à deux variables indépendantes, Annales de l'Ecole Normale Supérieure, XII, 1895.

Lévy P. Quelques aspects de la pensée d'un mathématicien, Blanchard, 1970.

Maggi G.-A. Sulla propagazione libera e perturbata delle onde luminose in mezzo isotropo. Annali di Matematica Pura ed Applicata. 16, 21–48.1888–1889.

Mazliak L. The Ghosts of the Ecole Normale: Life, death and destiny of René Gateaux. Statistical Science. 30 (3): 391–412, 2015a.

Mazliak L. Volterra and the journeys of French students to Italy in the 1910s. Revista Brasileira de História da Matemática, 14 (29), 2015b.

Mazliak L. and Tazzioli R. Mathematicians at war, Springer, 2009.

Maz'ya V. and Shaposhnikova T. Jacques Hadamard, A Universal Mathematician. American Mathematical Society-London Mathematical Society, 1998.

Nastasi P. and Sallent del Colombo E. Mittag-Leffler Volterra correspondence, to appear.

Padova E. Una nuova interpretazione dei fenomeni elettrici, magnetici e luminosi. Pisa, Tip. Pieraccini, 1891.

Padova E. Una nuova interpretazione dei fenomeni elettrici, magnetici e luminosi. Nuovo Cimento (3) XXIX, 225–241, 1891.

Peano G. Applicazioni geometriche del calcolo infinitesimale. Fratelli Bocca, Torino, 1887.

Peano G. Calcolo geometrico secondo l'Ausdehnungslehre di H. Grassmann preceduto dalle operazioni della logica deduttiva. Fratelli Bocca Editore, 1888.

Pincherle S. and Amaldi U. Le operazioni distributive e le loro applicazioni all' analisi. Zanichelli, 1901.

Plantade F. La diffusion des géométries non euclidiennes dans le bassin méditerranéen vers 1870 sous l'impulsion de Jules Houël (1823–1886), preprint, 2014.

Riemann B. Sur la propagation d'ondes aériennes planes. Académie Royale des Sciences de Göttingen, VIII, 1860.

Serret J.A. Cours de Calcul Différentiel et Intégral. Gauthier-Villars, 1868.

Siegmund-Schultze R. Die Anfänge der Funktionalanalysis und ihr Platz im UmwälzungsprozeSS der Mathematik um 1900, Archive for History of Exact Sciences Volume 26, Issue 1, pp 13–71.

Taylor A.E. A Study of Maurice Fréchet. I. His Early Work on Point Set Theory and the Theory of Functionals. *Archive for History of Exact Sciences*, 27(3), 1982.

Taylor A.E. A Study of Maurice Fréchet. II. Mainly about his Work on General Topology, 1909–1928. *Archive for History of Exact Sciences*, 34(4), 1985.

Taylor A.E. A Study of Maurice Fréchet. III. Fréchet as Analyst, 1909–1930. *Archive for History of Exact Sciences*, 37(1), 1987.

Taylor M.E. Partial Differential Equations I. Basic theory. 2nd Edition. Springer, 2011.

Tazzioli R. Ether and Theory of Elasticity in Beltrami's Work, Archive for History of Exact Science, 46, 1–38, 1993.

Tazzioli R. Construction engineering and natural philosophy: the work by Gabriel Lamé, in P. Radelet-de Grave, Edoardo Benvenuto : Between Mechanics and Architecture, Birkhäuser, 1995, 316–330.

Tazzioli R. Green's Function in Some Contributions of 19th Century Mathematicians. Historia Mathematica 28, 232–252, 2001.

Tedone O. Sull'integrazione dell'equazione $\frac{\partial^2 \varphi}{\partial t^2} - \sum_{1^m} \frac{\partial^2 \varphi}{\partial x_i^2} = 0$.Annali di Matematica, III, 1. 1–23, 1898.
Vezes M. Rapport. Cinquantenaire de la société des sciences physiques et naturelles de Bordeaux, Gauthier-Villars, 17–18. 1906.
Volterra V. Alcune osservazioni sulle funzioni punteggiate discontinue. Giornale di Matematiche, 19, 1881a, 76–87.
Volterra V. Sui principii del calcolo integrale. Giornale di Matematiche, 19, 1881b, 333–372.
Volterra V. Sui fondamenti della teoria delle equazioni differenziali lineari. Memorie della Società Italiana delle Scienze detta dei XL, III, 6, 1887a, 1–107.
Volterra V. Sulle equazioni differenziali lineari. Rendiconti dell'Accademia dei Lincei, IV, 3, 1887b, 393–396.
Volterra V. Sopra le funzioni che dipendono da altre funzioni (Nota I). Rendiconti dell'Accademia dei Lincei. IV, 3, 1887c, 97–105.
Volterra V. Sopra le funzioni che dipendono da altre funzioni (Nota II). Rendiconti dell'Accademia dei Lincei. IV, 3, 1887d, 141–146.
Volterra V. Sopra le funzioni che dipendono da altre funzioni (Nota III). Rendiconti dell'Accademia dei Lincei. IV, 3, 1887e, 153–158.
Volterra V. Sopra le funzioni dipendenti da linee (Nota I). Rendiconti dell'Accademia dei Lincei. IV, 3, 1887f, 225–230.
Volterra V. Sopra le funzioni dipendenti da linee (Nota II). Rendiconti dell'Accademia dei Lincei. IV, 3, 1887g, 274–281.
Volterra V. Sopra una estensione della teoria di Riemann sulle funzioni di variabili complesse. Nota I. Rendiconti dell'Accademia dei Lincei, IV, 3, 1887h, 281–287.
Volterra V. Sopra le funzioni che dipendono da altre funzioni. Rendiconti Accademia dei Lincei, 3, 97–105, 1887i.
Volterra V. Sulla teoria delle equazioni differenziali lineari. Rendiconti del Circolo matematico di Palermo. 2, 1888, 69–75.
Volterra V. Sopra una estensione della teoria di Riemann sulle funzioni di variabili complesse. Nota II. Rendiconti dell'Accademia dei Lincei, IV, 4, 1888a, 107–115.
Volterra V. Nota III. Sopra una estensione della teoria di Riemann sulle funzioni di variabili complesse. Rendiconti dell'Accademia dei Lincei, IV, 4, 1888b, 196–202.
Volterra V. Sur une généralisation de la théorie des fonctions d'une variable imaginaire. Acta Mathematica, 12, 1889a, 233–286.
Volterra V. Sur une Généralisation de la Théorie des Fonctions d'une Variable imaginaire. Acta. Math. t.12, pp. 233–286, 1889b.
Volterra V. Sopra le equazioni di Hertz. Nuovo Cimento, 39, 53–63. 1891.
Volterra V. Sopra un'estensione della teoria Jacobi-Hamilton del calcolo delle variazioni. IV, 6, 1891, 127–138.
Volterra V. Sur les vibrations lumineuses dans les milieux biréfringents. Acta Mathematica, 16, 1892, 153–215.
Volterra V. Sur les vibrations lumineuses dans les milieux biréfringents. Acta Mathematica, t. XVI, 153–215, 1892a.
Volterra V. Sulle vibrazioni luminose nei mezzi isotropi. Rendiconti della R. Accademia dei Lincei, s. V, vol. I2. 161–170. 1892b.
Volterra V. Un teorema sugli integrali multipli. Atti dell'Accademia delle Scienze di Torino. 32, 1896–97, 859–868.
Volterra V. Sur les équations aux dérivées partielles. Compte Rendu du Deuxième Congrès International des Mathématiciens, tenu à Paris du 6 au 12 août 1900. Gauthier-Villars, 1902, 377–378.
Volterra V. Leçons sur les fonctions de ligne, rédigées par Joseph Pérès. Gauthier-Villars, Paris, 1913.
Volterra V. and Pérès J. Théorie générale des fonctionnelles, Gauthier-Villars, Paris, 1936.

The French 'Analysts' and Peano's Mathematical Logic: Couturat's *Remarques* to Borel, Baire and Lebesgue

Erika Luciano

The historiographical categories 'School of Peano' and 'Italian Mathematical Logic' will be discussed, before illustrating the logical and foundational contributions by the Peanians included in the collective work Formulario di Matematica. *The reception in France of these results will then be analysed, both at the level of research activity and in the context of teaching of real analysis (1883–1890). Particular attention will be paid to the role of L. Couturat in popularizing Peanian logic in the years 1896–1910, and to an emblematic episode: the debate in 1904 between Couturat, E. Borel, R. Baire and H. Lebesgue, regarding the possible use of Peano's ideography within the* Collection de monographies sur la théorie des fonctions. *The implications of this story will be examined, focusing on how the destiny of the diffusion of mathematical logic in France was linked to Couturat's cultural policies, which significantly conditioned the evolution of French research in the fields of philosophy and foundations of mathematics.*

This research was performed as part of the Project *PRIN 2009 Scuole Matematiche e Identità Nazionale nell'età moderna e contemporanea*, Unity of Turin University.
RMM: Revue de Métaphysique et de Morale
FBP: Fonds Émile Borel, Archives de l'Académie des Sciences, Paris

E. Luciano (✉)
Department of Mathematics 'G. Peano', University of Turin, Turin, Italy
e-mail: erika.luciano@unito.it

F. Brechenmacher et al. (eds.), *Images of Italian Mathematics in France*, Trends in the History of Science, DOI 10.1007/978-3-319-40082-2_7

1 The 'School of Peano' and 'Italian Mathematical Logic'

From the 1890s onwards, recurrent references to the '*School of Peano*'and '*Italian Mathematical Logic*' have been made in international mathematics literature to identify—respectively—a community of scholars, and a set of mathematical contents and meta-mathematical stances constructed, transmitted and socialized by this equip.

It can be affirmed that the tendency was to refer the term '*School of Peano*' to a team of mathematicians, teachers, philosophers and engineers (A. Padoa, C. Burali-Forti, G. Vailati, G. Vacca, M. Pieri, R. Bettazzi,...) linked to the milieu of the University of Turin and collectively engaged in developing a common research program, under Peano's direction. Neither the members of this '*School*', nor scholars such as E. Schröder, L. Couturat, E.H. Moore, J. Pierpont, E. Wilson, T. Shearman, F. Klein and P. Jourdain,[1] who commented on the works of the Peanians from an external perspective, were able to univocally define either the pattern of this group or the line of studies it carried out. However, in general terms, it may be said that the image of the '*School of Peano*' which was affirmed was that of a closely-knit team of Turin mathematicians linked among them and to the *Maestro* Peano by a ramified network of scientific exchanges and friendships. Recent studies (Luciano 2016), based on unpublished sources, have shown how this image was in fact quite contrived. Yet, despite differences in opinion, rivalries, and occasional arguments between the Peanians, there is no doubt that they were closely interrelated through their research, publishing and educational activities in the realm of mathematical logic.

In turn, the label '*Italian Mathematical Logic*' not only comprised an obvious allusion to the national dimension in which the '*School of Peano*' was active, but also characterized the line of studies developed by this group with special (indeed almost unique) reference to the symbolic language (namely 'ideography') that they formulated, and then used in most of their publications and teaching practice. This ideography, which first appeared in Peano's writings of 1888–89, consisted of around twenty symbols to denote logical connectives and quantifiers (ε, $\subset$, $\sim$, ι, $\cup$, $\cap$,...), and around two hundreds signs to translate classical theories into a formal language: rational arithmetic, elementary geometry, general algebra and the first elements of number theory, infinitesimal and vector calculus, differential geometry, etc. Over time, this ideographic language was expanded by the members of the '*School of Peano*' (the so-called 'first generation' including the aforementioned scholars and many more such as T. Boggio, M. Bottasso, M. Gramegna, M. Peyroleri, V. Mago, P. Quarra and U. Cassina) in order to render it more flexible and suitable for expressing advanced mathematical disciplines such as functional analysis, finite difference calculus, measure theory and numeric calculus.

[1] Cf. Schröder (1897, p. 154), Couturat (1900, p. 400–401), Moore (1903, p. 402–405), Pierpont (1904, p. 147), Wilson (1904, p. 76, 78, 93), Shearman (1906, p. 200), Klein (1909, p. 484), Jourdain (1912a, p. 287, 288, 289, 295, 303, 304, 311, 313).

The various chapters of mathematics translated into ideographic language by Peano and his *protégés* were brought together in the *Formulario di Matematica*, an encyclopaedic treatise published in five editions between 1895 and 1908.[2] Designed specifically as a collective work, for almost 20 years the *Formulario* represented the pivot of the ‘*School of Peano*’ research programme. Peano himself, in particular, dedicated all his efforts to the coordination and compilation of the *Formulario* with contributions from himself and his collaborators, also at the risk of condemning his and his ‘*School*’ to be marginalized from the more promising scientific trends. It was Peano himself who described the work in progress connected with the edition of the *Formulario* as follows:

> Each of the sections must contain all the propositions, theorems and definitions, to which reference is made. In that way, once the *Formulario* will be sufficiently complete, anyone who wishes to get up to date on science, on a given point already dealt with in the *Formulario*, will only need to check the book, and will find all the known propositions there. [...] Quotations and historical information allow the reader to compare the original texts where the questions are discussed at greater depth. [...] Now I am going in for the composition of this *Formulario*, and each day brings a new part to translate into symbols. [...] I am working on the publication of the *Formulario* and I am happy to have the collaboration of a number of my colleagues, and of several young graduates, who are enthusiastically designing the various parts.[3]

In its complete definitive version, the *Formulario* included over 5,000 mathematical statements and proofs. The language used for *all* mathematical contents was logical ideography, whilst the meta-mathematical comments (i.e. the references to literature and sources, or historical, philosophical and methodological annotations) were in French for the first four editions and in *latino sine flexione* for the last one, in honour of those ideals of internationalization that marked all of Peano’s cultural activity.

The *Formulario* had a paratactic structure, so much so that at first glance it seemed to be a simple collection of formulas. To remedy this, Peano published some ad hoc works to explain how to consult this encyclopaedia, and also took advantage of some editorial tricks. As well as the meta-representations (♣, §, <), which refer to the conceptual key points throughout, the *Formulario* included special meta-textual devices, consisting of indexes of nouns and subjects, occurrence prospects for the symbols and concordance tables between the five editions.

This encyclopaedic work was published in Turin by Bocca, an education-focused publishing house, as annexe to the *Rivista di Matematica*, the journal founded and directed by Peano himself in the years 1891–1906. It was a mother-child text, i.e. mathematical sections were printed in larger font size, while

---

[2]Peano (1895aa, 1899b, 1901b, 1903f, 1908a). All the published and unpublished works by Peano are now available in Roero (2008a) (ed.). Peano’s papers are quoted according to the classification in this DVD. On the history of the *Formulario* cf. Lolli (1985, p. 49–83), Grattan-Guinness (2000, p. 219–267), Luciano (2008b, p. 65–81), Roero (2011, p. 83–132).

[3]G. Peano to F. Klein, 29.8.1894, in Luciano and Roero (2012, p. 169–170).

pedagogical, philosophical and historical remarks were printed in a smaller font. This choice of layout well reflected the dual goal of the *Formulario*, which was conceived from the beginning not only as a catalogue or syllabus of mathematical contents, but also as an auxiliary text for teaching and teacher training.

Aware of the distrust and hostility that every new formal language encounters at its birth, the 'phalanx' of Peanians—and, above all, their leader—devoted great energy to the socialization of ideography and the *Formulario*. Though failing to formulate a wholly efficient popularization strategy, they intertwined a broad web of partnerships and correspondence with European and American colleagues. Furthermore, they participated in the International Congresses of Mathematicians in Zürich (1897), Cambridge (1912) and Toronto (1924), in the International Conferences of Philosophy (Paris 1900, Geneva 1904 and Bologna 1911) and in other symposia, such as the Congrès International de Philosophie Scientifique, held in Paris in 1935. On many of these occasions, Peano, Burali-Forti, Padoa and Pieri presented the progress being made by ideography and its most recent applications in the *Formulario*. Some supplementary meetings, expressly devoted to clarifying the meaning and use of the symbolic notations, were also organized by the '*School of Peano*', for example at the second International Congress of Philosophy in Geneva.[4] Moreover, the Peanians proposed cycles of lectures and seminars in Italy, France, Belgium, Switzerland and Argentina, in order to present the signs,[5] illustrate their *style* of research (i.e. their practices of translation of classical mathematics into logical language) and explain how to 'read' the *Formulario*.

2 The Reception in France of Peano's First Results in Foundations of Analysis (1883–1890)

For reasons of geographical proximity and cultural affinity, French mathematicians were among the referential interlocutors of the '*School of Peano*', even before the logical and formalistic landmark of its research program was definitely assessed in 1889–90.

In fact, from 1883 onwards, C. Jordan, C. Hermite, J. Tannery, and H. Laurent began to follow Peano's work on the foundations of infinitesimal analysis, and particularly his contributions to the theory of real functions, which were not yet expressed with ideographic symbols. His name became renowned, among other things, for his results on the integrability of the functions (1882b), on interpolar functions (1882c), and for the debate regarding the mean value theorem, subject of a polemic with C. Jordan, J. Serret, and P. Gilbert in the *Nouvelles Annales de Mathématiques* (1884a, 1884b).

[4]Cf. Vailati (1904k, p. 127–128).

[5]These experiences of vulgarization of mathematical logic are illustrated in Luciano (2010, p. 279–315).

Hence it is not surprising that the volume *Calcolo differenziale e principii di calcolo integrale* by A. Genocchi, published by the young lecturer Peano with his critical notes and additions (Torino, Bocca, 1884c and 1884c*), was favourably received. In a report that appeared in the *Bulletin des Sciences Mathématiques*, Amédée Paraf (1861–1911), a teacher of infinitesimal calculus and rational mechanics, firstly in Bordeaux and then in Toulouse, underlined "the care for philosophical aspects" in this textbook, structured according the "German spirit", and written "in a very modern and scientific style".[6] Paraf particularly appreciated the introduction of real numbers constructed as Dedekind cuts of rational numbers, and the concepts of limit superior and limit inferior. He also greatly admired the unusual precision of several definitions (such as that of a definite integral given with reference to the infimum and supremum of the integral upper and lower sums), the rigour of the statements and proofs, and the fact that the hypotheses and the conditions for validity of each theorem were explicitly stated throughout. The community of French analysts mostly agreed with Paraf's positive report. There were just a few exceptions, on the part of mathematicians with a more traditional view of courses in infinitesimal analysis. For example, Hermite looked on this new overcritical current of treatise writing with puzzlement, and doubted the utility for teaching of 'philosophical studies'.[7]

The public of French mathematicians proved to be equally attentive in perceiving the originality of the treatise on *Applicazioni geometriche del calcolo infinitesimale* (Torino, Bocca, 1887b), a text which can be considered the 'second part' of the *Genocchi-Peano* and which devoted peculiar attention to the results of the German mathematicians (G. Cantor, R. Dedekind, K. Weierstrass, C. Harnack, P. Du Bois-Reymond,…) in the realm of the foundations of analysis and set theory. The positive reception of this volume was also favoured by an appreciative review from Jules Tannery, who found a singular precision in Peano's *style* of exposition and

> a lot of new contents, including definitions regarding sets of points, exterior, interior and limit points of a set, distributive functions, outer/inner/proper length (area, volume) of a set, and the notion of integral extended to a set. All these definitions are presented very clearly and precisely, in a very abstract form.[8]

As a consequence, in the space of just a few years, some chapters of *Genocchi-Peano* and *Applicazioni geometriche del calcolo infinitesimale* became the subject of quotations and extensions by the French mathematicians, or they were used in teaching. For example—as Gispert, Gandon and Perrin amply documented—Jordan's measure theory was strongly influenced by Peano's approach, as outlined in the 5th chapter of the 1887 treatise.[9] Furthermore—as Medvedev proved—the hints

[6]Paraf (1885, p. 170, 172).

[7]Cf. C. Hermite to A. Genocchi, 31.10.1884, in Michelacci (2005, p. 180–181).

[8]J. Tannery (1887, p. 237–239).

[9]Cf. Gispert (1983, p. 46–47, 55, 61–62, 64–65, 69, 79–80), Gandon and Perrin (2009, p. 665–704, 2009, p. 415–430), Perrin (2010, p. 317–326).

at *funzioni distributive o coesistenti secondo Cauchy* (distributive or coexisting functions according to Cauchy), included by Peano in the same section of *Applicazioni geometriche*, constituted a major historical precedent for H. Lebesgue's results on set functions.[10] Regarding the teaching of analysis in France, we need only mention the legacy of Peano's 1884 and 1887 volumes on the treatises of differential and integral calculus by C. Jordan, J. Tannery and H. Laurent. Peano himself recorded, in his handwritten notes on the pages of *Genocchi-Peano*, the errors that these colleagues had recognized in their papers and the benefits they had taken of his remarks and suggestions.[11]

This surge of international successes included Peano's subsequent book, *Lezioni di Analisi infinitesimale* (Torino, Candeletti, 1893h). This work, despite lacking the fine critical rigour which marked the *Genocchi-Peano* and *Applicazioni geometriche* volumes, actually constituted the ideal completion of those texts and stood out for its treatment of the theory of linear operators in the sense of functional analysis, for a new expression of the complementary term in Taylor's formula, for a theorem on the existence and derivability of the implicit function, and for the extensive discussion of geometric calculus. This was the first book in which Peano used (although still sparingly) the symbols of mathematical logic. With respect to this choice C. Somigliana commented:

> [Peano] imprinted on the teaching methods, by introducing the algorithms of mathematical logic wherever possible. His *Lezioni di Analisi infinitesimale* (Torino, 1893) are a noteworthy example of the union of rigorous modern analytical methods with traditional simplicity of presentation. The later developments in Peano's activity [...] are, on the other hand, the subject of much discussion as it is thought that the priority given to logical formalism and its symbolic code may confuse students as to the purpose of calculus, its origins and applicative value as a tool to investigate natural phenomena.[12]

Thanks to the international ripple prompted by Peano's treatises of 1884, 1887 and 1893, the key instance of his pedagogic and 'philosophic' thought was soon being diffused abroad, i.e. the conviction of the *practical utility* of studies on logic and the foundations of mathematics, for a through revision and renewal of the teaching of real analysis.[13]

[10]Cf. Medvedev (1983, p. 112–117).

[11]Cf. Luciano (2008a, vol. II, p. 67–68, 68, 69, 70, 77, 87–88, 95–96).

[12]Somigliana (1911, p. 21). Similar appraisals arrived from P. Mansion, F. Gomes Teixeira, Z. García de Galdeano and J. Tannery: cf. P. Mansion, *Lezioni di analisi infinitesimale del Prof. G. Peano*, Mathesis, 2, IV, 1894, p. 192–193; [F. Gomes Teixeira], *G. Peano: Lezioni di Analisi infinitesimale, Torino, 1893*, Jornal de Sciencias Mathematicas e Astronomicas, XII, 1894, p. 11–12; [Zoel García de Galdeano], *G. Peano: Lezioni di Analisi infinitesimale* ..., El Progreso Matemático, III, 1893, p. 319–320.

[13]Cf. for example G. Peano to C. Jordan, 6.11.1894, in Borgato (1991, p. 95–96).

3 After 1890: The Ideographic Turn and the Project of the Formulario

Around 1890, Peano's activity underwent a radical evolution, resulting from the introduction—in *all* of his writings and lectures—of the system of logical signs constructed by his '*School*'.

For Peano, there was no fracture between his early production on the foundations of calculus and this 'new season' of work. In effect, the project of translating the various mathematical theories into signs required a preliminary form of 'microscopy of thought', i.e. a careful and complete examination of each definition and proof. In other words, it was the same type of iper-detail-oriented analysis of the ideas and theorems of infinitesimal calculus, which—as long as it has been carried out with a non-ideographic approach—had brought Peano prestige and praise from colleagues both at home and abroad.

At the same time, from February 1891 (even more so after the summer of 1900), the '*School of Peano*' turned its attention to re-examining the results achieved in the previous years in order to insert them in the *Formulario*, rather than undertaking new lines of study. As a result, the group's scientific output changed in nature. The publications by the Peanians in the period 1892–1908 became—in many cases—'preparatory' or 'comparative' papers, instrumental for the implementation of the *Formulario*. They were, in essence, erudite 'exercises', whose value lay not so much in the specific contents as in the original meld of mathematical culture, historical critique, and educational awareness.

Gradually, as the ideographic language was enriched and the various hypothetical-deductive systems transformed into symbols found a place in the *Formulario*, Peano moved forward by boosting the role of logic in his teaching. In 1897, when the *Formulario* was considered well advanced, he introduced it as a textbook for his lectures on infinitesimal calculus, with excellent results in his own opinion.[14] The syllabus for the calculus course held by Peano and three volumes of handouts from 1898, 1902 and 1904,[15] show how Peano changed his teaching practice in that period. Entirely reshaped with respect to his youthful courses (1884–1887), Peano's lessons now began with the theory of limits, carried out in parallel with that of series, before moving on to the rules of differentiation and theorems on derivatives. After dwelling time on Taylor's series Peano then tackled integration theory. Interpolation functions, functions of several real variables, implicit functions and the theory of ordinary differential equations had all disappeared from his teaching program. By contrast, Peano had increased the applications of calculus and thus the lectures on limits of variable figures, tangents, normal planes, osculation, curvature, measure theory, etc. Every year, at the beginning of the course, Peano provided his students with a synopsis of the ideographic symbols and the sections of the *Formulario* that would be explained during the lessons and

[14]Peano (1896b, p. 85–86).

[15]Cf. G. Peano, *Programma di Calcolo infinitesimale per l'anno 1902–03*, Historical Archive of the University of Turin; Peano (1898, 1902–03G, 1904d).

would be the topic of the final exam. He dedicated five whole lectures to logic, in order to “give a clearer idea of the name of symbols, skipping almost all their properties, the alternative definitions, and their applications in arithmetic and geometry”.[16] In order to develop the aforesaid analytical contents, Peano introduced 9 logic signs, 16 arithmetic signs, 7 for algebra and 11 for function theory. For each theorem, he first provided the symbolic statement, such as:

$$a,\, b\,\varepsilon\,\bar{q}\cdot a{<}b\cdot f\,\varepsilon\,\bar{q}Fa\bar{\ }b\cdot x\,\varepsilon\,a\bar{\ }b\cdot fx = \max f'a\bar{\ }b\cdot Dfx\,\varepsilon\,\bar{q}\cdot\supset\cdot Dfx = 0$$

followed by the explanation in natural language: “if a and b are two quantities and $a < b$, f is a function defined on the interval $a^{-}b$, x is a value of the interval, then every time that the function f by that value of x reaches a maximum or a minimum, for that same value of x must be $Dfx = 0$”.[17]

Peano’s teaching style thus adopted ideographic logic but only in an instrumental and expressive way. After all, Peano himself repeated on many circumstances that the introduction of logical tools into teaching would have to be a gradual process, commensurate to their actual utility at different school levels.[18] On the other hand, he had an unwavering trust in the advantages of adopting the *Formulario* as a textbook for university teaching and as an auxiliary handbook for middle and high maths education.[19]

The promotion of a work like the *Formulario*, unusual from the point of view of contents and form, demanded total commitment. It was Peano himself who orchestrated its popularization on an international scale, starting with a lecture he gave at the 23rd meeting of the *Association Française pour l’Avancement des Sciences,* held in Caen in 1894.[20] At the same time, he intensified his relationships with his French colleagues, writing to C. Meray, C. Jordan, R. Montessus de Ballore, A. Gérard, etc. to outline the virtues of ideography, and the structure, merits and aims of the *Formulario.* To all of them he reaffirmed that he was interested only in the *applications of mathematical logic,* which was therefore a *tool* springing from foundational research and placed at the service of mathematical research and teaching:

[16]Cf. G. Peano to G. Vacca, 15.11.1906, in Osimo (1992), letter 90.

[17]Peano (1904d, p. 72).

[18]Cf. for example G. Peano to E. Catalan, 25.1.1892, in E. Jongmans, *Quelques pièces choisies dans la correspondance d’Eugène Catalan,* Bulletin de la Société Royale des Sciences de Liège, 50, 9–10, 1981, p. 307–308: “When these theories are well developed, they may substitute partially or entirely the traditional teaching. But we must not, however, exaggerate, and assume that one can straight away explain these definitions and theorems in school, in the current form as published. They will turn out to be simply incomprehensible”.

[19]G. Peano to G. Vacca, 8.11.1897 and 28.12.1902 in Osimo (1992), letters 7 and 40. Despite the trust of Peano, and notwithstanding the presence of some enthusiastic witnesses, it cannot be denied that many criticisms arrived from his colleagues, and the students themselves were most unhappy at having to learn a formal language which (however patient and careful the relevant explanations might be) would prove of no further use or application in other teachings.

[20]Peano (1894h, p. 222–226).

> algebra has been *applied* to logic and we have succeeded in expressing a large collection of complete propositions, belonging to different fields of mathematical sciences. [...] Initially I set down only those propositions in which I was most interested, and last year I was able to adopt the *Formulario* in my calculus course at university.[21]

In his attempt to attract the attention of the French scholars to the *practical utility* of logic, Peano went so far as to propose reserving the name *mathematical logic* to applied research conducted by his '*School*', ceding that of *logistics* to the more abstract, and general, studies carried out by B. Russell, D. Hilbert and their collaborators.[22]

As information on the ideographic trait of Peano and his *School*'s activity began to circulate, reactions of perplexity at first, and later of open hostility, towards the logical symbols became widespread. Peano was substantially 'accused' of having forgotten what doing research in mathematics means, and of having devoted himself in a Manichean way to editing the *Formulario* and promoting his symbolic language. In the international *parterre*, the first to state the sense of disorientation experienced by several foreign colleagues when faced with the formalistic orientation of Peanian studies, was perhaps E. Borel. He attended Peano's plenary lecture at the first International Congress of Mathematicians in Zürich (1897) and released this report:

> Wednesday's session began with Mr. Peano's talk on mathematical logic. It is well known that the aim of this new discipline is that of completely eliminating natural language from mathematical texts, expressing all ideas solely with symbols. Mr. Peano explained the fundamentals of Logic; he eased the task of his audience by supplying them with the first issue of his work. Maybe his audience would have been more interested in his lecture if he had outlined the advantages of mathematical logic instead of simply teaching us the basic notions. [...] Mr. Peano could thus have possibly convinced his many critics of the *usefulness* of his symbols, those critics who jeer at these symbols without even knowing them; in turn they may have convinced Peano on the dangers which may arise when one plunges into the chasm between mathematicians and other scholars rather than attempting to build a bridge between them.[23]

In the meantime, a philosophical journal, the *Revue de Métaphysique et de Morale*, edited by Xavier Léon and André Lalande, was establishing itself as the principal vehicle of promotion of *Italian Mathematical Logic* in France, thanks to

---

[21]G. Peano to C. Méray, 25.7.1900, in Luciano and Roero (2005, p. 204). Analogous statements in G. Peano to C. Jordan, 6.11.1894 in Borgato (1991, p. 95–97); G. Peano to L. Couturat, 1.6.1899, in Luciano and Roero (2005, p. 18–21); G. Peano to R. Montessus de Ballore, 3.10.1896, 16.5.1897, 5.8.1897, in Le Ferrand (2012, p. 9–10).

[22]Cf. Couturat (1904, p. 1042) and, in the Peano-Vacca Archive (preserved in the Library of the Department of Mathematics, University of Turin) the manuscripts by G. Vacca, *Logica Matematica*, 1939, f. 1r; *Logica Matematica*, 1939a, f. 1r; *La Logica Matematica negli ultimi cinquant'anni*, 1939b, f. 1; *Nel Congresso Internazionale di Filosofia a Parigi 1900*, [post 1939], f. 1r.

[23]Borel (1897 p. 787).

the presence within the editorial board of Louis Couturat, a philosopher and logician who enjoyed a strong friendship and collaborative working relationship with Peano and his '*School*'.

4 Louis Couturat, Promoter of the Italian Mathematical Logic in France

A multifaceted intellectual, with interests ranging from philosophy to the history of linguistics, from set theory to the propaganda of Esperanto, Louis-Alexandre Couturat (1868–1914) maintained an active correspondence with Peano during the years 1896–1910, often involving him in mathematical and interlinguistic initiatives.[24] Over one hundred letters between the two show how fertile this period was for the '*Italian School of Mathematical Logic*'. These letters allow us to evaluate the entrepreneurial skills of Couturat in his capacity of advocate of the logistic movement in France; they document his strategies for the vulgarization of the ideography and *Formulario* via essays and reviews; and they illustrate his involving of the Italian colleagues in international congresses in Paris and Geneva, and his defence of their work from P. Boutroux, H. Poincaré, J. Hadamard, E. Borel and R. Baire.

Trained in philosophy and mathematics at the *École Normale Supérieure*, under the guidance of J. Tannery and Louis Liard, from as early as 1893, Couturat was taking part in the debates on philosophy of mathematics triggered by the discovery of non-Euclidean geometries and fostered by discussions on the nature of the continuum. However, it was the writing of the PhD thesis in mathematics, entitled *De l'Infini mathématique* and completed in June, 1896, which led him to read the most recent studies in algorithmic logic from Austria, Germany, Great Britain and Italy, and to open a correspondence with B. Russell. To the same period dates back the first contact between Couturat and Peano: in fact, the editor in chief of the *Bulletin des Sciences mathématiques et physiques*, Albert Léon Gérard, informed Peano about the publication of the book *De l'Infini mathématique* (Paris, Alcan, 1896) and, at the same time, introduced Couturat to the works of the '*Turinese School*', underlining their philosophical significance. Couturat thus discovered the *Italian Mathematical Logic,* and above all the advancements of ideography and its applications to the theory of real functions.

[24]On Couturat's biography and cultural activities cf. Schnippenkötter (1910, p. 447–468), Lalande (1914, p. 644–688), Dassen (1934, p. 136–143, 1939, p. 73–204); Loi (1976, p. 683–699); Schlaudt and Sakhri (2010); Luciano (2012, p. 41–64); Luciano and Roero (2005, p. VI-XXIV). From 1900 onwards, Peano and Couturat actively engaged in promoting the international language, advocating two different projects: Peano, the so-called Latino sine flexione, and Couturat, a form of artificial language derived from Esperanto and called Ido. The French philosopher however, employed increasingly dogmatic tones, proving intransigent in the dissemination of this project. This attitude caused rifts and disagreements with Peano, until they completely ceased all contact in the winter of 1910.

Concerning the symbolic language, Couturat vacillated in his evaluation from the outset. He admired the neat and compact expression of deductive processes which the Peanian ideography allowed, but he seemed to consider this system of signs as a mere form of tachigraphy. As he gradually delved deeper into his studies however, Couturat began to feel the need to explore further the most recent literature. Such purpose strengthened following his appointment to the faculty of liberal arts at the university of Caen. There he held some philosophy courses (1897–1900), which were aimed at illustrating the foundations of pure mathematics and algorithmic logic. During this period, Couturat's letters to Peano were dotted with requests of clarifications and of bibliographic references: Peano suggested to Couturat the reading of some of his own articles on the foundations of geometry, whose analysis was conducted for the first time treasuring the ideographic language. Meanwhile, Couturat and Peano entered into a discussion about the role of notations, paying particular attention to their expediency and to the problem of their polisemy if the same sign be used in both class theory and predicative logic.

The debate on the *status* of logic, the true focus of interest for the French philosopher, gradually began to dominate their correspondence and from 1898 onwards Couturat asked Peano on different occasions whether his mathematical logic should be interpreted solely as a language or if it were—properly speaking—a kind of algebraic algorithm. To answer this question, Couturat (on Peano's advice) undertook an in-depth examination of the work of G. Boole, H. MacColl, E. Schröder and A.N. Whitehead, comparing them with the Italian system. This comparative analysis resulted in the essay *La logique mathématique de M. Peano*, which appeared in *Revue de Métaphysique et de Morale* in October 1899. Couturat here advanced some comments on the system proposed by Peano and his co-workers. His objections concerned the consistence of certain notations (for example the symbol for negation was criticised as "ambiguous", due to its close resemblance to the algebraic sign for subtraction and negative numbers), and the axioms of logic classified as minimal by the '*Italian School*'. The French philosopher observed, on this point, a lack of economy in Peano's system, and held that he had over-developed the ideography whilst neglecting the algorithmic aspect (that is the rules of inference):

> At first, I was amazed by the great number of axioms that had been admitted as independent and irreducible. Undoubtedly, the author has already eliminated some, as we can see whilst flicking through his papers; but there are still a lot, maybe too many. [...] If we have a look at this system in its entirety, to give an overall evaluation of it, we ascertain that between the two elements of logical calculus, notation is much more developed than algorithm.[25]

Despite having proven his detailed knowledge of the first two editions of the *Formulario* (1894–95, 1897–99), at this time Couturat had not yet reached a full understanding of some of the most distinctive features of *Italian Mathematical Logic*. His critique in fact revealed to be partially unjust. For example, he equated Peano's ideography with a simple form of abbreviated writing (stenography), even

[25]Couturat (1899, p. 636, 643).

claiming the use of a specific copula ε to be redundant in expressing the predication *x*εF, for F single judgement.[26] On his part, Peano proved to be brimming with explanations, and despite renouncing to offer any elucidation on the inspirations behind his project, he underlined on many occasions in his letters to his French colleague the methodological aims he had set himself:

> As you well notice, my objective is to *apply logic* to mathematical sciences. I perfectly understand the importance of the studies on theoretical logic but, due to the extent of these studies, I prefer to direct my efforts towards its *application.*[27]

This goal—as Peano explained to Couturat—had materialized fully in the compiling of the *Formulario,* even if it had not always been possible to 'free' mathematics from the obstacle of the natural language, as the ideography was still too 'weak' to encode certain fields of mathematics.[28]

In the light of his correspondence with the Italian colleague, Couturat began to re-evaluate some of his opinions as expressed in the essay *La logique mathématique de M. Peano,* and he soon became aware of his initial misjudgement of the contributions of the '*Italian School*':

> I admit in my article that it is a little unfair to judge your work (as is my duty) from the point of view of (pure) formal logic, when it is mainly destined to the analysis and the control of mathematical proofs. There will be time to study your contributions and criticise them from the point of the *application* to sciences, but at the moment this would divert me too much from my purpose.[29]

The analysis of mathematical logic in its applicative uses was in fact developed, almost at the same time, by Giovanni Vailati. In the article *La logique mathématique et sa nouvelle phase de développement dans les écrits de M. J. Peano*, the outline of which was suggested by Peano himself,[30] Vailati expressly affirmed that deductive logic was strictly connected to mathematics in two ways: firstly because in exact sciences logical calculus best revealed its potential both as a tool for demonstrating and proving intuitive truths, and as a support of mathematical creativeness; and secondly because, thanks to the analogy between logical and algebraic operations, it allowed the replacing of reasoning with calculus, avoiding the errors and misunderstandings caused by the use of common language in proofs. The protagonists of the modern and contemporary phase of algebra of logic—according to Vailati—had been G.W. Leibniz, J.-H. Lambert, J. Segner, G. Boole, as well as G. Peacock, D. Gregory and A. De Morgan, who had all contributed to the development of this discipline by introducing new symbols or establishing their properties. However, Vailati also remarked that Peano had been the first to have created a logical-symbolic language powerful enough to be *applied* to arithmetic and geometry. Fascinated by these research studies, the members of his '*School*'

---

[26]Couturat (1899, p. 628–630).

[27]G. Peano to L. Couturat, 1.6.1899, in Luciano and Roero (2005, p. 19).

[28]*Ibidem.*

[29]L. Couturat to G. Peano, 4.6.1899, in Luciano and Roero (2005, p. 23).

[30]On the events behind the writing of this article see Roero (2009–2011, p. 169–184).

had continued to pursue them, with Burali-Forti using the ideography to investigate the theory of magnitudes (i.e. real numbers), and Pieri and Padoa applying it to the study of foundations of geometry and arithmetic.[31]

In the months that passed between the publication of the essays by Vailati and Couturat, the latter also started to study the work by A.N. Whitehead (*L'Algèbre Universelle de M. Whitehead*, RMM, VIII, 1900, p. 323–362). In his comparative analysis of the systems of Peano and Whitehead, Couturat awarded greater recognition to the contributions of Whitehead, approaching them to those of Leibniz in their range and generality. Peano and his '*School*' were instead charged with having neglected to highlight the formal analogy between logic and algebra, having been obliged instead to work on different signs in order to avoid ambiguity in the translation of various theories.

Thanks to the network of exchanges and through the manifold inputs arriving from Peano and Russell, in the last years of the 19th century, Couturat reached to elaborate an original position as regards the epistemological status of formal languages and their applicability to mathematics, in the dual role of descriptive and cognitive tools. Aware of the deplorable French isolation on the philosophico-foundational front, from this period onwards Couturat increased his activity of vulgarization of works on logic carried out abroad, particularly those by the '*School of Peano*'. Incited by feelings of true "scientific and philosophical fraternity", Couturat collaborated on the draft revisions for the second and third editions of the *Formulario*. Moreover, charged with organizing the section on *Logic and history of the sciences* at the first International Congress of Philosophy (Paris, 1–5 August 1900), he contacted Peano, inviting him to participate as a member of the *Comité de patronage* of the Congress, and encouraged the leading members of the '*School of Peano*' to present talks. Burali-Forti, Pieri, Padoa, Vailati and Vacca took up the challenge and the first three outlined the results achieved by applying ideography to the theory of real numbers, algebra and neutral geometry, whilst Peano himself chose to present in that event his ideas on definitions.[32]

The Paris Congress marked a milestone in Couturat's story and career, winning him fame within the French intellectual milieu. His position on the organizing committee of the conference put him alongside other outstanding scholars such as Poincaré, Jules and Paul Tannery. Couturat also took responsibility for the edition of the third volume of the Congress *Proceedings*, concerning the session he himself had led, and he wrote a report for *L'Enseignement Mathématique*. That was a perfect opportunity to permanently crystallize in the eyes of the French mathematical community the 'portrait' of the '*School of Peano*' and of its cultural project:

---

[31]Vailati (1899b, p. 100, 102).

[32]In the third volume of the Congress Proceedings, entitled *Logique et histoire des sciences*, Paris, A. Colin, 1901 the following articles appear: G. Peano, *Les définitions mathématiques*, p. 279–288; C. Burali–Forti, *Sur les différentes méthodes logiques pour la définition du nombre réel*, p. 288–308; A. Padoa, *Essai d'une théorie algébrique des nombres entiers, précédé d'une introduction logique à une théorie déductive quelconque*, p. 309–365; M. Pieri, *Sur la Géométrie envisagée comme un système purement logique*, p. 367–404.

> The Italian School has offered significant examples of its works on Mathematical Logic. Whatever one thinks about the usefulness or convenience of logical symbolism as used by Peano and his collaborators in the *Rivista di Matematica* or in the *Formulario*, we must recognize that they have lead to great progress in the analysis of concepts and principles of mathematics, and that thanks to this symbolism (and particularly the insight skills of those practitioners who manage it) they have reached wonderful results of rigor and subtlety. The four papers presented at the Congress by Italian mathematicians provide a general but consistent overview of their works in the main domains of mathematics, and nothing does more honour to the Italian School and its Head, Peano, than this spontaneous entente of his collaborators (which in no way diminishes their independence or individual originality).[33]

In the meantime, in order to give an exhaustive overview of the research tradition recently established by the '*Italian School*', "rendering more legible"[34] the *Formulario*, Couturat published a detailed review in the *Bulletin des Sciences Mathématiques* of the third edition of this encyclopaedia.[35] Here he expressly oriented the attention of French mathematicians towards the aspect of the scientific work by the '*Italian School*' that Peano himself had asked him to emphasize. Thus, Couturat showed his fellow-countrymen the *utility of the language* in which the *Formulario* was written for *application* to mathematical studies:

> We will not dwell on Peano's logical calculus, which after all occupies relatively little space in his *Formulario*, and we will instead study its *application* to mathematics. It should be noted that some modern mathematical theories (for example set theory) are simply offshoots of logical calculus, and would benefit from being expressed by the corresponding symbolism. […] So, thanks to this formal symbolism the reconstruction and re-foundation of the entire system of mathematical truths can be accomplished.[36]

With his 1901 review, Couturat introduced a new chapter in the reception of Peano's production in France, in that it clearly focused the attention on *Peanian logical language*, a future target of irony and contempt, even ostentatious, on the part of many mathematicians.

In the period following the first International Congress of Philosophy, right up to the summer of 1904, Couturat continued undeterred with his popularizing in France of the international studies of mathematical logic. To this end, he undertook an ambitious editorial project: the publication in the *Revue de Métaphysique et de Morale* of a series of five articles on the principles of mathematics, later collected in the volume (*Les Principes des Mathématiques*, RMM, XII, 1904, p. 19–50, 211–240, 664–698, 810–844, XIII, 1905, p. 224–256 and Paris, Alcan, 1905). His intention was to comment on and expand (in an educational perspective) the content of B. Russell's famous work *The Principles of Mathematics*, completing a study

[33]Couturat (1900, p. 401).

[34]L. Couturat to G. Peano, 1.10.1899, in Luciano and Roero (2005, p. 31).

[35]Couturat (1901, p. 141–159). Couturat had published an anonymous review of the second edition of the *Formulario* in the 1899 issue of the *Revue de Métaphysique et de Morale* (RMM, VII, Supplément p. 5–6). He also appears to have analysed the last edition of the *Formulario Mathematico*, in another anonymous review, which was published in the 1906 issue of the same journal (RMM, XIV, Supplément Mai 1906, p. 11).

[36]Couturat (1901, p. 141, 149, 159).

which he described to Peano as: "something midway between a review and a translation; in other words, a free commentary."[37] Hoping to address a varied readership of mathematicians, philosophers, humanists and intellectuals from different professionalizations, Couturat chose to frame Russell's contributions within the general context of studies carried out by Frege, Peano and Whitehead. In addition to presenting propositional calculus, class logic and the theory of relations, Couturat also tackled the constructions of real numbers, outlined the principal aspects of the debates over the continuum, and investigated the various axiomatizations of geometry. This led him to return to the works of the '*School of Peano*', particularly those of M. Pieri which he compared with the papers of O. Veblen. At the same time, the French philosopher continued his discussion with Peano on the ideographic notations introduced and used by the '*Italian School*' and by Russell. Examining the legacy of the Italian logic system on Russell, Couturat expressed on many occasions his conviction that the book *The Principles of Mathematics* was the 'capstone' of the studies launched by the School of Peano:

> I studied Russell's fine book [...]; I find every moment an opportunity to resume and quote your *Formulario* and other essays. This book, inspired by your method, does you the greatest honour; it shows the philosophical implications of your work.[38]

In the framework of the same strategy of promoting logic in France, in January 1904 Couturat drafted two handbooks, one of a historic slant—l'*Histoire de la Logistique*—closely linked to his course at Collège de France in 1905–1906 (when he substituted H. Bergson), and a second more specialistic volume, entitled *Manuel* or *Traité de Logistique*. This second text, particularly, should have offered a balanced summary of the systems by Peano, Russell and Frege, including a presentation of sentential and relational logic, and ending with a section dedicated to methodology. The *Manuel* or *Traité de Logistique* would have been accompanied by a wide bibliography and catalogue raisonne of sources, compiled with the assistance of Peano. This bibliography would have included all the recent works written in Peanian ideographic language, adopted and expanded by Russell. The plan to publish the *Manuel* or *Traité de Logistique*, repeatedly postponed, was dropped in autumn/winter 1904 due to two intense polemics involving Couturat, which saw him confronting firstly Poincaré's and then Borel's entourage.

5 The Controversy 'on Rigour and Intuition' (1904–1907)

In September 1904, the second International Congress of Philosophy was held in Geneva. Both Peano and Couturat took part in the section of *Logic and philosophy of the sciences* with Couturat not only speaking about the movement in favour of the international language, but also expressing his fervent belief in the usefulness of

[37]L. Couturat to G. Peano, 6.12.1903, in Luciano and Roero (2005, p. 54).

[38]L. Couturat to G. Peano, 6.12.1903, in Luciano and Roero (2005, p. 54).

mathematical logic (*Sur l'utilité de la logique algorithmique*, RMM, XII, 1904, p. 706–713). On the contrary, dealing with the notion of 'correspondence' in infinitesimal analysis, the philosopher Pierre Boutroux, son of mathematician Émile and nephew of H. Poincaré, expressed strong reservations on the possibility that logic could ground mathematics, and made harsh criticisms on the expectation of reducing mathematics to logic, isolating the minimum number of axioms and eliminating (thanks to ideography) any resort whatsoever to intuition.[39] With his speech, Boutroux thus shifted the debate from the question of the utility of logic *applied* to mathematics, to that of the (hierarchical) relationships between logic and mathematics. A lively discussion ensued, in which Peano maintained that logic and mathematics were strictly linked and mutually at the service of the other, despite the distinction of their own fields of research:

> G. Peano of Turin, whose doctrine is being questioned, answers [...]: the two theories are distinct, but nevertheless they generate each other. [...] There is neither separation nor opposition between logic and mathematics, but logical transition from one to the other.[40]

Charged with writing a report on the Congress and the publication of the *Proceedings* of the section *Logic and Philosophy of the Sciences,*[41] Couturat turned to Peano and Russell, for assistance in constructing a complete refutation of Boutroux's "unjust accusations" against logic. As Couturat wrote to Peano, Boutroux had in fact shown himself to have a shallow knowledge of this area of research and yet aspired to judge its usefulness and productiveness. However, all of this was not surprising, as his case was emblematic of the overall French *Weltanschauung* and showed how: "theories pertaining to logic still face a lot of prejudice amongst philosophers".[42]

At this point, Boutroux decided to contact Peano directly, in a brief period of correspondence in the autumn of 1904, to hold a long-distance discussion continuing the conversation that had taken place after the reading of his essay at the Geneva Congress.[43] Excerpts from Peano's letters were published by Boutrox as part of the article *Correspondance mathématique et relation logique.*[44] Despite Peano's explanations, Boutroux did not alter his opinion, vindicating a clear distinction between mathematics and logic, since logic renounced its creative

[39]Boutroux (1904, p. 909–920).

[40]Couturat (1904, p. 1046).

[41]Couturat (1904, p. 1037–1077).

[42]L. Couturat to G. Peano, 2.10.1904, in Luciano and Roero (2005, p. 77). Hence, from this date onwards, there were many complaints in the correspondence between Couturat and Peano regarding Boutroux, Poincaré, Hadamard, Laurent, Borel, who did not know the topic of Logic indepth and yet criticized it, "speaking of what they ignore". Cf. L. Couturat to G. Peano, 22.9.1904, 8.11.1904, 5.1.1905, 18.7.1905, 2.8.1905, 21.11.1905, 15.3.1906, 17.1.1906, 3.5.1906, 14.6.1907, in Luciano and Roero (2005, p. 75, 77–78, 79, 85, 89–90, 93, 98, 101–102, 106–107, 132).

[43]Cf. P. Boutroux to G. Peano, 21.11.1904 and G. Peano to P. Boutroux, [November–December 1904] in Luciano and Roero (2005, p. 218–219, 219–220).

[44]Boutroux (1905, p. 620–637).

dimension in science in favour of a merely expositive role. In his second essay, strongly criticized by Couturat, Boutroux held that, if describing logic and mathematics metaphorically, the first was a "backward path", the second a "forward gear".

This second article by Bourtroux deeply embittered Couturat, who accused his colleague of improper behaviour, having taken advantage of some explanations received confidentially from Peano. The publication of the summary of the Geneva congress and the two reports by Boutroux unleashed a series of interventions by Couturat, Poincaré, Hadamard, Russell, Borel, Winter and, from the Italian ranks, Pieri, Vacca, Padoa and Peano.[45]

The so-called 'debate on rigour and intuition' was exacerbated by the publication of Poincaré's essay, *Les mathématiques et la logique* (RMM, XIII, 1905, p. 815–835; XIV, 1906, p. 17–34 and p. 294–317). Resuming the stances of Boutroux, Poincaré here carried out a trenchant attack on the logicist theories vulgarized by Couturat. In particular, he expressed some doubts on the possibility of isolating the logical pieces from mathematical reasoning, and accused logicians of trying to banish intuition from mathematics, reducing the principle of induction to a mere definition of the natural number. He also questioned the usefulness of symbolism and its efficacy to overcome the problem of the paradoxes of naive set theory. His critique was specially harsh regarding the ideography constructed by the '*School of Peano*', to which Couturat had attached exaggerated importance (according to Poincaré), so much so that "Peano himself should have been surprised". The ideography—claimed Poincaré—could only cause a pointless weighing down of the mathematical discourse and obstruct the free unravelling of the intuition in the phase of "discovery". Questioning the utility of the symbolic language, Poincaré wrote:

> On the question of fertility, it seems M. Couturat has naive illusions. Logistic, according to him, lends invention "stilts and wings", and on the next page: "Ten years ago, Peano published the first edition of his *Formulaire*". How is that, ten years of wings and not to have flown! I have the highest esteem for Peano, who has done very pretty things (for instance his "space-filling curve"); but after all he has not gone further nor higher nor quicker than the majority of wingless mathematicians, and would have done just as well with his legs. On the contrary, I see in logistic only shackles for the inventor. It is no aid to conciseness -far from it, and if twenty-seven equations were necessary to establish that 1 is a number, how many would be needed to prove a real theorem?[46]

Poincaré also stated that the application of symbolism did not constitute in itself a guarantee of infallibility, i.e. the use of ideography did not prevent the emergence of paradoxes, as had been proven by the experience of C. Burali-Forti.[47]

---

[45] On this debate cf. Jourdain (1912b, p. 481–483), Sanzo (1975, p. 369–418), ENS (1983, p. 97–111), Heinzmann (1986), Luciano and Roero (2005, p. XXIX–LX).

[46] Poincaré (1906b, p. 295). Regarding the hostility to Peano's ideography cf. also Poincaré (1903, p. 5) and L. Couturat to G. Peano, 15.3.1906 and 3.5.1906, in Luciano and Roero (2005, p. 101, 107).

[47] Poincaré (1905a, p. 825).

Along the same line of thinking, J. Hadamard observed a short time later in the *Revue générale des sciences pures et appliquées*, that the Peanian logic—a collection of 'Kabbalistic-looking' signs, to be combined according to a limited arsenal of rules—had failed in its goal, proving itself incapable of ensuring the complete exactness of deductive processes.[48]

Couturat responded to these criticisms with the pamphlet *Pour la logistique*, published in July, 1906. Aware of Poincaré's reputation in the international scientific arena, the French philosopher decided to adopt an ironic, quasi-apologetic, tone in his response. Particularly, concerning the work of the '*Italian School*', he published a short extract of one of Peano's letters in which he affirmed to have proven often "the importance of symbolic notation, its great utility in difficult and delicate questions and its indispensability, both in foundational studies and in his teaching". On the advice of Peano, Russell, Pieri and Burali-Forti, Couturat then discussed other criticisms made by Poincaré regarding the principle of induction, the antinomies, the axiom of choice, etc.

To complete the defence of Peano's logic, two other interventions arrived: one by J. Richard and the other by B. Russell. J. Richard commented:

> Peano imagined a kind of algebra of logic; he employs this 'algebra' for writing propositions and deducting these propositions one from another, according to fixed rules. He never had, I could prove it by many quotes, the idea of using that algebra to supply the foundational notions of science, such as the concept of number. So if others are adapting Peano's system at this latter need and thus diverted it from its proper objective, it is not the fault of Peano.[49]

Meanwhile, Russell pleaded with Poincaré to abandon the provocative tone which had marked his essays and defended with energy the *Italian School*'s cultural project:

> One point on which I respectfully beg to differ from Poincaré's view regards his appraisal of Peano; especially since he assumes that my work, if valid, would destroy those of Peano, a belief which is in fact the opposite of my own opinion. Poincaré said: [...] that most of Peano's work is of no interest for his studies. But Peano forged a very powerful tool for some lines of research. Some of us are interested in these lines of research, and hence honour Peano, who has gone (we believe) so much farther and higher than "wingless" mathematicians so that they have lost sight of his, and do not know how much he foreruns them.[50]

Despite constant pressure from Couturat, Peano never intervened in defence of his ideography in the French context. It was only in the spring/summer of 1906 that he explained his point of view on some aspects of this debate 'on rigour and intuition', publishing two articles: *Super theorema de Cantor-Bernstein* (1906b) and *Additione* (1906e).

---

[48]Hadamard (1906, p. 161–162).

[49]Richard (1906, p. 957–958). As far as Richard's appreciation of Peano's logical language is concerned cf. also Richard (1935, p. 18, 20).

[50]Russell (1906, p. 628).

The controversy continued unabated until 1907. In that year, E. Borel and M. Winter dedicated two significant works to the question. Although they conceded that mathematical logic could be useful for teaching, they both underlined the mere instrumental and 'grammatical' purpose of the *Italian Mathematical Logic*; they denounced the verbosity of the *Formulario*, and expressed the conviction that a research project of this nature could not be 'original' in any respect:

> Mathematics makes its own science; and I am not afraid of being contradicted by a living mathematician; one should resist by referring to texts in which eminent mathematicians speak of mathematics as if reduced to its elementary parts, not wanting to flaunt their knowledge; I should possibly imitate their reserve; but it does not seem possible to me to admit that it is sufficient to dissect some elementary reasoning in order to find out what mathematics is; it's a bit like a Chinese landing on our shores and carefully examining a bathing cabana should pretend to have formed in his mind an overall idea of European architecture, from the Parthenon to the Eiffel Tower via Notre Dame of Paris.[51]

> Maybe Peano and his disciples felt that when limiting their work to the individuation of grammatical-logical principles of mathematics, logistics constitute only a limited specialty and, as the idea of a new general method probably haunted them, they have introduced into the *Formulario* a number of pages that should not, strictly speaking, be included. [...] On page 35 we read, in the usual arrangement, the developments of the first powers of binomials $(a + b)$ trinomials $(a + b + c)$ and other algebraic expressions such as $(a+b)^4 + a^4 + b^4 = 2(a^2 + ab + b^2)^2$. All these expressions might also occur in an ordinary handbook of algebra, and we see no improvement in the method except at the beginning of paragraph 2 with the expression: $a, b \in N_0 \cdot \supset \cdot$ and at the beginning of paragraph 3 with the analogous expression $a, b, c \in N_0 \cdot \supset$. This means that in the following propositions a, b, c must be considered as numbers: we had little doubt... We do not see, in general, what algebra hopes to accomplish with such obvious remarks.[52]

The 'debate on rigour and intuition' finally ended in 1908, with a general consensus between Russell and Poincaré. In fact, in the end, they both agreed that logic is an auxiliary of intuition and that, like mathematics, it is marked out by an inextricable weaving of intuition and rigour, induction and deduction. For Couturat, reduced to an inferior position, the epilogue of the debate marked the beginning of a new phase in his life and work. Deeply offended, he abandoned all his studies on logic and dedicated himself exclusively to interlinguistic issues. From 1907 to 1912 he wrote not a single line on the topic of logic, not even in his correspondence with Peano, and it is noteworthy that he returned to the question on only one more occasion, as a consequence of the original debate. In an essay that appeared in the *Revue du Mois* Boutroux attributed in fact to Poincaré, recently passed away, the merit for having halt the diffusion of logic in France.[53] Couturat then decided to write a short text for the *Revue de Métaphysique et de Morale* in which, in a calm tone and an attempt to play down the disagreements of the past, he recalled some of the most salient aspects of the argument which had inflamed the French cultural world from 1904 to 1907.[54]

---

[51]Borel (1907, p. 282).

[52]Winter (1907, p. 192–193, 195–196).

[53]Cf. Boutroux (1913, p. 181).

[54]Cf. Couturat (1913, p. 260).

6 A Second 'Controversy': Couturat Versus Baire, Borel and Lebesgue

It is in this same *Weltanschauung* that a second episode may be inscribed, namely the fruitless attempt by Couturat to arouse the interest of Baire, Borel and Lebesgue in Peano's ideography.

In 1904, Borel inaugurated a collection of monographs on the theory of functions, which he himself supervised.[55] In December of that same year, he sent the first two volumes of the collection to the editorial board of the *Revue de Métaphysique et de Morale*: his *Leçons sur les fonctions de variables réelles* and R. Baire's *Leçons sur les fonctions discontinues,*[56] asking Couturat to announce them in the next supplement.

Couturat wrote an anonymous review of the two volumes, which was to appear in the January issue of the *Revue de Métaphysique et de Morale.*[57] At first sight, it might seem strange to find, in such a journal, a review of two such technical books on advanced mathematics, like those of Baire and of Borel. Nevertheless, it must be remembered that, from 1899, Couturat had taken upon himself the task of indicating in this journal the mathematical works that could most interest a public of philosophers, and above all those which highlighted the increasing importance of set theory and mathematical logic, as a basis of analysis.[58]

So, in his review, the French philosopher stressed that, on one hand, Baire had rigorously shown the (logical) existence of transfinite numbers, "by pointing out carefully-chosen examples: derived sets of infinite order, sets of points of a segment, integer sequences",[59] and—on the other hand—he had proven how the study of the ordinal properties of perfect, dense and nowhere-dense sets was essential to determining the distribution of the discontinuities of the functions.[60] Furthermore, Couturat underlined that in his *Leçons*, Baire had generalized his fundamental theorem on pointwise discontinuous functions to finite or infinite value functions of several variables, thanks to an extension of the concepts of limit made by Borel.[61]

---

[55]On the history of Borel's *Collection* cf. Erhardt (2011, p. 111–139).

[56]Borel (1905), Baire (1905).

[57]Couturat (1905c, p. 6). Couturat's review focused mainly on Baires' text. In fact, as Borel had announced beforehand to Couturat, it was from *Leçons sur les fonctions discontinues* that he drew the greatest food for thought, both because it was simpler than Borel's book, and because it provided the necessary conceptual tools for a full understanding of some sections of Borel's own text.

[58]See in particular Couturat's reviews of G. Peano, G. Cantor, P. Poretsky, E. Schröder works, published anonymously in the *Revue de Métaphysique et de Morale.* Cf. in RMM: Suppl. Janvier 1899, p. 5–6; Suppl. Novembre 1899, p. 4; XIV, Suppl. Mai 1906, p. 10.

[59]Couturat (1905c, p. 6); Attachment from L. Couturat to E. Borel, 13.12.1904, FBP, ms. 053, p. 3.

[60]See in particular the proof of Baire's theorem (1905, p. 67–98): "the necessary and sufficient condition for such a function be the limit of continuous functions is that it is punctually discontinue on any perfect set".

[61]Cf. Baire (1905, p. 121–124), Borel (1905, p. 4).

Couturat therefore concluded that, in the light of Baire's work and his book *De l'Infini mathématique* (1896), infinity could be considered not only a value of the variable, but also a point of discontinuity:

> What do the finitists *à la* Renouvier think about? Nothing, of course, and for good reason; they will continue for a century or more to rehash as gospel truths the platitudes of a pseudo-critic of sciences, which has contributed so much to producing the anti-intellectual and anti-scientific reaction which abounds today.[62]

Couturat sent Borel, Baire and Lebesgue the draft of his review, together with his report of the 2nd International Congress of Philosophy before mentioned, and a short manuscript entitled *Remarques d'un logicien.*[63] His intention was twofold. On one hand, by illustrating to colleagues how he had reacted to Boutroux's speech on the relationships between logic and function theory, Couturat aimed at 'recruiting' Borel and Baire in a battle supporting Russell's line of thought:

> As I said to Baire, regarding his footnote on p. 121, Russell has highlighted the fact that pure Mathematics is based on the notion of order, on the "ordinal" relationships of numbers, and consequently on the Logic of relations, because any order is defined through relations. As much as I was upset to see a mathematician such as Boutroux oppose this doctrine, which seems consistent with the spirit of modern mathematics, I was equally happy to acquire from Baire confirmation on a specific and important detail, such as the definition of limit.[64]

On the other hand, Couturat's *Remarques*—a sort of memorandum, expressly written for private use—underlined the reflections that Baire's and Borel's *Leçons* might suggest to a logician, not to a mathematician, nor to a philosopher. The opening of these *Remarques* was a cutting one:

> Both authors used (without knowing) almost all of Peano's fundamental notions of mathematical Logic.[65]

The term *notions* must not deceive us. In effect, encouraged by the conviction that the socializing of Peano's Mathematical Logic meant, above all, spreading knowledge of his ideography (particularly suitable for *applications* to mathematics), Couturat here did only speak of Peano's *notations*: the ideographic signs for implication, logical sum, product etc., which Baire and Borel had indicated

---

[62]Couturat (1905c, p. 6); Attachment from L. Couturat to E. Borel, 13.12.1904, FBP, ms. 053, p. 4.

[63]Attachment from L. Couturat to E. Borel, 13.12.1904, FBP, ms. 053, p. 5.

[64]L. Couturat to E. Borel, 13.12.1904, FBP, ms. 053, p. 1: "Comme je le dis à Baire, à propos de sa note de la p. 121, Russell a bien mis en lumière ce fait que la Mathématique pure repose sur la notion d'ordre, sur les relations "ordinales" des nombres, et par suite sur la Logique des relations, puisque tout ordre se définit au moyen de relations. Autant j'ai été contrarié de voir un mathématicien comme Boutroux combattre cette doctrine qui me semble conforme à l'esprit des mathématiques modernes, autant j'ai été heureux de trouver chez Baire une confirmation, sur un détail précis et important, comme la définition de la limite".

[65]Attachment from L. Couturat to E. Borel, 13.12.1904, FBP, ms. 053, p. 5: "Les deux auteurs ont employé, sans le savoir, presque toutes les notions fondamentales de la Logique mathématique de Peano".

"without knowing" with new symbols, since they ignored the works of the '*Italian School*':

> The inclusion between two sets that Mr. Baire (p. 16) indicates with the sign $\leq$, is indicated as € by Schröder, and $\supset$ by Mr. Peano. Logical addition, which Mr. Borel (p. 17) marks as + (like Schröder), and Mr. Peano as U. For an infinite sum, Mr. Baire uses the symbol $\sum$; Mr. Peano, U'. For logical product, Mr. Borel uses (,) (p. 19) and Mr. Peano $\cap$. For an infinite product, Mr. Baire adopts the symbol: D(P_1, P_2, ...), which he calls (like Mr. Dedekind) the maximum common divisor of the sets P_1, P_2, P_3, ... Elsewhere (p. 100), he refers to logic product as a common set. For an infinite (or simply undefined) product, (p. 105) he refers to the section of P as determined by $\sum$; it is the logical product of P by $\sum$. For the logical zero (empty set), Mr. Baire adopts the notation = 0 (p. 15) as in arithmetic. Mr. Peano uses Λ. Furthermore, M. Baire (loc. cit.) states: "P is null in the ab interval". He intends to say: "$P \cap ab = \Lambda$": the logical product of P and of ab is a null set. [...].[66]

Couturat concluded by saying:

> I hope that my articles and "Remarks" attract your attention to the mathematical Logic (Logistics) of Peano and co. [...] But it is the Peano system, which is most interesting for the logic <u>of</u> mathematics.[67]

With his letter to Borel and Baire, and with these three attachments, Couturat abandoned the 'caution' that had marked his previous works of vulgarization of Italian logic (1899, 1900, 1901) and even mixed currents of studies very distant from one another (set theory, Peano's *mathematical logic* and Russell's *logistics*).

Unfortunately, Borel's and Baire's answers have not been passed down to us. Nevertheless, we can perceive their reactions through a network of allusions to Couturat's *crusades*, present in Baire's and Lebesgue's letters to Borel, and in the epistolary exchange between Couturat, Peano and Russell. Borel was evidently very sceptical, especially in relation to the possibility of 'fighting' finitist philosophy with mathematical arguments of any kind, and to the nominalist aspect of 'modern' logic. Baire and Lebesgue, on the other hand, appeared annoyed, both at Couturat's style and at his surreptitious attempt to engage them in his *crusades*: one against F.

[66] Attachment from L. Couturat to E. Borel, 13.12.1904, FBP, ms. 053, p. 5: "La relation d'inclusion entre deux ensembles, que M. Baire (p. 16) figure par le signe $\leq$, Schröder par le signe €, et M. Peano par le signe $\supset$. L'addition logique, que M. Borel (p. 17) figure par + (comme Schröder), et M. Peano par le signe $\cup$. Pour une somme infinie, M. Baire emploie le signe $\sum$; M. Peano, le signe $\cup$'. La multiplication logique, que M. Borel figure par le signe (,) (p. 19) et M. Peano par le signe $\cap$. Pour un produit infini, M. Baire emploie le symbole: D(P_1, P_2, ...), qu'il appelle (comme M. Dedekind) plus grand commun diviseur des ensembles P_1, P_2, P_3, ... Ailleurs (p. 100), il appelle le produit logique: ensemble commun. Pour un produit infini (ou simplement indéfini), (p. 105) parle de la portion de P déterminée par $\sum$; c'est le produit logique de P par $\sum$. Le zero logique (la classe nulle): M. Baire le représente par = 0 (p. 15) comme le zero arithmétique. M. Peano le figure par Λ. En outre, M. Baire (loc. cit.) emploie la locution: "P est nul dans l'intervalle ab". Il veut dire: "$P \cap ab = \Lambda$": le produit logique de P et de ab est nul. [...]".

[67] L. Couturat to E. Borel, 13.12.1904, FBP, ms. 053, p. 2: "Je souhaite que mes articles et mes "Remarques" attirent ton attention sur les théories de Logique mathématique (de Logistique) de Peano et cie. [...] Mais c'est le système Peano qui est le plus intéressant pour la Logique <u>des</u> mathématiques".

Evellin, E. Le Roy and the finitists, and one in favour of the 'modern' logic (both in the sense of Russell's theory of relations and in that of Peano's ideography). Thus Lebesgue did not hesitate to define Couturat's *Remarques* as "stupid as far as contents, presumptuous for their tone".[68] For his part, Baire did not intend to be enlisted "in the bataillon of those who share the bed with the infinite concept" and vented his feelings with Borel, commenting:

> I have received Couturat's stuff. Oh, yes! He is a raconteur, so much so that I'm going to write to him myself, rather sharply, because truth comes first. On the other hand, he is really quite irritating with his stock-phrase 'without knowing'.[69]

No less negative were the reactions of the three mathematicians regarding Peano's ideography. In this connection, Baire was pithy in his affirmations:

> The fact of translating deductions expressed in ordinary language into formulas may have some advantages but it has at least two serious detriments: 1. these people seem to believe that they will build frameworks that will comprise all future science whether you like it or not; 2. on the contrary to what Couturat said in one of his analyses of Peano, Schröder or Russell's systems, the algebraic language holds less guarantees of infallibility for research than a language which is adapted to each particular subject. Whence, I see striking proof in the calculation error committed by König. It is absurd for me to perform calculus in set theory. I realize that my misunderstanding of Zermelo's sentence on the products of sets was a direct consequence of the fact that he himself speaks "the language of logic": so, the intersection of sets is, in his view, the logical product of these sets? That is far-fetched![70]

In relation to Baire's view of ideography (very close to that stated slightly later by Poincaré), it must be remembered that he was one of the few French mathematicians to have had the opportunity of direct contact with the '*School of Peano*'. In fact, in the winter semester of 1898, he had spent a stay of studies in Turin, conducting research on analysis with V. Volterra.[71] He had also come into contact with C. Segre and V. Volterra who strongly criticized the ideographic turn of Peano's activity. There does not, however, seem to have been any interaction between Baire and Peano, busy in those months completing the second edition of the *Formulario*.

In the space of three days, Couturat wrote twice to Borel,[72] who among other things was his *ancien camarade* at the *École Normale Supérieure* in Paris, to clarify his positions. As regards the tone of his *Remarques*, he complained in turn about having received from Lebesgue a letter "of prankster and humbug nature, in which

[68]H. Lebesgue to E. Borel, 17.12.1904, in Bru and Dugac (2004, p. 98).

[69]R. Baire to E. Borel, 15.12.1904, in Dugac (1990, p. 80–81).

[70]R. Baire to E. Borel, 15.12.1904, in Dugac (1990, p. 80).

[71]Cf. Dugac (1976, p. 305, 318–329, 341–363, 1990, p. 33–48, 93, 95–96).

[72]L. Couturat to E. Borel, 16.12.1904, FBP, ms. 054, p. 1–4; L. Couturat to E. Borel, 19.12.1904, FBP, ms. 054, p. 1r-2v.

he does not answer because there is nothing to answer, proving only his levity"[73] and confessed that he did not understand what could have inspired such acrimony in his colleague. It was not at all his intention—Couturat added—to make personal criticisms of anyone, but quite the contrary:

> I wish to say only this: *You know*, you use concepts that logicians (who were also not bad mathematicians: Boole, Schröder, Peano) have managed for the past 50 years; they designate otherwise. Would it not be better for you to agree with them on vocabulary and notations? This is what happens every day in the cross fields between various sciences, where scholars from opposite sides come together, *often without knowing*.[74]

Faced with the philosophical objections by Borel and Baire, Couturat on one side emphasised that he did not fall into the same incongruities as the finitist philosophers—since through logical arguments he did not claim to show the coherence of a metaphysical construction of any kind—and on the other side warned his friends against over-insisting on the nominalist character of the 'new' logic:

> By aligning words we can prove something, because words mean ideas and judgments, and any deduction consists of the connections of ideas and judgments. Mathematicians harm them, or rather harm their science, by claiming to just align words. You talk about language conventions but a language means anything, damn it![75]

However, Couturat's reply focused above all on Peano's ideography. Regarding the symbols, he had never thought of proposing to Baire and Borel that they write books or articles entirely in symbolic language, but only to

> point out some expressions and notations of Logic, and ask you if it would be appropriate to use them instead of phrases and signs you borrow from other branches of science, in consideration of actual analogies, analogies that may deceive the reader. Instead of using, for example, the + sign for the logical sum, it would cost you nothing to adopt Peano's sign, since in any case you are forced to define the + sign in your new meaning.[76]

[73]L. Couturat to E. Borel, 19.12.1904, FBP, ms. 055, p. 1.

[74]L. Couturat to E. Borel, 19.12.1904, FBP, ms. 055, p. 1–2: "Je vous dis simplement ceci: *Vous savez*, vous employez certains concepts que les logiciens (qui étaient aussi des mathématiciens non méprisables: Boole, Schröder, Peano) manient depuis 50 ans; ils les désignent autrement. Ne serait-il pas bon de vous entendre avec eux sur le vocabulaire et les notations? C'est ce qui arrive tous les jours dans les domaines entre diverses sciences, ou des chercheurs venus de côtés opposés se rencontrent *souvent sans le savoir*".

[75]L. Couturat to E. Borel, 19.12.1904, FBP, ms. 055, p. 1: "en alignant des mots on peut prouver quelque chose, car les mots signifient des idées et des jugements, et toute deduction consiste dans des liaisons d'idées et de jugements. Les mathématiciens se font tort, ainsi qu'à leur science, en prétendant ne faire qu'aligner des mots.Vous parlez de convention de langage mais un langage signifie quelque chose, que diable!".

[76]L. Couturat to E. Borel, 16.12.1904, FBP, ms. 054, p. 2: "signaller quelques expressions et notations de Logique, et vous demander s'il ne conviendrait pas de les employer au lieu des locutions et des signes que vous empruntez à d'autres branches de la science, en vertu d'analogies réelles, mais qui peuvent tromper le lecteur. Au lieu d'employer, par exemple, le signe + pour la somme logique, il n'en coûterait pas plus d'employer le signe de Peano, puisque de toute façon vous êtes obligés de définir le signe + dans son nouvel emploi".

According to Couturat, the use of Peano's ideography by *applied mathematicians*[77] like Baire, Borel and Lebesgue not only would not cost them any effort, but also would present the advantage of proving the useful *applications* of 'new' logic to the theory of functions, bringing their works closer to those of *mathematicians-logicians.* The preference for Peano's system over, for example, those of E. Schröder and G. Boole, derived from the fact that the ideography worked out by the '*Italian School*' was the only one to ensure the absence of ambiguity between logical signs and algebraic ones.[78] Besides, Couturat noted, Peano's symbols were the best known in the mathematical world, and had already been employed in numerous articles and treatises published in Italy, Great Britain and the United States. This, therefore, made it urgent to diffuse them in France, so that:

> it would be good to know these symbols [...], and for this it is natural to rely mainly on young mathematicians. Moreover, I will soon publish a manual on Logistics to popularize the knowledge of this new Logic; because [...] although the symbols are due to Peano, the ideas date back half a century and are roughly the same for all modern logicians, whose agreement guarantees the objectivity of these theories.[79]

> As far as the "new Logic" is concerned, it was created [...] by mathematicians and was the result of mathematical studies, thus it is not appropriate for mathematicians to disregard it, it would be stupid to send us back to Aristotle! The secular divorce between Logic and Mathematics derives precisely from the fact that Aristotelian logic *does not apply to mathematics.* The new logic is—on the contrary—the Logic *of* mathematics, and that is why agreement between mathematicians and logicians is natural and desirable. But mathematicians must not confuse the new and old Logic, holding them in equal disdain, and ignore it a priori.[80]

---

[77]L. Couturat to E. Borel, 16.12.1904, FBP, ms. 054, p. 2. The title of "applied mathematicians", given to Borel, Baire and Lebesgue by Couturat, has a precise meaning. Just as Peano claimed to construct ideography purely for its *usefulness* (i.e. its *applications* to mathematics), thus did Borel, Baire and Lesbegue claimed to adopt set theory, exclusively for its *practical value,* i.e. its *applications* to other branches of mathematics, including function theory. For this reason, Couturat considered his French colleagues to be 'applied mathematicians' and not practitionners of abstract set theory. This stance of Borel, Baire and Lesbegue is fully analyzed and documented by Gispert (1995, p. 53, 58–59, 72–73, 75).

[78]L. Couturat to E. Borel, 16.12.1904, FBP, ms. 054, p. 2.

[79]L. Couturat to E. Borel, 16.12.1904, FBP, ms. 054, p. 2: "il serait bon de les faire connaître [...], et pour cela il est naturel de compter surtout sur les jeunes mathématiciens. Du reste, je vais publier bientôt un manuel de Logistique destiné à vulgariser la connaissance de cette Logique nouvelle; car [...] si les symbols sont propres à Peano, les idées datent d'un demi-siècle, et sont à peu près les mêmes chez tous les logiciens modernes, dont l'accord garantit l'objectivité de ces théories".

[80]L. Couturat to E. Borel, 19.12.1904, FBP, ms. 054, c. 2v: "Quant à la "Logique nouvelle", fondée [...] par des mathématiciens et issue de l'étude des mathématiques, ce n'est pas à des mathématiciens qu'il convient de la méconnaître; et ils sont malvenus à nous renvoyer à Aristote! Le divorce séculaire de la Logique et des Mathématiques vient justement de ce que la logique aristotélicienne ne s'appliquait pas aux mathématiques. La logique nouvelle est au contraire la Logique *des* mathématiques, et c'est pour quoi l'entente entre mathématiciens et logiciens est naturelle et désirable. Mais il ne faut pas que les mathématiciens confondent la nouvelle Logique avec l'ancienne dans un égal dédain, et l'ignorent de parti-pris".

In the meantime, the echo of this affair reached Peano and Russell, to whom Couturat expressed his bitterness at the reception of his *Remarques* amongst Borel's circle. Over a few days, he wrote to Russell and Peano:

> Mathematical logic still faces many prejudices and biases, even amongst mathematicians; I recently experienced this when speaking to several of my friends and comrades in favour of your system and work; they reacted in the most disdainful manner and, I dare say, the most foolish.[81]

> It is useful to popularize logistics, or even make it known, I see this daily. I had the following small experience: [...] It is clear that [Borel, Baire and Lebesgue] disdain and ignore the works of Peano and his School, and believe them to be absolutely useless and sterile. "There is no need of these signs to reason correctly, etc.". I fear this is the attitude of all mathematicians with regard to logistics; that people who live amongst symbols (to the point to reduce their entire science to a pure formal game) have a strange aversion to any symbols they do not understand. They do not, however, save the ancient algebraic symbolism.[82]

Russell's answer came in January 1905.[83] He limited himself to criticising the attitude of Baire, who found it reasonable to use notations invented by him or attributed to R. Dedekind, while at the same time accusing the mathematicians that invented ideographic languages of dealing with frivolities. Nevertheless, Russell showed he was confident about the fact that, occupying a chair, and having founded a '*School*', Peano would have an opportunity to spread the new logical-mathematical theories. No further mention of this story has been found in Couturat's correspondence with Peano. Moreover, soon afterwards Baire, Borel and Lebesgue would be active above all on the front of debates on set theory, following the publication of *Cinq lettres sur la théorie des ensembles*, while Couturat and the '*School of Peano*' were to be submerged by the 'polemic on rigour and intuition' which exploded in the *Revue de Métaphysique et de Morale*.

7 Conclusions

The forms, vehicles and dynamics through which the French mathematical community came into contact with *Italian Mathematical Logic* and with the output of the '*School of Peano*' were significantly affected by the work of cultural mediation carried out by Couturat. The images of the *Italian Mathematical Logic* and of the '*School of Peano*', as furnished by Couturat in his essays, only partly corresponded to the real *status* of the research by the Turin team, and in particular unnaturally glossed over the complex dynamics inside this work group, which preceded and accompanied their scientific teamwork activity. Of course it is natural that any practice of socialization, though entrusted to specialists like Couturat, simplifies and possibly banalizes the thought of an author or a group of authors, omitting the

[81]L. Couturat to G. Peano, 5.1.1905, in Luciano and Roero (2005, p. 83).

[82]L. Couturat to B. Russell, 18.12.1904, in Schmid (2001, vol. 2, p. 454).

[83]B. Russell to L. Couturat, 1.1.1905, in Schmid (2001, vol. 2, p. 461–462).

details, nuances and contradictions present *in* and *between* their writings. This process of levelling is all the more radical if—as happened for the '*School of Peano*'—it was its own members that piloted the choice of aspects on which to centre popularization activity, for the purpose of constructing and validating a well-defined and stable image of their collective work.

Following the directives of Peano himself, Couturat clearly oriented the perception and evaluation in France of the foundational studies conducted by the members of the Turin team. Specifically, in an attempt to attract to ideography the interest of the *applied mathematicians* Baire, Borel and Lebesgue, Couturat played precisely on the *applicative* aspect of Italian logic, and hence on the employment of ideography in the sphere of real analysis. In doing this, he wittingly caused a fusion between two separate planes: that of *applied set theory*, in accordance with Borel and his colleagues, and that of *applied logic*, in accordance with the '*School of Peano*', which refer to two very different ways of conceiving *application* and *utility*. The results of this strategy were not positive, particularly as this confusion of meanings and cultural orientations caused the protests of Baire, Borel and Lebesgue.

A second element that conditioned the approach of the French mathematicians to the logical-foundational studies of the '*School of Peano*' was the almost complete absence of references, in Couturat's essays, to the teaching implications of this research. In this connection, the unifying factor in Peano's scientific production, which accounts for the ideographic turn in his studies on the principles of analysis, can be identified in the interaction between research and teaching, and in particular in the desire to construct a symbolic language *useful* in bringing complete rigor into university teaching. The emphasis on the transposition of logical-foundational studies into teaching also linked Peano to many of his own *protégés*, from A. Padoa to C. Burali-Forti, who introduced the ideography in high schools, technical institutes, and even middle schools all over Italy.

In France, on the other hand, no attempt was made to experiment with the teaching of mathematical logic, either at the level of higher training or at the level of pre-university education, nor could the French mathematicians and philosophers draw on a publishing market specialized in textbooks of mathematical logic.[84] There was more: Couturat himself—uninterested in the teaching implications of ideography—almost entirely neglected to examine this component despite repeated warnings by Peano. It follows that very few French scholars (among them Borel Richard and Winter), looking at Italian mathematical logic through the 'lens' of the popularization work by Couturat, succeeded in perceiving the pedagogic *applications* that had led to the development of the ideography.

[84]Regarding this last aspect, we just recall that in Italy the publication of the first treatise on *mathematical logic* (*Logica Matematica*, published by C. Burali-Forti with Hoepli) dates from 1894. In France it was only in 1912 that Gauthier-Villars published Padoa's handbook *La logique deductive dans sa dérnière phase de developpement*. The preceding attempts by Couturat to publish a *Traité* or *Manuel de Logique algorithmique* with Alcan had come to nothing.

In the light of these two ‘vices’ of form and of substance, Couturat’s initiatives for the promotion of Peanian logic, though effective and important in many respects, also involved negative long-term effects, conditioned the view of Peanian scientific activity, and influenced the very evolution of this line of studies in France, right down to the nineteen-thirties.

References

Baire R. 1905. *Leçons sur les fonctions discontinues professées au Collège de France par René Baire, Maître de Conférences à la Faculté des Sciences de Montpellier, rédigées par A. Denjoy, élève de l’École Normale Supérieure*, Paris: Gauthier-Villars.

Borel E. 1897. Congrès International des Mathématiciens, Première session: Zürich, Août 1897, *Revue Générale des Sciences Pures et Appliquées*, 8, p. 783–789.

Borel E. 1905. *Leçons sur les fonctions de variables réelles et les développements en séries de polynomes professées à l’École Normale Supérieure par Emile Borel, et rédigées par Maurice Fréchet avec des notes par Paul Painlevé et Henri Lebesgue*, Paris: Gauthier-Villars.

Borel E. 1907. La logique et l’intuition en mathématiques, *RMM*, XV, p. 273–283.

Borgato M.T. 1991. Alcune lettere inedite di Peano a Genocchi e a Jordan sui fondamenti dell’analisi, in A. Conte, L. Giacardi (eds.), *Angelo Genocchi e i suoi interlocutori scientifici*, Torino: Dep. Sub. Storia Patria, p. 61–97.

Boutroux P. 1904. Sur la notion de correspondence dans l’analyse mathématique, *RMM*, XII, p. 909–920.

Boutroux P. 1905. Correspondance mathématique et relation logique, *RMM*, XIII, p. 620–637.

Boutroux P. 1913. L’œuvre philosophique de Henri Poincaré, *Revue du Mois*, 15, p. 155–183.

Bru B., Dugac P. 2004 (eds.). *H. Lebesgue. Les lendemains de l’intégrale. Lettres à Émile Borel*, Paris: Vuibert.

Couturat L. 1899. La logique mathématique de M. Peano, *RMM*, VII, p. 616–646.

Couturat L. 1900. Les mathématiques au Congrès de philosophie, *Ens. math.*, II, p. 397–410.

Couturat L. 1901. Peano G. professeur à l’Université de Turin. Formulaire de Mathématiques..., *Bull. Sci. Math.*, XXV (2), p. 141–159.

Couturat L. 1904. Logique et Philosophie des Sciences, Comptes rendus IIme Congrès de Philosophie, Genève 4–8 Septembre 1904, *RMM*, XII, p. 1037–1077.

Couturat L. 1905a. *Les principes des mathématiques*, Paris: Alcan.

[Couturat L.] 1905c. Leçons sur les fonctions de variables réelles et les développements en séries de polynomes, par Emile Borel, avec des notes par P. Painlevé et Lebesgue, 160 p. in-8; Leçons sur les fonctions discontinues, par René Baire, 128 p. in-8, Paris, Gauthier-Villars, 1905, *RMM*, Supplément N° de Janvier, p. 6.

Couturat L. 1906a. Pour la logistique, *RMM*, XIV, p. 208–250.

Couturat L. 1913. Logistique et Intuition, *RMM*, XXI, p. 260–268.

Dassen C.C. 1934. Luis Couturat (Síntesis de un trabajo del Dr. C.C. Dassen), *Anales de la S. C. Argentinia*, e. III, t. CXVIII, p. 136–143.

Dassen C.C. 1939. Vida y Obra de Luis Couturat, *Anales de la Academia Nacional de Ciencias Exactas, Físicas y Naturales de Buenos Aires*, IV, p. 73–204.

Decaillot A.M. 2008. *Cantor et la France*, Paris: Kimé.

Dugac P. 1976. Notes et documents sur la vie et l’oeuvre de René Baire, *Arch. Hist. Exact Sci.*, 15, 4, p. 297–383.

Dugac P. 1990 (ed.). Lettres de René Baire à Emile Borel, *Cahiers du séminaire d’histoire des mathématiques*, 11, p. 33–120.

ENS 1983. *L’oeuvre de Louis Couturat (1868–1914) ... de Leibniz à Russell ...*, Paris: Presses de l’École Normale Supérieure.

Erhardt C. 2011. Du cours magistral à l'entreprise éditoriale. La « collection Borel », publiée par Gauthier-Villars au début du XXe siècle, *Histoire de l'Education*, 130, p. 111–139.

Gandon S., Perrin Y. 2009. Le problème de la définition de l'aire d'une surface gauche: Peano et Lebesgue, *Arch. Hist. Exact Sci.*, 63, 6, p. 665–704.

Gispert H. 1983. Sur les fondements de l'analyse en France (à partir de lettres inédites de G. Darboux et de l'étude des différentes éditions du « Cours d'analyse » de C. Jordan), *Arch. Hist. Exact Sci.*, 28, 1, p. 37–106.

Gispert H. 1995. La théorie des ensembles en France avant la crise de 1905: Baire, Borel, Lebesgue ... et tous les autres, *Revue d'Histoire des Mathématiques*, 1, p. 39–81.

Gispert H. 2009. Les traités d'analyse et la rigueur en France dans la deuxième moitié du XIXe siècle, des questions, des choix et des contextes, in G.P. Brizzi (ed.) et al., *Dalla pecia all'e-book. Libri per l'Università: stampa, editoria, circolazione e lettura. Atti del convegno internazionale di studi. Bologna, 21–25 ottobre 2008*, Bologna: Clueb, p. 415–430.

Grattan-Guinness I. 2000. *The Search for Mathematical Roots 1870–1940, Logics, Set Theories and the Foundations of Mathematics from Cantor through Russell to Gödel*, Princeton: Univ. Press.

Hadamard J. 1906. La Logistique et l'Induction complète. *Revue Générale des Sciences Pures et Appliquées*, 17, p. 161–162.

Heinzmann G. 1986 (ed.). *Poincaré, Russell, Zermelo et Peano. Textes de la discussion (1906–1912) sur les fondements des mathématiques: des antinomies à la prédicativité*, Paris: Blanchard.

Jourdain P. 1912a. The development of the theories of mathematical logic and the principles of Mathematics, *Quaterly journal of pure and applied mathematics*, XLII (3–4), p. 219–314.

Jourdain P. 1912b. The philosophy of Mr. Bertrand Russell. With appendixes of leading passages from certain other works, *Monist*, 21, p. 481–508.

Klein F. 1909. *Elementarmathematik vom höheren Standpunkte aus*, v. II *Geometrie*, Leipzig: Teubner.

Lalande A. 1914. L'oeuvre de Louis Couturat, *RMM*, XXII, p. 644–688.

Le Ferrand H. 2012. *Six lettres de Giuseppe Peano à Robert de Montessus de Ballore*, preprint hal-00709080, version 1.

Loi M. 1976. Couturat méconnu/Couturat misconosciuto/Couturat unknown, *Scientia*, 111, p. 683–699.

Lolli G. 1985. Quasi alphabetum: logica ed enciclopedia in G. Peano, in G. Lolli, *Le ragioni fisiche e le dimostrazioni matematiche*, Bologna: Il Mulino, p. 49–83.

Luciano E. 2008a. *Giuseppe Peano docente e ricercatore di Analisi 1881–1919*, PhD thesis in Mathematics, Turin: Dept. of Mathematics, 2 vols.

Luciano E. 2008b. Un sessantennio di ricerca e di insegnamento dell'Analisi infinitesimale a Torino: da Genocchi a Peano, *Quaderni di Storia dell'Università di Torino*, 9, p. 27–150.

Luciano E. 2010. Sulla didattica della Logica Matematica: dalle conferenze di A. Padoa (1898) all'istituzione dei corsi ufficiali (1960), in C.S. Roero (ed.), *Peano e la sua Scuola fra matematica, logica e interlingua. Atti del Congresso internazionale di Studi* (Torino 6-7 ott. 2008), Torino: Dep. Sub. Storia Patria, p. 279–315.

Luciano E. 2012. Peano and his School between Leibniz and Couturat: the Influence in Mathematics and in International Language, in R. Krömer, Y. Chin-Drian (eds.), *New Essays on Leibniz Reception in Science and Philosophy of Science 1800–2000*, Basel: Birkhäuser, p. 41–64.

Luciano E. 2016. Defining a Mathematical School. Oral Knowledge in Peano's *Formulario*, *Revue d'Histoire des Mathématiques*, 42 p. in press.

Luciano E., Roero C.S. 2005 (eds.). *Giuseppe Peano-Louis Couturat Carteggio (1896–1914)*, Firenze: Olschki.

Luciano E., Roero C.S. 2012 (eds.). From Turin to Göttingen: Dialogues and Correspondence (1879–1923), *Boll. Sto. Sci. Mat.*, 32 (1), p. 7–232.

Medvedev F.A. 1983. Le funzioni d'insieme secondo G. Peano, *Archives Internationales d'Histoire des Sciences*, 33, p. 112–117.
Michelacci G. 2005 (ed.). Le lettere di Charles Hermite a A. Genocchi (1868–1887), *Boll. Sto. Sci. Mat.*, 25, p. 1–270.
Moore E.H. 1903. On the foundations of mathematics, presidential address, *American M. S. Bull.*, (2) 9, p. 402–424.
Osimo G. 1992 (ed.). *Lettere di Giuseppe Peano a Giovanni Vacca*, Milano: Università Bocconi, Quaderni Pristem, N. 3.
Paraf A. 1885. Angelo Genocchi. - Calcolo differenziale e principii di Calcolo integrale. - Pubblicato con aggiunte dal Dr Giuseppe Peano. - Torino 1884, *Bull. Sci. Math.*, IX (2), Ière partie, p. 170–172.
Peano G. 1882b. Sull'integrabilità delle funzioni, *Atti R. Acc. Sci. Torino*, 18, p. 439–446.
Peano G. 1882c. Sulle funzioni interpolari, *Atti R. Acc. Sci. Torino*, 18, p. 573–580.
Peano G. 1884a. Extrait d'une lettre, *Nouvelles Annales de Mathématiques*, s. 3, 3, p. 45-47.
Peano G. 1884b. Réponse à Ph. Gilbert, *Nouvelles Annales de Mathématiques*, s. 3, 3, p. 252–256.
Peano G. 1884c and 1884c*. *Angelo Genocchi, Calcolo differenziale e principii di calcolo integrale, pubblicato con aggiunte del Dr. G. Peano*, Torino: Bocca.
Peano G. 1887b. *Applicazioni geometriche del calcolo infinitesimale*, Torino: Bocca.
Peano G. 1893h. *Lezioni di Analisi infinitesimale*, Torino: Candeletti.
Peano G. 1894h. Notions de logique mathématique, in *Association française pour l'avancement des sciences, Congrès de Caen 1894, Séance du 11 Août 1894*, Paris, s. 11, v. 23, p. 222–226.
Peano G. 1895aa. *Formulaire de Mathématiques, tome 1 publié par la Rivista di matematica*, Torino: Bocca.
Peano G. 1896b. Introduction au tome II du "Formulaire de mathématiques", *Revue de Mathématiques*, 6, p. 1–4.
Peano G. 1898. *Appendici alle Lezioni di Analisi infinitesimale del Prof. G. Peano 1898, R. Università di Torino*, [Torino]: Litografia Tasca.
Peano G. 1899b. *Formulaire de Mathématiques, publié par la Revue de Mathématiques. t. II, n. 3. Logique mathématique. Arithmétique. Limites. Nombres complexes. Vecteurs. Dérivées. Intégrales*, Turin: Bocca - Ch. Clausen.
Peano G. 1901b. *Formulaire de Mathématiques*, t. III. Turin: Bocca - Ch. Clausen.
Peano G. 1902–03G. *Analisi infinitesimale, Dispense* 1–12, [Torino]: without editor.
Peano G. 1903f. *Formulaire mathématique, édition de l'an 1902–03 (tome IV de l'édition complète)*. Turin: Bocca - Ch Clausen.
Peano G. 1904d. *Lezioni di Calcolo infinitesimale tenute dal prof. G. Peano nella R. Università di Torino Stenografate da Igino De Finis*, Torino: Tipolitografia G. Paris.
Peano G. 1906b. Super theorema de Cantor-Bernstein. *Rendiconti del Circolo Matematico di Palermo*, 21, p. 360–366 and *Revista de Mathématica*, 8, p. 136–143.
Peano G. 1906e. Additione [a Super theorema de Cantor-Bernstein], *Revista de Mathématica*, 8, p. 143–157.
Peano G. 1908a. *Formulario Mathematico, Editio V. (tomo V de Formulario completo)*, Torino: Bocca.
Perrin Y. 2010. Le rôle précurseur de Peano dans la définition de l'aire d'une surface, in C.S. Roero (ed.), *Peano e la sua Scuola fra matematica, logica e interlingua. Atti del Congresso internazionale di Studi* (Torino 6-7 ott. 2008), Torino: Dep. Sub. Storia Patria, p. 317–326.
Pierpont J. 1904. The history of Mathematics in the Nineteenth Century, *Bull. Am. Math. Soc.*, 11 (1), p. 136–159.
Poincaré H. 1903. Review of Hilbert's Foundations of Geometry, *Bull. Am. Math. Soc.*, 10 (1), p. 1–23.
Poincaré H. 1905a. Les Mathématiques et la Logique, *RMM*, XIII, p. 815–835.
Poincaré H. 1906b. Les Mathématiques et la Logique, *RMM*, XIV, p. 294–317.
Richard J. 1906. A propos de la Logistique. *Revue Générale des Sciences Pures et Appliquées*, XVII, p. 957–958.

Richard J. 1935. Réflexions sur la logique, in *Actes du Congrès International de Philosophie Scientifique, vol. III. Langage et pseudo-problèmes*, Paris: Sorbonne, p. 18–20.
Roero C.S. 2008a (ed.). *L'Opera Omnia di Giuseppe Peano*, Torino: Dip. Mat., dvd-rom n. 3.
Roero C.S. 2009–11. Un manoscritto di G. Peano per G. Vailati 'Sulla storia della Logica matematica e suo stato presente', *Quaderni di Storia dell'Università di Torino*, 10, p. 169–184.
Roero C.S. 2011. The Formulario between Mathematics and History, in F. Skof (ed.). *Giuseppe Peano between Mathematics and Logic, Proceeding of the International Conference in honour of Giuseppe Peano on the 150th anniversary of his birth and the centennial of the Formulario Mathematico, Torino (Italy) October 2–3, 2008*, Milano: Springer, p. 83–132.
Russell B. 1906. Les Paradoxes de la Logique, *RMM*, XIV, p. 627–650.
Sanzo U. 1975. Significato epistemologico della polemica Poincaré-Couturat/Epistemological significance of the Poincaré-Couturat controversy, *Scientia*,110, p. 369–418.
Schnippenkötter J. 1910. Die Bedeutung der mathematischen Untersuchungen Couturats für die Logik, *Philosophisches Jahrbuch der Görresgesellschaft*, 23, p. 447–468.
Schlaudt O., Sakhri M. 2010 (eds.). *Louis Couturat - Traité de Logique algorithmique*, Basel: Birkhäuser.
Schmid A.-F. 2001 (ed.). *Bertrand Russell Correspondance sur la philosophie, la logique et la politique avec Louis Couturat (1897–1913)*, Paris: Kimé, 2 vols.
Schröder E. 1897. Ueber Pasigraphie, ihren gegenwärtigen Stand und die pasigraphische Bewegung in Italien, in F. Rudio (ed.), *Verhandlungen des ersten internationalen Mathematiker-Kongresses in Zürich vom 9 bis 11 August 1897*, Lepizig: Teubner, v. 1, p. 147–162.
Shearman A.T. 1906. *The development of symbolic logic. A critical-historical study of the logical calculus*, London: William and Norgate.
Somigliana C. 1911. Intorno all'ordinamento degli studi matematici nel primo biennio universitario in Italia, *Bollettino della Mathesis*, III, *Suppl. Atti della Sottocommissione italiana per l'insegnamento matematico*, p. 15–23.
Tannery J. 1887. Peano (G.). – Applicazioni geometriche del Calcolo infinitesimale. 1 vol. in -8°; XII-334 p. Turin, Bocca frères, 1887, *Bull. Sci. Math.*, XI (2), p. 237–239.
Tannery J. 1888. Peano (G.). - Calcolo geometrico secondo l' « Ausdehnungslehre di H. Grassmann », preceduto dale operazioni della logica deduttiva. 1 vol. in -8°; X-170 p. Turin, Bocca frères, 1888, *Bull. Sci. Math.*, XII (2), p. 261–262.
Vailati G. 1899b. La logique mathématique et sa nouvelle phase de développement dans les écrits de M. G. Peano, *RMM*, VII, p. 86–102.
Vailati G. 1904k. Relazione sul II Congresso di filosofia, *Boll. Bibl. e Storia Sci. Mat.*, 7, p. 127–128.
Wilson E.B. 1904. The Foundations of Mathematics, *American M. S. Bull.*, 11 (2), p. 74–93.
Winter M. 1907. Sur l'introduction logique à la théorie des fonctions, *RMM*, XV, p. 186–216.

Louis Rougier's Reception of the Peano School

Paola Cantù

1 Introduction

Among the numerous influences and reciprocal interactions between France and Italy at the beginning of the 20th century, it is interesting to investigate the complex case of Louis Rougier's reception of Italian mathematical logic (including in particular the contributions by some members of the Peano school: Giuseppe Peano, Giovanni Vailati, Alessandro Padoa, and Mario Pieri). This paper aims to investigate the role and the influence of the Peano school on the inversion of this French tendency of philosophers to ignore logic and mathematics. The influence held by the Peano school on Couturat's works is wellknown in the literature (Luciano 2012: 49–51; Grattan Guinness 2000: 255, Sect. 5.5)[1] as well as the negative reaction of

[1]Luciano (2012) is a detailed analysis of the common interests of Couturat and Peano in Leibnizian logic and universal language, as well as of the influence of Vacca on Couturat's investigations into Leibniz's unpublished materials. Luciano shows how Couturat emphasized the role of the geometrical characteristics in his book on Leibniz's logic and contributed some historical notes on Leibniz to the third edition of the *Formulary*, but also revised certain parts of his volume on Leibniz's unpublished writings under the influence of discussions with Vailati and Vacca (Luciano 2012: 49–51). Grattan Guinness (2000) claimed that Couturat "secured a more international public" for Peano's contributions, allowing the members of the Peano school to publish in the *Revue de Métaphysique et de Morale* (Sect. 5.4.6, p. 251), and being largely responsible for the section on "Logic and history of sciences" at the Second International Congress of Mathematicians which took place in Paris from 1 to 5 August 1900, and which saw the participation of Peano, Padoa, Pieri, Burali-Forti, and Vailati (Grattan Guinness 2000: 255, Sect. 5.5).

P. Cantù (✉)
Université Aix-Marseille and Centre National de Recherche Scientifique, CEPERC UMR7304, Aix-en-Provence, France
e-mail: paola.cantu@univ-amu.fr

F. Brechenmacher et al. (eds.), *Images of Italian Mathematics in France*,
Trends in the History of Science, DOI 10.1007/978-3-319-40082-2_8

Poincaré to Peano's 'logistics'[2] (Poincaré 1912; Heinzmann 1991). Yet the relation between the Peano school and Louis Rougier, a French philosopher who became seriously involved in symbolic logic, has not been adequately investigated.

Louis Rougier had a multidisciplinary education, having obtained a degree in Philosophy at the Faculty for Humanities and a degree in Physics at the Faculty of Sciences at Lyon University in 1909; in 1920 he obtained a Ph.D. in Philosophy at the Sorbonne University in Paris under the direction of André Lalande. He taught Philosophy as a Full Professor at the Besançon University from 1924 to 1948, and at the Caen University from 1954 to 1959. From the beginning of his career he investigated the relationships between philosophy, logic, geometry and physics, but he also developed many other different interests, ranging from history to linguistics, and from religion to political philosophy.[3] I will consider here the way he reacted to Italian 'logistics' and will therefore limit myself to an investigation of his writings on the philosophy of science and in particular on the philosophy of logic and mathematics.

Rougier's philosophy of logic has been investigated in the literature (see for example the special issue of *Revue d'Histoire des Sciences* edited by Bourdeau and Brenner (2014), and the two special issues of *Philosophia Scientiae* devoted to Rougier by Pont and Padovani (2006, 2007), but most essays only contain a generic reference to Peano as a source of Rougier's interest in logistics, without undertaking any deep investigation of the impact of the Peano school on Rougier's thinking nor any detailed analysis of Rougier's quotations from authors belonging to the Peano school. Apart from this gap in the literature, there are other reasons to be interested in the investigation into Rougier's reception of the writings of the Peano school, both from a historical and a theoretical perspective. From a historical perspective, it contributes to a better understanding of the reception of mathematical logic in France, but it also sheds some light on the role and the heritage of two multidisciplinary research groups: the Peano school and the *Revue de Métaphysique et de Morale*. From a theoretical perspective, it offers a common thread that might be useful in reconstructing Rougier's philosophy of logic throughout different stages of his intellectual development, but it also brings to the forefront some aspects of the epistemology of the Peano school that have been somewhat neglected in the literature.

My claim is that Rougier was not only personally interested in Italian mathematics and logic, and in particular in the innovations made by Peano and his school, but used it to support a project of renewal of French logic and scientific philosophy. Rougier recalls the contributions of the school to axiomatics and to mathematical logic, in order to criticize the standard teaching of logic in France, mainly based on

[2]'Logistics' translates the French expression 'logistique', which was first used by Louis Couturat, André Lalande and Gregorius Itelson to denote mathematical logic (Grattan Guinness 2000: 367).

[3]Fundamental sources for Rougier's biography are Rougier (1961, 1963), Marion (2004), Berndt and Marion (2006).

syllogistic theory, as well as the view, still defended by his teacher Goblot in (1907), that mathematical proofs are largely based on intuition. For this purpose, any reference to Hilbert's axiomatics would have sufficed. The reason why Rougier refers also to Peano and his entourage is related to a common interest in linguistics and in epistemology.

1.1 The Peano School

A point of contact between the Peano school and Rougier concerns the linguistic turn, and the interdisciplinary approach to it. Peano's works on universal language have long been considered in the literature (Grattan-Guinness 2000; Kennedy 2006) as completely independent from his results as a mathematician and logician. Yet Peano's interest in 'latino sine flexione' as a universal language is 'closely connected' to his search for an adequate ideographic symbolism to represent all fundamental logical and mathematical ideas, as he had himself declared (Kennedy 2002: 79). The interrelation between the study of formalized languages and the construction of a universal language that might be used for international scientific communication is often understated in the literature about the Peano school: on the contrary, Peano conceived it as an essential part of the new kind of scientific collaboration that he wanted to realize in the edition of the *Formulary*, a dictionary of mathematics that should evolve like a modern wiki, thanks to the suggestions sent by the editors in chief and by the readers themselves. In this project, the universal language (*latino sine flexione*) is developed on the basis of scientific analyses of natural languages as a tool to improve international scientific communication: often the same reasons advocated in defense of his mathematical symbolism are also used to defend the efficiency of the *adopted* universal language: e.g. conciseness is praised in mathematical logic (Peano 1889: 114, 1891a: 87, 1891b: 267, 1900: 284, 288) as well as in the *Interlingua* (Peano 1904: 279–280).

In a letter to Russell, Peano claimed that the *Interlingua* is "apparently quite different, and judged to be absurd by a great many persons, but in reality closely connected with mathematical logic" (Kennedy 2002: 79). There are several aspects of Peano's approach to the *Interlingua* that bear a considerable similarity with Peano's approach to mathematical logic. Firstly Peano took a language in use as a starting point for the development of the universal language: very formal Latin scientific language; similarly, he adopted the symbols effectively used by mathematicians in the development of symbolic logic, sometimes tolerating—notwithstanding his well known crusade in defense of rigor—imprecise expressions used in mathematical practice, as for example that "3/4 is equal to 6/8", whereby Padoa rightly remarked that one should rather say that the two fractions belong to the same equivalence class but are not identical (Peano 1894: 29, 1918: 197; Padoa 1909: 68–72). Secondly, he selected a group of collaborators to the Interlingua project and the journal *Discussiones de Academia pro Interlingua*), just as he did with the group of authors who collaborated on the project of the *Formulary* and the *Révue Mathématique* (Kennedy 2006: 185). Thirdly, he considered the problem of a universal language as a heritage of Leibniz's idea (Peano 1903: 9), as he also believed of the symbolic calculus (Peano 1894: 3): he also combined logical and universal language in the last editions of the *Formulary*, using the latter as an informal language (Peano 1906).

There are three other important aspects of the *Interlingua* that I claim bear a considerable similarity with aspects of Peano's mathematical logic. Firstly, there is a search for naturality (Minnaja and Paccagnella 2010: 576, 600) and for the possibility of mirroring concepts by words in the *Interlingua* that can be recognized in the *Formulary* too (Cantù 2014: 171). Secondly, Peano preferred a universal language having a minimal grammar but a larger lexicon (Minnaja and Paccagnella 2010: 595); analogously, having built symbolic logic in order to avoid the syntax errors of ordinary language, he preferred a symbolic calculus that is based on a minimum number of formation rules, but admits possible nominal definitions and thus some arbitrariness in the choice of synonymous terms. Finally, investigations developed by other members of the Peano school reveal certain interactions between the analysis of ordinary language and the analysis of mathematical language: it is for example the case of several remarks by Padoa on the notions of pre- and post-function (Peano 1894: 27; Padoa 1912a: 59–60, 64).

Another point of contact between the Peano school and Rougier is the epistemological approach to the analysis of scientific language. Giovanni Vailati investigated the role of argumentative strategies, metaphors, and questions of words in science, thus making the relation between language, logic, philosophy and science fully explicit. Like Rougier, Vailati associated an epistemological aim to the linguistic investigation:

> To this tendency to recognize the identity of theories, beyond or under differences of expression, symbolism, language, representative conventions and the rest, is to be attributed also the constant interest of the mathematical logicians in linguistic questions –from Grassmann […] to Nagy […]; from Couturat, joint author with Leau of a History of the Project of 'Universal Language', to Peano, inventor and propagandist of one of the most practical among them: the 'latino non flexo' (Vailati 1906: 166).

A further point of contact between Rougier and the Peano school is the effort to go beyond the institutional separation between hard and human sciences by means of an interdisciplinary or multidisciplinary approach. The group of mathematicians who studied or worked in Turin under the supervision of Giuseppe Peano was highly specialized in logic and the foundation of mathematics, but did not neglect the study of linguistics, psychology, history of sciences, philosophy, philology, and economics. In particular, Giovanni Vailati, who was considered as the "philosopher" of the group, applied a truly interdisciplinary approach to the investigation of scientific definitions, language, and justification—linking the historical study of mathematics to language analysis, psychology, and philosophical pragmatism—and made several efforts to promote a multidisciplinary reform of university faculties.[4]

By interdisciplinary research I mean an investigation that is not only conducted from different disciplinary points of view, but that is based on the interactions between different disciplines. Peano's work on logic and linguistics is truly interdisciplinary, because we have seen that the development of an ideography and the search for a universal language influence each other. The same holds for Vailati's inquiries in history and philosophy of science, because his central idea of considering scientific assertions as predictions is developed in strict relation to his historical investigations into mechanics, his pragmatistic understanding of philosophy and his ideas on education as an active process of students (Vailati 1905; Cantù and de Zan 2010: ii–iii). Another example is Vailati's investigation into definitions, which involves contributions from philosophy, history, logic, psychology, and theory of education that interact with each other. The topic of definitions received a huge attention from all members of the Peano school, but Peano, Padoa, Pieri and Burali-Forti (1901) discussed it mainly from a logical point of view, or with respect to the determination of the primitives of geometry. Vailati gave the issue an interdisciplinary twist, remarking that although two definitions can be equivalent from a logical point of view, they may differ from an epistemological perspective, because, given the pragmatistic correlation of meaning with the consequences that can be derived from a given proposition, one definition might be heuristically more fruitful than another (Vailati 1899a: 77).[5]

The specialization of the university system, which had been introduced by the 1859 Casati Reform at the university level and later extended to secondary schools

---

[4]Vailati promoted the idea of a unique secondary school curriculum, and a specialization at the higher education level, but he also suggested that Philosophy should be taught in all university faculties, to guarantee the possibility of a fruitful communication between experts of different disciplines. This position differed from Enriques' conception insofar as the latter suggested mitigating the specialization of university faculties by the introduction of a common course in History of Sciences rather than Philosophy.

[5]Peano claimed that mathematical definitions are particular kinds of equalities that play the role of typographical abbreviations, and discussed the conventional nature of definitions and the notion of primitive concept, the distinction between nominal and real definitions, the conditions of completeness and homogeneity, and the occurrence of definitions by abstraction in mathematics (Peano 1900, 1911, 1915, 1921). Padoa introduced a criterion to determine independent results in the case of definitions (Padoa 1900, partially translated in Van Heijenhoort 1967). Burali Forti (1901) had compared definitions by abstractions, by assumptions and nominal definitions. Vailati combined an epistemological analysis of the role of definitions with an historical investigation of their origin (Vailati 1903b, 1904c), an investigation of the contribution of definitions to the predictive efficiency of scientific theories with a linguistic inquiry on the metaphorical use of scientific expressions (Cantù and De Zan 2010). But definitions are investigated also with respect to their psychological nature (Vailati 1901), to their fallacious use (Vailati 1903a), and to their ambivalent role in education (Vailati 1899b, 1900, 1905), where they can be used either as the object of a mnemonic study or as the starting point of an interrogative method that invites students to discover for themselves the predictive consequences of a theory.

thanks to the 1922 Gentile Reform, was probably not so favorable to the development and to the valorization of interdisciplinary research. This might have been a supplementary reason, among many others, that might explain why several members of the Peano school never obtained an academic position (Vailati, Burali-Forti and Padoa, who finally secured one but only at the very end of his career). Unfortunately, the scholars who managed to get a position, were not appointed in Turin (Pieri got a chair in geometry in Sicily, and Vacca became professor in Chinese language and literature at the University of Rome), thereby abandoning the effort to build a multidisciplinary research group.[6]

If the separation between the two cultures was strong in Italy at the beginning of the century, it was even stronger in France, where it took place already at the end of primary schools, which explain why philosophers were not usually interested in the investigation of mathematical and logical results (Guillaume 2009: 179).[7] Commonly quoted examples are Leon Brunschvicg, who was highly skeptical towards logicism and Russell's empiricist realism (Brunschvicg 1911: 146; Grattan Guinness 2000: 459–460; Berndt and Marion 2006: 16), and Edmond Goblot, who ignored the contributions by Peano and Russell in his *Traité de logique* (Goblot 1918; Grattan Guinness 2000: 461–462; Lalande 1919; Kergomard and Salzi 1937). Yet, as a consequence of Rougier's criticism, at least Goblot offered an apology for this omission, and quoted Rougier's article (1916) for integration (Goblot 1918: xix–xi).

Rougier's success in promoting the interaction between philosophy and science is certainly related to his efforts to make the results of the Vienna Circle better known in France—a result he obtained by organizing a conference in 1935 on scientific philosophy—, but was also an effect of a specifically French

[6]The 1859 Casati reform had separated the faculty for literature and philosophy from the faculty for mathematical, physical and natural sciences, thus introducing for the first time specialized curricula in the Italian high school education system. The huge debate about the 1882 Baccelli reform, aimed at reintroducing a unique Faculty for Philosophy and Science, involved political, scientific, and financial issues, and raised the question of the relations between state centralism and local universities, both at an administrative and a didactic level. Under the influence of positivism, many mathematicians, like Vailati and Enriques, appreciated the specialization introduced by the Casati reform, but suggested reintroducing a bridge between the two cultures by activating common courses in all faculties. The tension between a specialized and an interdisciplinary approach to knowledge was not limited to the discussion on the didactic reform of universities (and secondary schools), but emerged in scientific research too. The Gentile Reform brought the debate to an end, establishing a quite rigid separation of the curricula at the secondary school level, and the supremacy of a humanistic education, because a certification in classical studies gave access to all university faculties, whereas a certification in scientific studies gave access to scientific faculties only.

[7]The image of a French philosophical tradition that was almost completely unaffected by the developments of mathematical logic is also inaccurate, as recent investigations have shown. See e.g. Brenner (2014), Bourdeau (2014), Mazliak and Sage (2014). Notable exceptions were Couturat, Nicod, Herbrand, Cavaillès, Lautman, authors whose contributions to the philosophy of logic were strongly limited by the early end of their lives (van Heijenoort 1968, p. 1).

phenomenon: the editorial line inaugurated by the *Revue de Métaphysique et de Morale.*[8] In some sense, Rougier himself is a byproduct of this editorial enterprise, because, as with many other French philosophers of his time, he learned about recent scientific developments not through academic courses, but through his assiduous reading of the journal. An investigation of Rougier's quotations, as well as of his own autobiographical remarks, shows that his vast knowledge in science and scientific philosophy was mainly derived from the articles published in the *Revue de Métaphysique et de Morale.* Concerning symbolic logic, in particular, Rougier claimed that between 1917 and 1921 he had read the papers published by Peano, Padoa, Russell, Couturat, and Whitehead (Rougier 1961: 12).[9]

Yet, Rougier's interdisciplinary methodology was quite different from the editorial line of the journal, which aimed at a rigorous multidisciplinary inquiry that could open up new questions or new approaches to a specific domain, whereas Rougier sometimes made an uncritical use of interdisciplinary investigation to justify his own philosophical perspective. For example his interest in history and language is often subordinated to the urge to justify a philosophical anti-rationalism that is developed independently (see infra Sects. 3 and 5). For this reason, Rougier's encyclopedic knowledge was appreciated, but also gave rise to considerable mistrust in his readers, who were afraid it might fail to be accompanied by a rigorous scientific methodology.

In a review of the essay *Du Paradis à l'utopie* (Rougier 1979) Laforest praises Rougier's encyclopedic mind and defines him as "epistemologist, linguist, philosopher, economist, constitutionalist, expert in politics, historian of civilizations, and witness of his time" (Laforest 1981: 342). Some reviewers tried to classify Rougier by selecting one discipline as his main topic of interest: e.g. Cohen and Cohen (1962, p. 542) remarked at the very beginning of his review that "The author is a logician", so as to specify that he was an outsider in linguistics; Pap (1956, p. 148) qualified Rougier as a French (i.e. continental versus analytical) philosopher, belonging to a minority group that opposes both existential philosophy and rationalism. Other reviewers raised explicit doubts about the scientificity of Rougier's methodology, either by questioning his references and his tendency to plagiarize (Gilson 1926: 313, Pont 2007: 16–17), or by underlining the rhetorical impetus of his approach (B. 1937: 134).

[8]See in particular the contributions by Maximilien Winter, recently underlined by Charles Alunni (2015: 117), who also presents a clear and illuminating prospect of the scientific articles published between 1893 and 1947 in the *Revue de Métaphysique et de Morale*.

[9]Peano was never actually published in the *Revue*, but Rougier probably had in mind two articles explicitly devoted to Peano's work: Couturat (1899) and Vailati (1899b) (Cassina 1933 appeared later, after the death of Peano). Padoa, on the contrary, is often quoted by Rougier: among the articles that he probably read are Padoa (1911a, b, c, 1912b, 1917) (and maybe even a posthumous paper: Padoa 1938). Pieri published only one article in 1906, which Rougier quotes in *La structure des théories déductives* (1921a). Even if Rougier mostly mentions other articles by Vailati, he probably also read the articles published in the *Revue de Métaphysique et de Morale* (Vailati 1898, 1904b), together with Lalande's review of Vailati's work which appeared in 1907. On the contrary, as far as I know, Rougier never mentions Vacca (1911a) or (1911b). The mentioned papers by Whitehead are presumably Whitehead 1907, 1916 and Couturat's review (1900).

1.2 The Evolution of Rougier's Philosophy of Logic

Apart from the historical interest of a comparative investigation of Vailati's and Rougier's interdisciplinary approaches, the comparison with the Peano school is also useful to trace an evolution in Rougier's own philosophy of logic. Although the division is not strictly chronological, four main phases of Rougier's reception of the Peano's school might be detected: 1916–1921, 1920–1935, 1936–1955, and 1955–1960.

In a first phase (Sect. 2) Rougier used Italian symbolic logic and Hilbert's axiomatic logic as a tool to renew the French teaching of traditional syllogistic logic (Rougier 1916, 1917, 1921a) but was also influenced by several French authors. Rougier's early essays on geometrical proofs and on the teaching of logic, published in the *Revue de Métaphysique et de Morale* (1916, 1917), underwent the influence of Goblot's traditional approach to logic, of Poincaré's writings, of Comte's positivism, and of Renan's theories on language. The influence of Goblot explains Rougier's interest in the syllogistic analysis of geometrical proofs.[10] Rougier was Goblot's student, and even if he criticized some of the claims of his master (Rougier 1916, 1919), he always declared his high esteem for such a 'remarkable logician' (Rougier 1921a: xi), who first showed the incompatibility between traditional syllogistic deduction, which goes from the general to the particular, and deductive sciences, which always generalize (Rougier 1916: 810, footnote 1, 1921a: xi). The influence of Comte mainly concerned a general positivistic understanding of the history of philosophy (based on what has later been called 'whig history'), and an interest in a multidisciplinary approach to science. The influence of Comte is not just present in Rougier's early writings, but remains a dominant trait of his epistemology also in later writings, when Rougier develops an increasingly strong interest in linguistics, sociology, anthropology, and history.[11] The influence of the Peano school can be seen especially in the analysis of definitions (see below Sect. 2.1).

In a second phase (Sect. 3) references to the Peano school become more precise, and Rougier combined them with his interpretation of Poincaré's conventionalism, using symbolic logic and history of logic as a razor against French classical rationalism, mathematical realism and the Scholastics' reprisal of Aristotelian metaphysics (Rougier 1913, 1920a, b, 1925, 1935). The influence of Poincaré is very strong, because it had already begun during Rougier's university studies and is from the outset related to an epistemological interest in logic, mathematics, and physics. Rougier's conception of the relativity of logic is a development of

[10] I think that Goblot's influence explains what puzzled Grattan-Guinness (2000: 461), i.e. that "Rougier relied upon syllogistic logic, if possible".

[11] The influence of Comte concerns also certain specific philosophical questions such as the role of intuition in mathematics, the nature of proofs, the relation between mathematics and logic, and scientific empiricism (Rougier 1931: 752, cit. in Pont 2006: 311).

Poincaré's conventionalism that is linked not only to geometry and physics but also to logic (Heinzmann 2010: 4–5). As in the case of Peano, one should not assume that Rougier was influenced by Poincaré in the sense that he adhered to his ideas in an historically faithful perspective, but rather that he used certain insights of Poincaré, understood in the light of successive developments of mathematical logic, and in the light of his own epistemological project. Even if Rougier's work as a translator and as a commentator of Poincaré's writings gave a more decidedly conventionalist turn to axiomatics, a comparison with Vailati and an analysis of several remarks on the economizing role of symbolism, and on Russell's definition of mathematics, suggests a certain proximity between Vailati's pragmatistic interpretation and Peano's logic.

In a third phase (Sect. 4) Rougier's interest in the epistemological role of logic becomes stronger. Symbolic logic is conceived both as a tool for the analysis of science—in so far as Rougier adheres to the main tenets of logical empiricism—and as a result of the development of new scientific theories, to the extent that Rougier considers the belief in the relativity of logic as a consequence of quantum physics (Rougier 1936a, b, c, 1939, 1941, 1955). Even if references to Peano are maintained in the *Treatise on Knowledge* (1955),[12] the influence of the Vienna Circle is prevalent, and Rougier does not show any more interest for the topics developed by the Peano school at the turn of the century (definitions, choice of primitive notions and primitive axioms, economy of symbolism, …). Rougier's increasing interest in epistemology and in the criticism of rationalism leads him to a relativist position in logic, which is quite far away from Peano's own conception: it is particularly difficult to reconcile Rougier's strong criticism of Leibniz's rationalism and panlogism (Rougier 1920b: 190–198; 1939: 211) with Peano's belief that symbolic logic might realize Leibniz's project of a universal characteristic (Peano 1894: 3, 59). The absence of relevant new quotations from Peano or other members of the school is probably related to the decline of the school itself on the one hand, and on the other hand to the fact that Rougier's interest is shifting more and more from the criticism of traditional logic and syllogism to the criticism of scholastic metaphysics and rationalism. Rougier's interest in epistemology, already evident in his first writings and strictly related to Poincaré's philosophy (Rougier 1913) now becomes absolutely preeminent, and other Italian authors that were more deeply involved in the discussion of epistemological questions come to the forefront. For example, Rougier develops a growing interest in Enriques' works, which begins with the 1913 translation of the book on the problems of science, becomes more significant in the volume on *The paralogisms of rationalism*, and is then maintained right through to later writings, such as the *Treatise of knowledge*, where one finds no

[12]This is partly due to the fact that Rougier's books often contain elaborations or literal reproductions of previous texts, as we will show below, and partly due, as Philippe Nabonnand has claimed (2006: 222) to the fact that Rougier uses history of science as a source of confirmation for his own claims, especially when he lacks solid arguments.

more references to Peano, Padoa and Pieri, but several references to Enriques (Alunni and André 2015; Lolli 2012).

Yet, certain specific traits of the Peano school, such as the interdisciplinary approach inherited from positivism, the interest in the analysis of ordinary language and in some scientifically fruitful pseudoproblems, as well as a pragmatistic background for the thesis on the relativity of logic, all underpin Rougier's epistemology right up to his last works on the theory of knowledge and on the metaphysics of language. This last phase (Sect. 5) shows how Rougier combines symbolic logic with linguistics and history of ideas to analyze scientific language (pseudo-propositions, and latent metaphysical conceptions), and scientific knowledge (pseudo-problems, and pseudo-scientific conceptions) (Rougier 1936d, 1955, 1960). Rougier's treatment of pseudo-problems, which differs from the approach that became standard in the Vienna Circle, reveals several points of contact with Vailati's writings on the history and language of science.

1.3 Methodological Remarks

One way to investigate Rougier's reception of the Peano school is to detect and analyze all references to Peano, Pieri, Padoa, and Vailati in Rougier's philosophical production. This approach is not without difficulty, because of Rougier's sloppy use of sources, and often of second-hand sources: for example a quotation by Benjamin Peirce is erroneously attributed to Pieri, an excerpt by Padoa is used to justify an idea that Padoa would not subscribe to (see below Sect. 2.2), and references to Giuseppe Peano's works are usually made by means of second hand articles published in the *Revue de Métaphysique et de Morale* (see footnote 9). Besides, Rougier's tendency to quote entire chapters of his earlier works in later writings makes it difficult to evaluate the evolution of the influence of the Peano school on his work. Finally, Rougier's habit of extrapolating quotes from their original context makes it extremely difficult to verify the exact impact of other scholars on his own epistemology.

For all these reasons, one cannot limit oneself to the investigation of explicit references of Rougier to the authors of the Peano school, but one should also compare several topics that Rougier analyzed without any explicit mention of the Peano school, but that reveal, to a detailed analysis, interesting points of contact with several views shared by some member of the group. Particular attention will be devoted to the nature of deduction, the analysis of scientific and ordinary language, the role of pseudo-problems in science, the interdisciplinary approach to knowledge, and the idea that logic is strictly related to epistemology.

Another problem arises from the fact that Rougier was a very prolific author, who published on several different topics, ranging from the history of philosophy to the history of ideas, from politics to epistemology.[13] For this reason, a selection of Rougier's writings will be made, based on their proximity to the interests of the Peano school. I will divide them into four groups, according to the prevailing research interest and the polemical use that Rougier makes of mathematical logic.

Finally, Peano is an author who was often quoted in 20th century essays on logic and philosophy of science without specific references to his work, but just as one of the fathers of the axiomatic method. This makes it difficult sometimes to separate Peano's own influence on Rougier from the influence of other authors, such as Pasch and Hilbert. This is another reason why a detailed analysis of Rougier's writings is essential to distinguish general tributes to the fathers of modern logic from specific references to some of the members of the Peano school.

2 Logistics and the Renewal of French Philosophy of Logic

2.1 Alternative Systems of Geometrical Primitives: Peano, Pieri and Padoa

In the article *La demonstration géométrique et le raisonnement déductif*, published in 1916 in the *Revue de Métaphysique et de Morale*, Rougier discusses a question raised by his teacher Goblot: whether mathematical demonstrations can be necessary and deductive, and at the same time innovative and capable of fruitful generalizations. Goblot claimed that this is not the case, because either proofs are reducible to one or more syllogisms, and cannot thus derive anything that is not already contained in the premises, or they rely on some empirical intuition, and cannot thus be merely deductive and necessary (Goblot 1907: 270, 274). Proofs cannot be reduced to syllogisms, because the syllogism (due to the principle *de omni et de nihilo*) derives particular truths from general truths, while proofs go from the particular to the general (e.g. from triangles to polygons). A proof consists, according to Goblot, in an operation and in an act of acknowledgement of the result based on the observation of some figure, thus guaranteeing intuitive evidence but not apodictic certainty (Rougier 1916: 809–811).

[13]For a complete bibliography of Rougier's writings see Padovani (2006b, 2007). For an outline of some of the papers that failed to be included in our discussion see Rougier (1961, 1963), Pont (2007), Berndt and Marion (2006). For an analysis of Rougier's writings on physics, see especially Lacki (2006). In particular, no mention will be made here of Rougier's political writings, for it would open up a new domain of research. Yet, it would be very interesting to compare different political positions held by the members of the Peano school with Rougier's political ideas, and especially the Pragmatism of the *Leonardo* group, who influenced Vailati, with the antidemocratically counterpart of Rougier's criticism of rationalism. For recent literature on Rougier's political engagement see Pont and Padovani (2007), Berndt and Marion (2006), and in particular Marion (2007), who tries to make a connection between Rougier's conventionalist epistemology and his defense of economic liberalism.

Responding to Goblot's criticism of Euclid's implicit use of intuition in proofs, Rougier mentions the recent developments of geometry contained in Pasch, Poincaré, and Klein, who eliminate all intuition from geometrical proof, and account for different geometrical theories by means of the purely abstract concept of group of transformations, and in Hilbert's *Grundlagen der Geometrie*, which Rougier does not quote in the original but based rather on the French translation of Halsted's book on *Rational geometry* (1904). Rougier's objection to Goblot is a defense of working mathematicians, who want to eliminate any recourse to intuition from proofs (Bourdeau 2014), and especially a defense of Poincaré's conception of mathematical induction as a proof by recurrence that goes from the particular to the general (Poincaré 1902: 10–11). Yet, unlike Poincaré, who claimed that mathematics cannot be based only on syllogisms, because there is a limited number of syllogistic applications, Rougier follows Goblot in the belief that the variety of mathematical proofs might be grounded in syllogistic theory, provided one allows a variety of syllogistic premises—formation principles, postulates of existence, and axioms of relation, and that one admits complex sets of syllogisms to explain a single mathematical proof (Rougier 1916; 844–845).[14]

Couturat and Padoa are explicitly mentioned as the sources for Rougier's treatment of logical notions and of the conception of a deductive theory (Rougier 1916: 814, footnote 1). One reason why Rougier quotes Padoa's book and not Peano's *Formulary* might be its greater accessibility (it was published in France by Gauthier-Villars), but one should not forget that Padoa dedicated a whole section of his book to syllogisms, and in particular to the reduction of all syllogisms to the form Barbara (expressed in mathematical logic by the transitivity of the implication), and to the syllogism in individual form (expressed in mathematical logic by the singleton class: the class of the y that are equal to x) (Padoa 1912a: 67, 75).[15] Anyway, Rougier often quotes by memory, and never refers to specific passages. In this case, he explicitly says that he is 'freely' presenting the content of Padoa and Couturat's work. Actually, a detailed analysis of the definition of equality reveals substantial differences both in respect to Padoa's and to Couturat's treatment, and

[14]Rougier's confutation of Goblot is based on the idea that there is no passage from singular to general, because the objects mentioned in the axioms only have a logical existence, and are thus logical variables that can "represent an infinity of particular cases": so, in geometrical proofs, the inference is always from general to general (Rougier 1916: 845). Secondly, Rougier argues that a mathematical proof is able to connect heterogeneous facts, exactly because it is not based on a single syllogism (in that case the conclusion should be contained in the premises), but on several hypothetical syllogisms whose major premise "asserts the formal implication of two heterogeneous facts", whereas the minor premise asserts "the logical possibility of the first of these two facts" (Rougier 1916: 844). All translations from Rougier's works are mine, if not otherwise stated.

[15]Peano explicitly refers to Padoa's work in Peano (1921: 181), as he makes a remark about syllogistic logic, claiming that the conversion from a universal to a particular proposition does not hold because the universal proposition does not necessarily have existential import.

interesting analogies with the procedure used in Peano's *Formulary* (Peano 1901), which is not explicitly quoted by Rougier.[16]

An indirect echo of the Peano school's debate on definitions can be traced in the analysis of the axioms of geometry as incomplete definitions or characterizations (Rougier 1916: 834), identified with Poincaré's 'disguised definitions', with Peano's 'definitions by postulates', and with Hilbert's 'implicit definitions'.[17] But the three concepts are different. Nominal definitions are based on the introduction of an equality between a new sign and a group of signs that has a known meaning. Definitions by postulates, as Burali-Forti called them, or under hypotheses, occur in mathematics when it is necessary to introduce an equality between a group of signs that include new signs and a group of known signs. The definition is introduced according to a certain hypothesis: for example, if a and b are positive integers (i.e. they satisfy the axioms of the system of positive integers), *quot (a, b)* is the biggest positive number x (including 0) such that the product $x0$ is smaller than a (Peano 1894: 44). It is evident from this example that the Peano school's definition by postulates does not coincide with Hilbert's notion of implicit definition. This is also due to the fact that Vailati used 'implicit definition' as being synonymous with contextual definition: "that which is defined is not an isolated word, but a group of words or phrases in which this word appears" (Vailati 1906: 167), so that definitions by abstraction are an example of implicit definition.

When Rougier refers to definitions by postulates, he might have had in mind Russell's references to "the three kinds of definition admitted by Peano—the nominal definition, the definition by postulates, and the definition by abstraction" (Russell 1903: 112) or he might have read Burali-Forti's essay on real numbers to which Russell refers, but this is less probable, because Rougier does not mention Burali-Forti's further claim that everything that cannot be defined nominally—including the notions of number defined by Peano, Cantor and Dedekind –is an intuition but not a concept (Burali-Forti 1901: 296).

An explicit reference to the Peano school concerns on the contrary the investigation of the primitives of geometry, its minimum number and the role of duality in determining economy of thought. Rougier claims that all different ways of presenting Euclid's metric geometry in a deductive way are equivalent: "Mr. Peano takes as primitive the notions of point and segment; Mr. Pieri, those of point and

[16]Padoa used reflexivity, symmetry, transitivity and substitutivity to define equality, and then considered the relation between the 'equaliform' relation of equality and the symbol $\supset$, either interpreted as inclusion (so that the bi-conditional expresses the equality of two classes) or as implication (so that the bi-conditional expresses the equality of two propositions) (Padoa 1912a: 61–66). Couturat first introduced equality as the bi-conditional in the calculus of propositions, then derived certain properties of this relation, then defined it again in the case of classes, and finally in the case of relations (Couturat 1905: 10, 19, 31). Rougier, on the contrary, defines equality by means of the symbol $\supset$ and then shows that it is an equaliform relation satisfying a condition of substitutivity.

[17]Heinzmann has showed how Rougier tends to identify Poincaré's 'disguised definitions' with Hilbert's implicit definitions, which are on the contrary quite different concepts (Heinzmann 2010: 13), whereas Hyder, quoting (Stumpf 1991), has shown that Rougier reads Poincaré in the light of epistemological remarks by Russell and Whitehead (Hyder 2006: 251, 254).

movement; B. Veblen those of point and order; Mr. Padoa those of point and distance between two points; Mr. Hilbert, those of point, straight line, plane, being on, being between, parallel line" (Rougier 1916: 830). The choice of one of the various systems is described as conventional but not arbitrary, because it has to satisfy the conditions of consistence (non-contradiction, compatibility or logical existence) and sufficiency. Rougier then claims that there are two other desired, even if not indispensable properties: the independence of primitive notions and axioms (which Rougier calls necessity), and the maximum economy in the choice of the primitive notions and axioms (Rougier 1916: 830, 1917: 592).[18]

Whereas all other metatheoretical properties (consistency, sufficiency and independence) might be taken from Peano but also from Hilbert, the insistence on the importance of choosing the smallest number of primitives was typical of Peano's geometrical axiomatization. The insistence on the virtue of economizing primitive notions is at the center of Vailati's connection between Peano's logic and pragmatism: "Pragmatists and mathematicians find themselves in agreement, too, in their efforts toward the maximum of conciseness and rapidity of expression—in their tendency to eliminate all superfluity and redundancy both of wording and of concept" (Vailati 1906: 604)[19].

Duality is praised not because of the new results it can prove, but because it helps us economize thoughts (Rougier 1916: 834): this is evidently a relevant difference with respect to Padoa's ideas, who considers duality as one of the main advantages of Peano's ideography, and describes the law of duality as "a rigorous tool for proving theorems as well as an easy and powerful tool for discovering theorems", and even as something that "explains the why of certain facts that had not been explained otherwise" (Padoa 1912a: 89, 86). Padoa even describes the duality principle as one of the two fundamental logical laws that become evident in Peano's system, the other being the substantial identity between the theory of classes and the theory of conditions, i.e. the propositional calculus (Padoa 1912a: 104).[20]

Finally, there is an interesting reference to mathematical formalism, based on Russell's definition of mathematics as a science where we do not need to know what we are talking about (Rougier 1916: 832, 850). Rougier mentions here only

[18]Rougier's attention might have been driven to this particular issue by Enriques' *Principles of Science*, where it is claimed that Peano made appeal to reasons of economy of thought to justify the use of the logical principle of permanence (Enriques 1906: 139).

[19]Vailati (1906, 1904a) were first published in the journal *Leonardo* and then reprinted in Vailati's *Scritti* (1911). *Rougier* knew the journal *Leonardo*, which he mentions in the *Paralogismes du rationalisme* as a negative example of the consequences of certain forms of pragmatism (Rougier 1920b: viii), and had read at least three articles from Vailati (1903a, b, 1904b) concerning the history of logic (Rougier 1920b: 145). See below Sect. 3.2.

[20]A different and subtler understanding of duality will emerge in Rougier (1955), where duality is still presented as a tool to economize thought but is also praised because it enables the discovery of unexpected properties, especially when the analogies are applied to different domains. Duality is associated to similitudes of mathematical structures that "do not only lead to an economy of thought, but are also fruitful" (Rougier 1955: 89).

the original paper by Russell (1901: 84), but this definition was quoted in Couturat (1904) and discussed at length in Vailati (1904a). Rougier's formulation is actually nearer to Vailati's formulation than to Russell's own presentation, which is centered on truth,[21] whereas both Vailati and Rougier insist on the absence of any intuitive meaning. Vailati had formulated formalism as a way to emancipate mathematics from intuition:

> A common characteristic of both these branches [Peirce's theory of relations and Peano's mathematical logic] is precisely the tendency to emancipate mathematical deductions from any appeal to facts or intuitions with reference to the meaning of the operations, or relations, under consideration. These are defined by the mere simple enunciation of a certain number of fundamental properties that, being able to be common to relations or operations with the most disparate and heterogeneous meanings, are compatible with the most various interpretations of the symbols appearing in their enunciation (Vailati 1904a: 141).

Rougier puts more emphasis on the fact that no intuitive meaning occurs either in the enunciation or in the demonstration of a theorem, which is coherent with his aim to contest Goblot's claim that intuition is needed in mathematical proofs:

> In a deductive theory, the intuitive, concrete, psychological or real meaning of particular notions denoting singular objects, classes of objects or relations, never occurs in the enunciation of theorems or in the demonstration of theorems (Rougier 1916: 832).

This reveals once more that the main aim of the paper is internal to French philosophy, and that references to Russell, Hilbert and the Peano school are subordinated to the aim of renewing French philosophy of logic and mathematics.

2.2 From Comte's Formalism to Padoa's Logicism

In 1917 Rougier published a second article in the *Revue de Métaphysique et de Morale* entitled *De la nécessité d'une réforme dans l'enseignement de la logique*. After introducing the distinction between formal logic—used in formal sciences—and material logic—used in natural sciences—, Rougier analyzes the main features of formal logic, recalling some of the topics already discussed in Rougier 1916: (1) the properties of judgments (simple and composite; judgments of existence, inherence and relation; truth judgments and hypothetical judgments; particular, singular and universal judgments; categorical and hypothetical judgments; apodictic, assertoric and problematic judgments); (2) the nature of deductive and inductive reasoning (the latter being reduced to the former), and of analogical reasoning (identified with a categorical syllogism whose major premise is synthetic); (3) the nature of modern axiomatics (the conventional choice of axioms

[21]"If such and such a proposition is true of anything, then such and such another proposition is true of that thing. It is essential not to discuss whether the first proposition is really true, and not to mention what the anything is, of which it is supposed to be true" (Russell 1901: 84).

intended as functional propositions that implicitly define the objects of the theory; the proofs of consistency of a theory either by recourse to intuition and to concrete models or by reduction to other consistent theories; the repartition of primitive propositions in postulates of existence, principles of formation, and axioms of relations). Insisting on the formal nature of all proofs (including mathematical proofs), Rougier criticizes the widespread use of intuition in the teaching of mathematics and logic.

The paper begins by a reference to the Italian, English and French '*logisticiens*', i.e. Peano, Padoa, Russell and Couturat, "who have taken formal logic to a higher level of precision and exactness" (Rougier 1917: 569), and ends with a quotation from Comte, presented as a precursor of the formalistic understanding of mathematics (Rougier 1917: 591). Here logicism is explicitly associated to formalism, which is considered as an interpretation that makes mathematics hardly distinguishable from logic. Rougier finds a precursor of logicism in Comte's understanding of mathematics, considered as "a huge and admirable extension of natural logic to a certain order of deductions" (Comte 1830: 63). Rougier is here making a non rigorous use of the history of science, identifying Comte's idea of abstract mathematics (*calculus*) with the formalism developed by modern axiomatics at the end of the century and even with logicism! In the same passage there is a reference to other authors "who came to confound the two disciplines": Pieri, Russell, and Padoa.

An attentive examination of Pieri's quotation and of the bibliographic reference in the footnote reveals a mistake. One might well take it for a typographical error, had it not been repeated as such in *La structure des théorie deductives* (1921). The quotation "la mathématique est la science qui tire des consequences nécessaires" is correctly referenced to an article which appeared in the *American Mathematical Journal* in 1881, but the author is not Mario Pieri, but Benjamin Peirce, who wrote in his paper on linear associative algebra that "mathematics is the science that draws necessary consequences" (Peirce 1881: 97). Yet, in this very paper, Peirce also remarked that "mathematics, under this definition, belongs to every inquiry, moral as well as physical. Even the rules of logic, by which it is rigidly bound, could not be deduced without its aid" (Peirce 1881: 97). Peirce's conception is thus misinterpreted by Rougier, who conceives mathematics as "the study of formally necessary reasoning that can be applied to any kind of material", whereas Benjamin Peirce, on the contrary, insists on the 'transmutation' that the material of any science needs to undergo in order to be compared with observation, and claims that even the laws of argument and the syllogisms need to undergo such a mathematical process of 'transmutation' in order to be compared with experience (Peirce 1881: 97–98). It is thus hard to determine to which author and to which claim Rougier is referring.

Another example of Rougier's sloppy use of quotations concerns the reference to Russell, which is taken, I suggest, from second hand literature (Couturat 1904). Russell's claim—pure mathematics is "a set of formal implications that are independent from all content"—is presented without any exact reference: Rougier is probably referring to Russell's claim in *The Principles of Mathematics* that "Pure mathematics is the class of all propositions of the form '*p* implies *q*', where *p* and *q* are propositions containing one or more variables, the same in the two propositions, and neither *p* nor *q* contains any constants excepts logical constants" (Russell 1903: 3). Rougier's formulation is more similar to Couturat's own formulation: "En d'autres termes, la Mathématique pure est un ensemble d'implications formelles indépendantes de tout contenu" (1904: 4).

The reference to Padoa on the contrary, is more precise, but Rougier forces the interpretation of Padoa's text in order to obtain a corroboration of his own thesis, i.e. of the claim that logicism amounts to a difficulty in distinguishing between logic and mathematics, or as he will define it with a slight pragmatistic turn in *The paralogismes*, "a need for unity, for simplicity, and for economy of our thought" (Rougier 1920b: 271). Rougier rightly claims that Padoa associated the question of the distinction between mathematics and logic with the question of the notion of couple being included in logic or in arithmetic, which is a matter of convention (Rougier 1917: 591). But Rougier omits some further remarks by Padoa, where he gives several arguments in defense of the claim that the notion of couple should be included in arithmetic rather than in logic, and the two sciences be kept separate (Padoa 1911a: 553–554).

These examples are quite paradigmatic of Rougier's non-scientific use of history (Nabonnand 2006), and explain why it is so difficult to give a historically adequate account of the influence of the Peano school on Rougier's thinking. This is also the reason why in the present paper I have decided to reconstruct this influence not only on the basis of exact quotations and references, which are, as we have seen, quite loose, but also on the basis of similarity of content.

2.3 Pieri's Logicism

Rougier's volume on *The structure of deductive theories* (1921a) is dedicated to Goblot, and includes a first chapter on the philosophy of logic, considering different types of judgments, the difference between induction and deduction, the notions of formal and empirical sciences, and different ways of reasoning, such as analogy and syllogism. The second chapter is an introduction to axiomatic theories (logical constants, relations, operations, definition, deduction). The third chapter concerns metalogical issues, as for example the independence of notions and axioms, the fruitfulness of axiomatics, and the definition by postulates. This chapter, entitled "The economy of deductive theories" (Preface, p. vi), studies the primitive propositions of a theory: the conventional but not arbitrary choice of uninterpreted notions and propositions (Rougier 1921a: 65–69), the distinction between logical axioms and axioms of the theory, the distinction between postulates of existence and unicity, constructive principles and axioms of relation (Rougier 1921a: 76–77). It is interesting to remark that consistency and sufficiency are necessary conditions, whereas independency, irreducibility, and economy of thought—the meta-theoretical notions that were deeply investigated in the Peano school—are "quasi-aesthetical desiderata" (Rougier 1921a: 69–71). The fourth chapter concerns proof: it presents Hilbert's view and considers the advantages of formal logic with respect to the traditional logic of syllogisms. The book is a literal synthesis of Rougier's writings up to this stage: it is built as a cut-and-paste essay from Rougier's previous publications.[22]

The references to Peano, Padoa, Russell and Couturat mentioned in the preface are exactly the same as in 1917, but they are used to support a further claim that marks Rougier's definitive separation from Goblot's perspective: symbolic logic has shown that no recourse to intuition is necessary in mathematical proofs, but also that mathematical deduction cannot be reduced to syllogistics, and certainly not to traditional syllogistics, which ignores relations.[23]

Concerning the influence of the Peano school on Rougier, there is thus not much more to say, except for an interesting reference to Pieri's logicism—and this time Rougier actually quotes from Mario Pieri's article which appeared in the *Revue de Métaphysique et de Morale* in 1906. Discussing a meta-theoretical property, i.e. consistency as a necessary condition to be respected in the non-arbitrary choice of a set of axioms, Rougier remarks that there are two possible ways to show the consistency of axioms. One relies on intuition, because it needs the empirical intuition of a concrete model that satisfies the axioms; the other method is purely logical and relative to a theory, because the consistency of a theory is proved by reduction to another theory that is taken to be consistent—e.g. the consistency of

[22]The first chapter on judgments is an extension of the last part of the second section of Rougier (1916: 827–829). The second chapter on logical constants, logical relations, logical operations and the logic of relations is almost identical to Sect. 2 of Rougier (1916: 823–827). The third chapter on the "economy" of deductive theories is also derived from Sect. 3 of Rougier (1916), with some small changes. The fourth chapter on mathematical proofs is already contained in Rougier (1916: 835–854). The conclusion discusses the topic of logicism, which was already presented in Rougier (1917: 591).

[23]Criticism of traditional logic and syllogistics is based on the following four claims: (1) traditional logic ignores relations, but inclusion and predication are relations: once relations are admitted, both Leibniz's monadology and Kant's transcendental aesthetic are consequently knocked down (Rougier 1921a: vii–viii); (2) traditional logic identifies syllogism with deductive reasoning *tout court*, but reasoning is either deductive or it is not reasoning at all (Rougier 1921a: vii–viii); (3) induction is opposed to deduction as a different form of reasoning, because it goes from particular to general, but if one does not limit deductive reasoning to syllogistics, there is no need to admit that reasoning from particular to general should not be deductive (Rougier 1921a: viii–ix); (4) mathematical proofs must contain intuition or some other non-deductive elements, because logic is considered as non-informative, whereas in mathematics new objects are built on the basis of certain constructive principles, without any recourse to graphical or physical operations (Rougier 1921a: x–xi).

geometry is reduced to the consistency of analysis, and the consistency of the latter is reduced to the consistency of arithmetic. At this point Rougier mentions Pieri, Russell and Whitehead as supporters of logicism, intended as the belief in the possibility of giving a logical interpretation of arithmetical principles, and thus a relative proof of the consistency of arithmetic to the consistency of logic (Rougier 1921a: 71–72). In the above-mentioned paper on arithmetical axioms Pieri actually claims that a relative proof of the consistency of arithmetic might be given by reduction to logic. He considers the notions of class and function as two ordinary logical notions and thus proves that number and successor of a number can be expressed as a finite class and as the subsequent of a finite class respectively (Pieri 1906: 204–205).[24] This passage by Rougier, together with other passages from *The paralogisms of rationalism,* where Rougier associates Russell with Peano and with all '*logisticiens*' (Rougier 1920b: 136, 271) in the effort "to reduce the principles of mathematics to the common principles of logic", suggests that Rougier attributes a form of logicism to all scholars in symbolic logic. To summarize, in his early writings Rougier often quotes the members of the Peano school by memory—which allows for some inconsistencies such as a citation from Benjamin Peirce attributed to Mario Pieri—and extrapolates the author's point of view from its original context, in order to support his own claim. For example, in Rougier (1916, 1917) the objective is to show that mathematical reasoning need not include any recourse to intuition, given that not only syllogistic reasoning (from general to general) but also inductive reasoning (from particular to general) can be explained as a kind of deduction. Deductive theories are presented as independent from any content, and formalism—whose origin is traced back through Russell and Hilbert to Comte—is taken to be what makes it so difficult to distinguish mathematics from logic. This indistinction, which Rougier misleadingly attributes to Padoa, is taken as an essential feature of logicism. Another relevant aspect is a pragmatic need for unity, simplicity and economy of thought, an issue that Vailati himself praised in his writings on Peano's symbolic logic. Having retained a very general notion of logicism, Rougier can apply it to almost all researchers in 'logistics'. It is only in his 1921 book on deductive theories that Rougier offers a more precise definition of logicism, quoting Pieri's metatheoretical remarks on the possibility of giving a logical interpretation of arithmetical principles and a relative proof of consistency based on the proof of the consistency of logic.

[24]I will not enter here in the discussion as to whether Pieri was actually a logicist or a supporter of a weak logicism. Cf. Marchisotto and Smith (2007: 58–60, 307), Grattan-Guinness (2000: 371), Francisco Rodriguez-Consuegra (1991: 129), Cassina (1949: 145).

3 From Poincaré's Conventionalism to a Pragmatistic Criticism of Rationalism

3.1 From Poincaré's Conventionalism to James' Pragmatism

The volume on *The Paralogisms of rationalism* (Rougier 1920b) is the first systematic work by Rougier, and is guided by a clear and precise project: a detailed criticism of rationalism based on the recent results of symbolic logic, on Poincaré's conventionalist understanding of geometry, but also on an analysis of the history of philosophy (especially ancient and medieval philosophy) and of logic. The interest in language is combined with a multidisciplinary treatment of scientific philosophy that Rougier will later describe as based on "the comparative results of anthropology, philology, history of religions, history of philosophy and history of sciences" (Rougier 1961: 34).

The book makes reference to previous articles by Rougier on physics, on medieval philosophy, and on the theory of knowledge, and also recalls many topics that had already been introduced in the articles on geometrical deduction and on the teaching of logic, and in the book on the geometrical philosophy of Henri Poincaré (Rougier 1920a).[25] Rougier's preeminent interest in epistemology is evident in the analysis of Poincaré's geometric conventionalism, because the idea that the principles of geometry might be conventions (i.e. propositions that are neither true nor false but convenient, not empirical but suggested by experience) is used to criticize rationalism (both in the Leibnizian and in the Kantian version) and empiricism (Rougier 1920a: 117). Poincaré's third way allows us to "reject the classic dilemma in which traditional logic tried to lock epistemology: the principles of the sciences are either apodictic truths (analytic according to Leibniz or synthetic a priori according to Kant), or assertoric truths (empirical or intuitive judgments)" (Rougier 1920a: 199).[26]

The aim of Rougier is to defend a form of anti-rationalist intellectualism, i.e. a conception based on the idea that knowledge can be acquired by discursive thinking rather than by intuition, even if there are no a priori and necessary truths that might be independent from mind and experience, and there is no faculty of reason that might be common to the human kind, but distinct from sensible perception and

[25]Actually, several topics discussed in the essay on Poincaré are literally reprinted or expanded from the previous essays. The first chapter is an almost literal transcription of Sect. 3 of the essay on geometrical proof (Rougier 1916), while the third chapter of the first part of Rougier (1920a), devoted to the systematization of geometry by means of the theory of groups of transformations, is an extended treatment of a topic that was already discussed in the last section of Rougier (1916). The quotations of Padoa, Pieri and Peano's choice of primitive notions for geometry do not substantially differ from the references made in Rougier (1916).

[26]Rougier shares Poincaré's conventionalism, because he considers it corroborated by the use of different geometries made by modern physics (Rougier 1920a: 146), and defines a convention as "a decree, stemming from a free decision of our mind, presenting itself to the acknowledgment of our will, which is free to accept or reject it" (Rougier 1920a: 121).

discursive intellect. The analysis of rationalism is indebted to Bergsonism[27] and Pragmatism. From Bergson Rougier inherits the criticism of some static representations of the world proposed by science, but unlike him Rougier condemns any form of intuition as a means of knowledge. Like the Pragmatist philosophers who were published in the journal *Leonardo*, Rougier uses logic as a means to criticize the rationalist theory of knowledge, but he condemns their assimilation of the truth of scientific laws to the fruitfulness of a juridical law or to the success of some practical recipe (Rougier 1920b: vii–ix).

Following Pragmatists, and in particular William James, Rougier extends conventionality from geometrical principles to logical laws, and uses James' arguments advanced in *The will to believe*, *The philosophy of experience*, and *Pragmatism* to attack rationalism and realism (Rougier 1920b: 394–96, 401–406, 458–460). In particular, Rougier praises James for having "ingeniously shown that the big philosophical systems are the expression of different temperaments" (1920b: 455), and depend on the structure of the language in which they are formulated.[28] Adapting James' theory of temperaments and its association to philosophical viewpoints, Rougier claims that different conceptions of logic, and different types of explanation—1. the anthropomorphic, animist or theological; 2. the ontological or metaphysical; 3. the symbolic, magic or mystical; 4. the positive or scientific—reveal different mental structures, contrary to the rationalistic claim that there is one and the same reason in every human being (Rougier 1920b: 445–447).[29] Rationalism is presented as the dangerous inertia of routine, as a conservative tendency that defends tradition against criticism, free thinking and doubt: the human mind is the result of a slow process of transformation and should be described without any recourse to intellectual intuition or a priori forms of knowledge (Rougier 1920b: 470–71).

3.2 From Enriques' Epistemology to Vailati's History of Logic

The epistemological interest of Rougier gets stronger and stronger and his interest shifts from Italian 'logistics' to French and Italian epistemology: there is not a single mention of Peano in the section devoted to the analysis of mathematical

[27]References to Bergson will be maintained also in later writings. Rougier (1960) uses an example taken from Bergson—the question "at what degree of organic complexity originated the first state of conscience in the chain of living organisms?"—to criticize the standard view in the Vienna Circle. This example already shows the intermediate nature of Rougier's conception, which cannot be reduced neither to the analytical nor to the continental philosophical tradition.

[28]"In this sense, as William James remarked, 'conjunctions, propositions, adverbs play a vital role in all philosophical theories'" (Rougier 1961: 32).

[29]Other sources of Rougier are here Comte (Rougier 1920b: 462), Le Roy (Rougier 1920b: 464), and Pareto (Rougier 1920b: 461).

realism,[30] but rather a significant use of articles by Brunschvicg and Enriques for analytical realism[31] and of Poincaré's *La science et l'hypothèse* for geometrical realism.[32] The absence of quotations from Peano does not mean that no reference is made to other members of the Peano school. On the contrary, Rougier makes some explicit references to Vailati and quotes authors that have been very influential on Vailati's version of pragmatism: the authors of the journal *Leonardo* and William James. Rougier's specific interest in Vailati's investigation into the history of logic probably arose during his translation of Enriques' *Principles of Science*, because Enriques makes several references to Vailati's remarks on the history of logic. What is common to Rougier and to Vailati's approach to the history of logical definitions is the search for logical fallacies, and in particular for one kind of paralogisms: the confusion between definitions *quid rei* and definitions *quid nominis*.

[30]Rougier distinguishes four kinds of realisms, which he considers as a kind of mentality based on the belief in the existence of eternal or necessary truths: mathematical (requiring logical existence), scientific (requiring empirical existence), Platonic (requiring a transcendent existence) and Aristotelian (requiring an immanent existence) (Rougier 1920b: 275–76).

[31]Analytical realism is realism about numbers and functions, and is presented in Russell's words as follows: "The philosophy I call analytical realism [...] shows that the exactitude of mathematics depends on its being concerned with an abstract world, a world of essences or Platonic ideas, which is not subject to the inaccuracies of sensory knowledge, and which is not constrained by the subjective variations one would have to fear in accepting a psychological theory such as that of Kant" (Russell 1911: 133). As in the case of geometrical realism, its genesis can be traced to a psychological tendency to objectify mathematical entities: to draw an analogy with a personality disorder called "alienation of the body's idea", the thought that mathematical objects have properties that we cannot recognize as derived by the postulates we have stated, induces us to consider them as extraneous to our creation, and thus to objectify them (Rougier 1920b: 289–290). Rougier here follows Brunschvicg's idea that the mathematician who "gave birth to a number is then forced to study it from the outside, just as a mother needs to have recourse to observation to understand the personality of her child" (Brunschvicg 1916: 332). The analysis of the history of analytical realism and its refutation (Rougier 1920b: 300–309) are entirely based on Enriques (1917), which is quoted twice. Enriques compares the collapse of the first phase of analytical realism associated with infinitesimal analyses, due to the discovery of paradoxes concerning limits and divergent series, with the collapse of the second phase of analytical realism associated with Cantorianism, due to the discovery of the paradoxes of set theory (Enriques 1917: 164). Rougier's refutation of analytical realism involves all kinds of numbers: natural, imaginary, hypercomplex, Kummer's ideals, and non-Archimedean numbers—wrongly attributed to Paolo instead of Giuseppe Veronese (Rougier 1920b: 292–293).

[32]According to Rougier, geometrical realism originated in a psychological tendency of the human mind, (1) to consider as existent some geometrical notions that appear to resist us, because once the postulates of the theory are chosen, we are no longer free to reject their logical consequences, and (2) to associate images endowed with intuitive evidence to the pure symbols of the theory (Rougier 1920b: 285–286). The refutation of geometrical realism is based on Poincaré's well-known thought experiment: beings who live in a spherical world, having maximal temperature in the center and zero temperature at the periphery, and who move so slowly as to always remain in a thermic equilibrium with their environment, would perceive their world as infinite, but their world is described by a Lobatscevkian rather than Euclidean geometry (Poincaré, 1902: 83–88). So, the refutation of geometrical realism is based on the idea that the apparent objectivity of geometrical postulates depends on the properties of our environment.

Discussing the volume on the paralogisms of rationalism in his *Itineraire philosophique*, Rougier praises symbolic logic for its usefulness in detecting four different kinds of paralogisms: (1) consider a proposition that is derived from a set of hypothetical axioms as a truth in itself; (2) infer from the necessity of a proof that the axioms from which a theorem is derived are necessary, whereas the necessity of the proof depends on the formal laws of deduction; (3) infer the empirical truth of a proposition from its formal truth; (4) consider axioms as analytical truths, i.e. as predicative judgments that are true *ex terminis*. But logic intervenes also in the criticism of pseudo-demonstrations, where it points out the implicit premises, for example when a nominal definition is taken as a real one, or when one makes recourse to intuition or to some metaphysical principle (Rougier 1961: 25–26). Vailati's analysis of the theory of definitions, and in particular of what Saccheri called fallacy of complex definition, is, according to Rougier, a good example of fallacies of reasoning that have been discovered by a clearer distinction between definitions *quid rei* and definitions *quid nominis* (Vailati 1903a, b).

Rougier uses the history of logic to criticize rationalism, to show that the principles of logic are not innate but evolve under historical and cultural influences, and to show that the validity of an argument might also evolve with time: several arguments that were considered valid in the past are considered fallacious now, and vice versa some forms of reasoning are considered valid now, although they were not in the past.[33] Vailati's investigations into the proof by absurdity (Vailati 1904b) are quoted by Rougier as an example of a proof that is now generally accepted, although it was not so in the past.

The use of symbolic logic to criticize philosophical systems will be maintained in the investigations into scholastics (Rougier 1925, 1935). In the case of logic, as in the case of the history of science in general, Rougier's historical methodology is guided by his philosophical views (Nabonnand 2006; De Libera 2006; Lacki 2006) and often undergoes the risk of building straw man fallacies in his epistemological criticism of scholastic realism (Cesalli 2006), and neopositivistic verificationism (Hyder 2006), because the claims he tries to confute are not always identical to the views advanced by his opponents.[34]

[33]In the conclusion to the volume, Rougier claims that logical principles are not innate in our mind, but have been progressively discovered in the history of mankind: the law of inherence was unknown to primitives (Levy-Bruhl 1910), the logic of relations was discovered at the end of the 19th century by symbolic logic, complete induction was unknown until the 16th century, the proof by absurdity was first made explicit only at the beginning of the 20th century (Rougier 1920b: 442). On the other hand, some operations have been long considered as logically valid, while they are not, such as subalternation, simple conversion, some principles of set theory (Rougier 1920b: 442). Finally, many principles of logic were formulated only thanks to a strong effort of abstraction, and are "partly stemming from experience" (Rougier 1920b: 444–6).

[34]On the reception of Rougier (1925) by contemporary historians of philosophy see also the literature quoted in Pont (2007), and especially Courcier (2006), who discusses why Rougier was accused of plagiarism. For a critical survey of Rougier's analysis of the scholastic mentality see Cesalli (2006) and for a criticism of Rougier's work as an historian of medieval philosophy see De Libera (2006).

4 The Interaction with the Vienna Circle and the Relativity of Logic

4.1 The 1935 Paris Conference and the Celebration of the Peano School

Rougier first met the members of the Vienna Circle in 1932, as he was invited by Reichenbach to Berlin, and then again in 1934, when he was invited by Schlick to Vienna. Yet, his most intensive relationship with the members of the Vienna Circle was during two international conferences: the International Philosophy Conference that took place in Prague in 1934 and the International Conference that Rougier co-organized in Paris in 1935 (Rougier 1961: 53–55, Bonnet 2006, Marion 2005, Stadler 2002). Rougier was in charge of the organization of the conference, but the scientific committee included Carnap, Frank, Jörgensen, Morris, Neurath, Reichenbach, and Stebbing. Rougier aimed to embrace different philosophical points of view in the program of the conference, including especially scholars from Latin language European countries—Belgium, France, Italy and Switzerland—, and exponents of empiricist rationalism and psychological empiricism. Rougier acknowledged his own adhesion to logical empiricism, which he described as being endowed with an "anti-metaphysician, pragmatistic, logicizing and empiricist spirit quite at the same time", especially as far as its basic claim is concerned: philosophical problems should be tackled with rigorous, scientific methods. Yet, he also remarked that the logical analysis of scientific language should always be accompanied by an explanation of the genesis, development and structure of scientific thought, that might take into account historical, psychological and sociological elements (Rougier 1936c: 4). The multidisciplinary approach to scientific philosophy that is evident here does not originate in Rougier's interactions with the Vienna Circle: on the contrary, it is rather an effect of Comte's influence, of Rougier's assiduous reading of the *Revue de Métaphysique et de Morale*, but also of Rougier's appreciation of Enriques and Vailati's approaches.

Given that Rougier's task was specifically that of finding speakers and participants in the Latin-language countries, he was responsible for the invitation of Alessandro Padoa, Federigo Enriques, Basilio Manià and Bruno De Finetti. Padoa was the only member of the Peano school participating at the conference, but it should be noted that at the time several members of the school were no longer alive, including Peano, Vailati, Burali-Forti and Pieri. Padoa was accorded the possibility to give not just one but three distinct papers, even if it was Federigo Enriques who had the honor of speaking at the introductory reception as a representative of the Italian school of scientific philosophy. Padoa's three papers—on the logical legacy of Peano (1936a), on a solution to Russell's antinomy of sets (1936b), and on larger systems of numbers as extensions of natural numbers (1936c)—did not include significantly new scientific results, but were devoted instead to a celebration of the Peano school. This celebration of the Peano school was reinforced by Russell, who recalled Leibniz's project and expressed his satisfaction for the due honors given to

Frege and Peano: "for to them ultimately, the movement which gave rise to the Congress is mainly due" (Russell 1936: 10).[35] What's more, Rougier himself had already celebrated the Peano school by declaring in his inaugural speech that celebrating logical empiricism implies celebrating the founders of 'logistics' (Frege, Peano, Hilbert and Russell). Yet, Rougier had also mentioned Mach, Poincaré and Duhem, bringing the epistemological element to the forefront: Rougier, in accordance with Enriques on this issue, explicitly remarked that the logical element alone would "reduce philosophers to the role of grammarians of science" (Rougier 1936b: 8). If the Peano school is praised as a fundamental precursor of modern logic, it is no longer to Peano that Rougier refers when he uses symbolic logic to criticize philosophical systems: in the article on *Scholastics and Logic*, where Rougier claims that the axioms of Scholastics are in disagreement with the principles of contemporary logic, and in particular with the theory of judgments of relations, with the theory of existential judgments, and with the theory of types (Rougier 1935: 108), the influence of Russell's theory is certainly predominant. Vailati, although he had died in 1910, was the only author who still had some influence on the history and philosophy of logic on an international scale, as was confirmed by Scholz, who recalled Frege, Russell and Bolzano as the fathers of symbolic logic—without even mentioning Peano—but quoted Vailati for his historical article on Plato and Theaetetus (Scholz 1936: 1), and by Lindenbaum, who recalls "the well known metaphor by Vailati" according to which "there is at present a democratic system among deductive theories" (Lindenbaum 1936: 32).

4.2 On the Relativity of Logic and Vailati's Contextualism

Soon after the publication of his main volumes on the paralogisms of rationalism and on the structure of deductive theories, Rougier fell under the influence of Wittgenstein's *Tractatus*. The idea that logical laws might be conventional was combined with Wittgenstein's claim that logical laws are tautologies expressing syntactical rules, and are thus entirely devoid of content (Rougier 1961: 44–46). Yet, Rougier's interest in epistemology induced him to give not only a syntactic explanation, but also an epistemological explanation of logical rules: he claimed that logical rules can all be reduced to disjunction, which derives from "the possibility of choice in our decisions that is due to ignorance of the event that will take

[35] On the contrary Enriques was rather more critical than celebrative—both in his inaugural speech and in his talk—underlying the risk of logicism (either in the ontological or in the nominalistic version) transforming itself into a new school of thought, and the risk of radical empiricism ending up as a dogmatic form of scientism (Enriques 1936a: 11, 1936b: 26–27). Analyzing the historical defeat of positivism, Enriques rightly remarked that one of its main weaknesses was the lack of assent among scientists: a truly scientific philosophy—he argued—"should be primarily accepted and supported by those who contribute to the progress of particular sciences […] who should acknowledge its accordance with the general spirit of their own research, or see in it a reflection of their ideals of knowledge" (Enriques 1936b: 24).

place among all possible eventualities", and negation, which "results from a deceived expectation" (Rougier 1961: 48).

The paper on the relativity of logic, first presented at the Harvard Conference in 1939 not only attests to the shift in Rougier's interests from the Peano school to Wittgenstein's *Tractatus*, but also reveals that by 1939 Rougier already held a critical position towards both panlogism and radical empiricism.[36] Wittgenstein's claim that logic is tautological is used to criticize the possibility of panlogism, while the relativism of logic is used to criticize the radical empiricism of the Vienna Circle,[37] as well as the idea that logical principles might be considered as analytic in se rather than as analytic with respect to a given logical system (Rougier 1941: 144).

> I should like to show how such a form of empiricism cannot be maintained. It supposes that the properties of tautological, synthetic, contradictory, and meaningless, applied to propositions, have an absolute meaning. Now this is not the case at all, because there exists an infinity of possible logics; and, according to the logic adopted, the same proposition is tautological or contradictory, analytic or synthetic, meaningful or meaningless. These properties, just like those of indemonstrable or demonstrable as applied to propositions, and of indefinable and definable as applied to concepts, are relative to the language we adopt (Rougier 1941: 144).

In particular, an early criticism of the idea that there could be propositions that are absolutely devoid of meaning (a recurrent theme in the Vienna Circle) is already mentioned here: "A statement can be called meaningless only within a given language and with reference to the logical structure of that language. The same statement can be meaningful in one language and meaningless in another" (Rougier 1941: 155).

The relative character of the notions of tautological, analytic, synthetic, contradictory and meaningless propositions is presented in analogy with the remarks of the Peano school on the notions of demonstrable or definable concepts, but the only explicit reference is cut and pasted from Rougier (1916: 830), and concerns alternative definitions of primitive elements in geometry (Rougier 1941: 139, 1939:

[36]For an earlier version of Rougier's criticism of certain aspects of the logical empiricism see Rougier 1938 and the correspondence with Reichenbach discussed in Padovani (2006a). The same criticism of panlogism (Leibniz and Hegel), and of radical empiricism is repeated without relevant changes also in Rougier (1955: 175–176).

[37]This might confirm Rougier's own assertion that he did indeed aim at diffusing the ideas of the Vienna Circle in France (he organized and edited the proceedings of an international conference in Paris that was perceived as the "first manifestation of logical empiricism in France"), but did not consider himself as a strict follower of the group (Rougier 1961: 55–56). Although he declared his agreement to the general thesis of the group, he disagreed with Neurath and Carnap on the unity of science and on physicalism. The argument against reductionism is the same that was already discussed in the paper on the relativity of logic: sentences devoid of meaning cannot be devoid of meaning per se, but only with reference to a given language, and there cannot be any reduction to a unitarian language, because between the ordinary language and the language of physics there are several different intersubjective languages of sensations (the language of music, the language of colors, and so on) that cannot be reduced to the language of physics (Rougier 1961: 57; Rougier 1938).

308). Actually, Vailati had suggested a relative understanding of definability, as he remarked that "to speak of the 'definability' or 'non-definability' of a given word or concept is to use a meaningless phrase, as long as no precise indication is given as to what other words or concepts may be used in the desired definition" (Vailati 1906: 167). As far as I know, Rougier does not explicitly mention this passage by Vailati. This might confirm the idea that the reference to the Peano school does not inform us about the genesis of Rougier's thesis on the relativity of logic, but rather tells us something about Rougier's tendency to refer to the Peano school to justify his theses. And in fact, the examples Rougier uses to present the urge for a relativistic conception of logic are rather taken from the analysis of the development of formal logic in the Thirties, including modal logics, *n*-values logics and Reichenbach's probabilistic logic. In particular, the many references to mechanics (e.g. to the opposition between corpuscular and wave-like theory or the Heisenberg indetermination principle in quantum theory) explain where Rougier got the idea of considering the choice between different logics as free but non arbitrary, given that it is dependent both on the facts investigated, and the aims of our inquiry (Rougier 1939: 322). At this time, the strongest influence on Rougier's philosophy of logic is certainly due to Jean-Louis Destouches and Paulette Février, who worked intensively on the relations between physics and logics, claiming that new logical systems, different from classical logic, should be used to axiomatize physics (see Lacki 2006: 297ff.).

5 An Epistemological Analysis of Scientific Language

5.1 Logic and the World: Padoa and Vailati

The *Traité de la connaissance* (1955) is an extended treatment of the topics introduced in the volume *La structure des théories deductives* (1921a). It is dedicated to Moritz Schlick and is deeply indebted to the epistemology of the Vienna Circle, especially to Hans Reichenbach, but also to Jean-Louis Destouches, to Paulette Février, and to the biologist Georges Matisse.[38] Rougier mentions the general bibliography on symbolic logic quoted in Church (1936), but also makes some specific references to Hahn and Gonseth.

The book is considered as part of a trilogy that includes *Les paralogismes du rationalisme* (1920b) and *Scolastique et Thomisme* (1925). As Granger ironically puts it, Rougier, although a strong critic of scholastics, publishes a "Summa totius logicae" (Granger 1956: 2), including four Chapters. (1) on the formal truth of

[38]The book got many positive reviews, including Granger (1956), Lenzen (1956), Champigny (1957), Beth (1959), and Blanché (1972). Together with a general appreciation, some strong criticism was expressed by Pap, who claimed that Rougier's treatment of logical empiricism was outdated, and that the analysis of necessary truth as conventions did not take into account the difference between sentence and proposition, terms and their interpretations (Pap 1956: 149, 158). A similar, even if more gentle remark, was repeated in Blanché (1972: 179).

propositions, (2) on the empirical truth of propositions, (3) on language and pseudopropositions and (4) on knowledge and on pseudoproblems.[39] I will consider in the following only some parts that can be fruitfully compared with Vailati's approach to analyticity and pseudoproblems. The only parts of Rougier (1955) which could be related to the Peano school on the basis of this topic are the introduction to the first book on formal and logical systems, and the third and fourth books on language. In the former, the influences of Wittgenstein's writings on the rules of logic and of Russell's notion of propositional function are preeminent, together with a strong influence of logical empiricism in general. A look at the table of contents, for example, already reveals that the opposition between formal and empirical truths is used as a criterion to organize content. Besides, the introduction is devoted to a discussion of the distinctions between a priori and a posteriori, analytic and synthetic, meaningful and meaningless propositions.[40]

The section on axiomatic systems shows that Rougier has updated his 1921 essay adding several recent developments of logic under the influence of Hilbert, Post, Gentzen, Whitehead, Russell, Huntington, Curry, and Gödel.[41] Rougier also discusses different philosophical approaches to mathematics, including logicism, intuitionism, formalism, and structuralism (Rougier 1955: 75–84, 102–115). Rougier introduces here a new definition of logicism, which is more similar to Carnap's formulation than to Pieri's: (1) mathematical notions should be reducible to logical notions; (2) proof methods in mathematics should be reducible to logical methods; (3) any mathematical proposition should be reducible to a logical proposition; (4) mathematical antinomies should be solved (Rougier 1955: 108).

Apparently, there is no influence of the Peano school on Rougier's philosophy of logic at this stage. The idea of a plurality of logics is justified by the fact that we are faced with different kinds of choices in nature and in life (alternatives, trilemmas, but also choices between several values) (Rougier 1955: 137–139), and this plurality of possible choices is grounded in the fact that logic is completely separated

---

[39]On Rougier's acceptance of the strong opposition between real and formal sciences suggested by the Vienna Circle, see Granger (1956: 16), who convincingly argues that formalism cannot be separated from the empirical aspect of knowledge, and mathematics cannot be considered as a mere language that plays no relevant role in physical theories.

[40]Rougier suggests a conceptual analysis of the words in question, because the dispute between philosophical sects is based on a "deep confusion of thought and a lack of precision of language" (Rougier 1955: 36). The objective of the analysis is epistemological: the distinction between a formal (or tautological) and a material (intuitive or empirical) notion of truth is necessary to develop a theory of knowledge that might avoid the mistakes of rationalists (classical rationalists, but also Kantians and German idealists), who claimed that scientific propositions are synthetic a priori, of empiricists (Mill and Spencer), who claimed that they are a posteriori, and neo-positivists (Neurath, Hempel), who claimed that consistence is the only criterion of truth both for formal and for real sciences (Rougier 1955: 43–44).

[41]In particular Rougier discusses the distinction between logical and non logical notions of a scientific theory; meta-theoretic properties such as consistence (relative and absolute), independence, completeness, saturation, categoricity, decidability; axiomatic formulations of propositional logic, arithmetic, and geometry; propositional functions; the distinction between a formal theory and its interpretation.

from the real in the sense that "no external reality corresponds to logical operations —such as negation, implication, equivalence, conjunction, disjunction, incompatibility—nor to logical values—such as true, false, probable, indeterminate—nor to modalities—such as necessary, contingent—nor to quantifiers—such as one, all, and so on." (Rougier 1955: 134). This separation between logical and real is strongly reminiscent of what Padoa has been reported to have said at the First International Conference of Philosophy that took place in Paris in 1900—a remark that Rougier probably read in the review of the conference that appeared in the *Revue de Métaphysique et de Morale* in 1900:

> Mr. Padoa says that the distinction between axioms and definitions has a mere logical and subjective value; in the real world, there are only facts that are given to us on the same level. There are no ideas that are simpler or more evident than others; undefined ideas and undemonstrated propositions subsist only in relation with the logical system that one adopts. And this logical system might be or not be verified by facts, depending on the interpretation given to the undefined ideas. In other words, Mr. Padoa derives a reciprocal independence, a complete separation of logical and real (Tardieu 1900: 560).

If Padoa had claimed that definability and provability are relative to a logical system, Vailati had made a further claim, as he asserted that also the distinctions between a priori and a posteriori, necessary and contingent, analytic and synthetic should be analyzed with respect to the relations between propositions and with respect to their context. Vailati remarked that "while philosophers from different schools of thought continue to discuss the nature of mathematical concepts more than ever, whether they are 'a priori' or 'a posteriori', whether they are 'necessary' or 'contingent truths', 'analytic' or 'synthetic', etc.", the mathematicians are not only uninterested by such questions, but believe that the question should concern the relations between such propositions (Vailati 1904a: 137–138). In a later pragmatistic piece of writing published together with Calderoni, Vailati added that

> "sometimes propositions, synthetic to begin with, become analytic, even though their exterior form is unaltered—that is, without this transformation being indicated by any special verbal sign which allows us to recognize it independently from the examination of the context of speech" (Vailati and Calderoni 1909: 139).

Having claimed in the paper on the relativity of logic that logic is completely separated from the real, Rougier now has to account for its aim; it is a way to compensate for our lacunose knowledge and for our imperfect language:

> To resume, logic relative to our conventions of language is not a theory on the objects of the world and on their behaviors. Even less is logic a set of a priori *rules* that can be applied to our mind or to nature. It is a set of tautologies that teach us to be consistent with the conventions of language that we have set, to recognize the equivalence of different sentences on the basis of the same conventions, to tautologically transform scientific statements that cannot be verified directly into statements that allow an empirical control. Logic results from a lack of univocity of our language and from the lacunose character of our knowledge. It is the crutch of our infirmity, the prerogative of our ignorance. Logic is a privilege of the deficiency of the creature—as some theologian would say: it proceeds from the fact that we are not omniscient. (Rougier 1955: 136).

The choice of a given logic is still considered as conventional, but a distinction is introduced between logical theories taken in isolation (as independent formal systems) and logical systems taken as parts of other scientific formal systems. Whereas in the first case the choice of a given logic is independent from experience and depends only on conventions of language, in the second case the choice is not arbitrary (Rougier 1955: 160). According to Rougier, a change in logic is dictated by the need to solve contradictions of a given scientific theory, whereas in ordinary life, on the contrary, one uses different logics without even noticing, depending on the action that one has in mind or on the context of our action (Rougier 1955: 172–173). Even if Rougier's remarks on the plurality of logics were developed independently from the Peano school, this reference to contexts, especially in the case of ordinary language, where language is often used at an unconscious level, shows a similarity with Vailati's remarks on the need to analyze the contextual role of propositions.

5.2 Pseudoproblems and Vailati's Questions of Words

The third and fourth books of Rougier (1955) are devoted respectively to linguistic pseudo-propositions and to philosophical pseudo-problems—topics that are central also in the last systematic work on metaphysics and language (Rougier 1960).[42] These books include issues that had been intensely debated in the Peano school: the analysis of logical language, the discussion of pseudopropositions and pseudolanguages, the investigation of language as a cause of pseudoproblems, and the discussion about the possibility of a universal language. Yet, in this section there is no mention of any of the members of the Peano school: linguists, Russell, Carnap, Tarski, Schlick, Neurath and Reichenbach are the most often quoted references.

Like Frege, Russell and the logical empiricists, Rougier remarked that the syntax of natural languages differs from logical syntax: "The progress of linguistics on the one hand, and the progress of logic on the other hand, have shown to what extent ordinary languages are ill conceived. They allow us to form words, to assemble them into statements, to combine these statements into arguments in a way that conforms to their grammatical syntax, but contrary to logic. They allow us to form,

[42] *La Métaphysique et le Langage* had a broad reception, even if not as positive as Rougier (1955). See the reviews by Malherbe (1973), Bardy (1975), Jacob (1960), van Haecht (1967), who remarked that the main claims were not new, Cohen and Cohen (1962), who underlined the influence of Cassirer (1946), Benveniste (1958), and suggested possible counterexamples to Rougier's main claim that grammatical structures shape thinking (Cohen and Cohen 1962: 543), Hallie (1962, p. 434), who remarked that "Rougier's linguistic hypothesis (borrowed from Freudenthal and others) leaves us in speculative linguistics, and is as thin as the claim that metaphysics is poetry". It should be noted that Rougier himself acknowledged his debt to Freudenthal (1958) (Rougier 1960: 231–232) and Benveniste (Rougier 1960: 12, 83, 1961: 73).

from the point of view of logic, pseudo-notions, pseudo-propositions, pseudo-arguments that give way to pseudo-problems. The gap between logic and grammar expresses the degree of illogicality of a language, its degree of incoherence" (Rougier 1960: 8).

The notion of pseudo-problem was first introduced in the 1921 volume on physics with reference to Comte's notion of a metaphysical problem:

> It has been customary since the time of Auguste Comte (1795–1857) to distinguish two entirely separate categories of problems: those which may be treated by the methods of science and which are, because of the perfection these methods have reached, capable of being solved sooner or later, and those which, being beyond these methods, lie outside of the limits of experience and on that account are called metaphysical problems. [...] It occurs much more often that metaphysical problems, supposedly incapable of solution, vanish, simply because the progress of ideas shows that they are fictitious problems or pseudo-problems, or problems which have been badly conceived (Rougier 1921b: ix–x).

Examples of pseudo-problems are "the action of imponderable energy on ponderable matter as it is investigated in classical physics, or Newton's notions of absolute space and universal time" (Rougier 1961: 22, 1921b: xi). Metaphysical problem and pseudoproblem are used almost as synonymous terms.

It was only in the essay presented at the Paris Conference (Rougier 1936a) that pseudoproblems were specifically investigated with respect to the grammar of a language.[43] Berndt and Marion (2006, pp. 32–33) suggested the influence of Brunschvicg, and remarked upon the connection made by Rougier himself with Carnap's *Scheinprobleme* (1928). I would like to highlight similarities and differences with respect to Vailati's remarks on the topic. Both were interested not only in ill-formed words and sentences, but also in ill-formed arguments, and considered it the objective of scientific philosophy to analyze not only the syntactical but also the epistemological aspects of scientific language. Vailati took his examples from the history of physics (Vailati 1899a), as Rougier himself had done in 1921b, but also analyzed Aristotle's logic (Vailati 1903b), as Rougier did in (1936d).

Rougier developed a treatment of pseudoproblems that differs from the one which became mainstream in the Vienna Circle. This is particularly evident in his book on metaphysics and language, where, unlike the logical empiricists, Rougier distinguished between two kinds of pseudo-problems: some pseudo-problems arise from syntax, and are insolvable because they are devoid of meaning, due to syntax errors (Rougier 1960: 14); other pseudo-problems cannot be verified and are thus unsolvable, though meaningful.

With respect to Vailati, Rougier insists more on the influence of specific languages on philosophical development: for example, he claims explicitly that metaphysics is a byproduct of the Greek language, because he claims that the

[43]Rougier remarked that Aristotle's logic had both contributed to eliminate pseudoproblems due to the desubstantialization of qualities (e.g. magical transformations), and created pseudoproblems due to the substantialization of relations and classes (Rougier 1936d: 37). So, according to Rougier, Aristotle's logic had both promoted and hindered the development of science: it favored the "development of natural sciences" and "contributed to hinder the impetus of the new astronomy and of modern physics at the time of Kepler and Galilei" (Rougier 1936d: 36).

particular features of Greek favored the tendencies to nominal realism and to conceptual realism (Rougier 1960: 7). In particular, the possibility to substantivize adjectives and verbs engenders, according to Rougier, a tendency towards undue abstractions, i.e. (1) to consider classes as existing independently of the individuals they collect, (2) to consider attributes as subsisting independently of the individuals they are attributed to, and (3) to consider relations as existing independently of the relata (Rougier 1961: 73–74). The most paradigmatic example is the possibility to substantivize the verb "to be", which is taken, following Freudenthal, as "a source of many mistakes, puns and metaphysical speculations" that have been clarified by symbolic logic, so that only those philosophers who care about the logical analysis of the copula might be considered as scientists, while all others are men of letters (Freudenthal 1958: 33–34 cit. in Rougier 1961: 76).

Rougier quotes just a part of Freudenthal's essay, and omits to quote other parts, where Freudenthal defends a thesis that is quite different from Rougier's own point of view. Freudenthal claims that the search for *Scheinprobleme* is interesting not to underline misuses of language (given that language, as Brouwer often recalled, is just a communication medium), but to show how certain problems that were ill-formulated at the beginning received a better formulation, and to distinguish certain pseudoproblems that "fertilized scientific thought" (Freudenthal 1958: 44–45). Omitting the last part of Freudenthal's essay, Rougier appears to be much more positivist than Vailati, because he considers the philosophical systems by Plato and Aristotle only as obstacles that hindered science (Rougier 1960: 96)[44] without ever assigning any positive value to pseudo-problems (Nabonnand 2006: 212). According to Rougier, metaphysical systems are legitimate only from a subjective point of view, because they are sentimental "poems that suggest reasons for living": different metaphysical systems just express different temperaments (Rougier 1960: 246). Vailati on the contrary admits that questions of words are a source of epistemological illusions but also remarks that they might have either negative or positive effects on the development of science, because they can either impose limits on science or favor the progress of knowledge (Cantù and De Zan 2010: xxxii; Vailati 1899a: 77).

Another point of contact between Rougier and Vailati is the influence of Renan and Müller. Rougier's thesis about the influence of language on philosophy is an extension of Max Müller's thesis on the influence of language on mythology (*nomina numina*) (Rougier 1936a: 35, 1960: 11). Rougier makes an explicit analogy between religion, considered as an illness of language, and metaphysics, which is also described as an illness of language, or rather as an illness of the grammar of language (Rougier 1936a: 35, 1961: 78). Another relevant influence is certainly Renan (1855), who is praised for his remark that linguistics might be a helpful tool for studying the history of philosophy (Rougier 1936d: 40). Both authors were well known to Vailati, who had read with close attention Renan (1876), (Cantù and de

[44]Rougier's description of the objects of physics as invariants that can be described only by a mathematical formalism based on relational judgments about structures and functions is strongly reminiscent of Cassirer's *Substance and Function* (1910), but Rougier does not quote it.

Zan 2010: xxii–xxiii), and quoted several passages by Müller on the origin of language.[45]

In 1960, Rougier analyzes all pseudo-problems as an effect of some confusion in the use and understanding of language, but also as the origin of specific metaphysical viewpoints (Rougier 1960: 29–37): materialism is a consequence of a confusion between different orders of language (formal or logical, physical, psychological, moral); nominal realism derives from a confusion between object language and meta-language (between the properties of a name and the characteristics of the thing designated by it, between symbol and the thing denoted by the symbol); conceptual realism is due to a confusion between different logical types (individual, class, class of classes). The confusion between different logical types is due to the insufficiency of grammatical categories, that do not correspond to logical types: for example—and here Rougier explicitly mentions Peano—the difference between membership of an individual to a class and inclusion of a class into another class is not explicit in the Greek syntax, where both logical types are expressed by the same grammatical structure: the apophantic form "Socrates is mortal" (Rougier 1960: 44).

Rougier's analysis of pseudoproblems undergoes an evolution that is similar to the evolution that one finds in Vailati's writings: when the emphasis is put on specific examples taken from the history of physics, the criticism of pseudoproblems is not so unilateral and trenchant as when the criticism is grounded on the analysis of philosophical pseudoproblems. Rougier's later works are animated by a 'purifying mission' (Pont 2007: 13) that reminds one of the pragmatism of the journal *Leonardo.* In his Vailati's writings (Vailati 1899a) the analysis of language was centered on the effort to make explicit certain unconscious aspects of language, as in the case of scientific metaphors, and to detect logical fallacies in scientific and philosophical writings. Yet, in later pragmatistic writings (Vailati and Calderoni 1909), where there was a philosophical issue at stake, the objective was rather the elimination of trifling matters of words and false questions, and any expression that could not be proved or refuted was taken to be meaningless, thus falling outside the field of linguistic analysis itself (Cantù and De Zan 2010: iiv–iv).

[45]Vailati (1904a: 139–140) quoted Müller's thesis that "language begins where interjections end" (Müller 1891: vol. 1, p. 507). Vailati mentioned Müller again in the paper "The Grammar of Algebra", taking it from a second hand source on linguistics: Zoppi, *Filosofia della grammatica*, 1880 (Vailati 1908: 208). Vailati used two ideas of Müller: (1) the assertion that language begins where interjections ends, i.e. where the meaning of words is not predetermined but depends on the way they occur in a sentence (claiming that the force of mathematical language depends on the amount of symbols that denote relations, operations, and functions), and (2) the claim that language is composed of words that do not have a meaning if taken in isolation (analyzing the distinction between verbs that require being completed by an object—i.e. 'generally transitive' verbs—and verbs that are strictly 'intransitive') (Cantù and de Zan 2010: xiv, xvii; Vailati 1904a: 140). Both ideas are used in (Vailati 1904a) to explain Russell's assertion that in mathematics one need not know what one is talking about. It is the very essay where Vailati quoted Russell's definition of mathematics that was discussed in Rougier (1916: 832).

6 Conclusion

Rougier, like some of the members of the Peano school (e.g. Vailati), cannot be ascribed either to analytic or to continental philosophy: on the one hand he underlined the linguistic turn and believed that the use of a plurality of logics was the only possible way to do scientific philosophy in a rigorous way; on the other hand he opposed certain tendencies of analytic philosophy to reduce all philosophical discourse to the analysis of language, and claimed that history, psychology and sociology of science are also necessary components of any investigation of scientific knowledge. This is due to the fact that he had an epistemological understanding of logic: on the one hand logic is necessary because our knowledge is lacunose and our language imprecise, but on the other hand it may help us to distinguish between different kinds of meaningless statements, depending on the rules they contravene (syntactic, semantic, type-theoretical, concerning the distinction between object language and meta-language, or between different orders of language: moral, psychological, and so on).

Rougier's conception, even if he is often presented as a mere divulgator, is on the contrary an original synthesis of the French tradition with Italian and Austrian sources. The French heritage is evident in his debt to Goblot, Comte, Poincaré, Renan, and Bergson. The interest in the Vienna Circle is attested by Rougier's efforts to promote the knowledge of the neo-positivist project of a scientific philosophy, but was not acritical. Rougier criticized physicalism, the belief in an absolute and definitive distinction between analytic and synthetic propositions, the idea that a proposition can be considered as truly scientific only if it's verifiable, and the characterization of the meaning of a proposition as its method of verification (Rougier 1960: 187–189).

The analysis of the influence of the Peano school on Rougier's philosophy of logic has shown that his interest gradually shifted from the writings of Peano and Padoa to the texts by Pieri and finally to the writings by Vailati. The changes in Rougier's choice of his references is not only explicable by an evolution in Rougier's readings (e.g. he probably read Pieri only in a second phase, which explains why Pieri's definition of logicism is mentioned in 1921 but not in 1917), but also by an evolution in Rougier's philosophy of logic, which became more and more involved in historical, epistemological and linguistic issues. The distinction of different phases of the reception of the Peano school has helped us to underline an evolution in Rougier's own philosophical analysis of science. In his early writings, Rougier used Padoa's presentation of the characteristics of deductive theories, together with several remarks on definitions, on the economy of thought, on formalism and logicism made by Peano and Pieri to answer Goblot's criticism of mathematical proofs. In later writings, an influence, or at least a commonality of views with the Peano school, and especially with Vailati's views, can be found in the analysis of the fallacies of complex definitions, in the remarks on the historical evolution of the logical notion of validity, in the consideration of analyticity as a relative notion, and finally in the evaluation of causes and effects of pseudoproblems in science.

Rougier's use of Italian symbolic logic (but also of Hilbert's axiomatics, of the Vienna Circle's scientific philosophy, and of Enriques' rationalism) is a spiral: Rougier starts from a French perspective (Goblot's logic, Poincaré's conventionalism, Destouches and Février's approach to physics and logic, Comte's positivism and Renan's linguistics), becomes interested in logical and philosophical tools developed outside France, and incorporates them in his own epistemological research project. Sometimes the effort to accommodate them either to his own 'ideological' conception of epistemology, or to questions that specifically belong to the French tradition, results in a betrayal or misrepresentation of the original texts (Padoa, Poincaré, Peirce, Freudenthal). Sometimes this historical and hermeneutical dislocation, together with the interdisciplinary displacement, is useful for avoiding a too-easy alignment with the standard view: this is the case of the early criticism of neo-positivist verificationism, as well as the interest in meaningful pseudo-statements and fruitful pseudoproblems.

Rougier's regard for Italy is the regard of a philosopher whose views were formed at the beginning of the 20th century by reading the *Revue de métaphysique et de morale*, which promoted a multidisciplinary and epistemological perspective that can be found also in the works of the Peano school and of Enriques. It is the regard of a scholar who did not read many fundamental texts of the school, but believed in them on behalf of Couturat's and Russell's praise. No matter what the fortunes of the Peano school in Italy had been, Rougier went on believing in them up until the Thirties, when he mentioned Peano as a fundamental contributor to the development of scientific philosophy,[46] and invited Padoa to give three papers at the International Conference on Scientific Philosophy. Beside several epistemological issues especially concerning the role of definitions in axiomatics and of linguistic fallacies and pseudoproblems in science, Rougier shared with Vailati and with Padoa the profile of the scientific 'engagé', even if on different fronts: with Vailati he shared the pragmatistic criticism of rationalism typical of the *Leonardo*, although not its irrationalistic tendencies; with Padoa he shared the belief that philosophers should act in the real world, and do not hesitate to take political stances (Berndt and Marion 2006; Cantù 2007; Tardieu 1900).

[46]See the letter to Peano dated Besançon, 2nd November 1930: "Dear Sir, The *Larousse mensuel* asked me to write an article on scientific philosophy from 1900 onwards. Given that I talked about you in the paper, the editorial board asked me for *a photo of yours and your birth date*. Could you, on the occasion, send me also a brochure that summarizes your life and your numerous writings? Please accept, Dear Colleague, the assurance of all my sentiments of admiration. Louis Rougier. [Cher Monsieur, Le *Larousse mensuel* m'a demandé de faire un article sur la philosophie scientifique depuis 1900. Comme j'ai en a parlé de vous, la direction me demande *votre photographie et votre date de naissance*. Pourriez-vous, à l'occasion, y joindre une brochure resumant votre vie et vos nombreaux travaux? Veuillez agréer, Cher Maitre, l'assurance de mes sentiments admiratifs. Louis Rougier]" Corrispondenze con Giuseppe Peano, in Peano (2002).

References

Alunni, C. and André, Y., editors (2015). Federigo Enriques o le armonie nascoste della cultura europea. Tra scienza e filosofia. Edizioni della Normale, Scuola Normale Superiore Pisa.

Alunni, C. (2015). Maximilien Winter et Federigo Enriques: des harmonies exhumées. In Alunni and André (2015), pages 101–147.

Arrighi, C., Cantù, P., de Zan, M., and Suppes, P., editors (2010). Logic and Pragmatism: Selected Essays by Giovanni Vailati. Center for the Study of Language and Information, Stanford.

B., C. D. (1937). Review. Les mystiques politiques contemporaines et leurs incidences internationales by Louis Rougier. International Affairs (Royal Institute of International Affairs 1931–1939), 16(1):134.

Bardy, Y. (1975). La métaphysique et le langage, essai by Louis Rougier. La Linguistique, 11(1):159.

Benveniste, E. (1958). Catégories de pensée et catégories de langue. Les études philosophiques, 13(1):419–429.

Berndt, C. and Marion, M. (2006). Vie et œuvre d'un rationaliste engagé. Louis Rougier (1889–1982). Philosophia Scientiæ, 10(2):11–90.

Beth, E. W. (1959). Review. Traité de la connaissance by Louis Rougier. Synthese, 11(1):84–85.

Blanché, R. (1972). Review. Traité de la connaissance by Louis Rougier. The Journal of Symbolic Logic, 37(1):178–179.

Bonnet, C. (2006). La première réception française du Cercle de Vienne: Ernest Vouillemin et Louis Rougier. Revue Austriaca, 63:71–84.

Bourdeau, M. and Brenner, A., editors (2014). Revue d'histoire des sciences, 67(2): De la logique philosophique à la logique mathématique.

Bourdeau, M. (2014). La logique à la croisée des chemins: la controverse Goblot-Rougier sur la nature du raisonnement déductif (1907–1921). In Bourdeau and Brenner 2014, pages 311–330.

Brenner, A. (2014). La réception du logicisme en France en réaction à la controverse Poincaré-Russell. In Bourdeau and Brenner 2014, pages 231–255.

Brunschvicg, L. (1911). La notion moderne de l'intuition et la philosophie des mathématiques. Revue de Métaphysique et de Morale, 19(2):145–176.

Brunschvicg, L. (1916). L'arithmétique et la théorie de la connaissance. Revue de Métaphysique et de Morale, 23(2):331–342.

Burali-Forti, C. (1901). Sur les différentes méthodes logiques pour la définition du nombre réel. In Bibliothèque du congrès international de philosophie, volume 3, Colin, Paris, pages 289–308.

Cantù, P. and De Zan, M. (2010). Life and works of Giovanni Vailati. In Arrighi et al. (2010), pages xxi–lxviii.

Cantù, P. (2007). Il carteggio Padoa-Vailati. Un'introduzione alle lettere inviate da Chioggia. Chioggia. Rivista di Studi e ricerche, 30:45–70.

Cantù, P. (2014). The right order of concepts: Graßmann, Peano, Gödel and the inheritance of Leibniz's characteristic. Philosophia Scientiæ, 18(1):157–182.

Carnap, R. (1928). Scheinprobleme in der Philosophie. Weltkreis, Berlin. Engl. transl. The logical structure of the world: And, Pseudoproblems in philosophy. Open Court Publishing, Chicago and La Salle (IL), 1969.

Cassina, U. (1933). L'oeuvre philosophique de G. Peano. Revue de Métaphysique et de Morale, 40(4):481–491.

Cassina U. (1949), Le dimostrazioni in matematica. Annali di matematica pura ed applicata, 29(4):131–146.

Cassirer, E. (1910). Substanzbegriff und Funktionbegriff. Cassirer Verlag, Berlin. Engl. transl. Substance and Function, and Einstein's Theory of Relativity. The Open court, Chicago/London, 1923.

Cassirer, E. (1946). L'influence du langage sur le développement de la pensée dans les sciences de la nature. Journal de Psychologie (normale et pathologique), 29:129–152.

Cesalli, L. (2006). Louis Rougier et la "mentalité réaliste". Philosophia Scientiæ, 10(2): 157–175.

Champigny, R. (1957). Review. Traité de la connaissance by Louis Rougier. The Journal of Philosophy, 54(4):107–109.
Church, A. (1936). A bibliography of symbolic logic. Journal of Symbolic Logic, 1(4):1–547.
Cohen, M. and Cohen, D. (1962). La metaphysique et le langage by Louis Rougier. L'Année sociologique (1940/1948-), 13:542–543.
Comte, A. (1830). Cours de philosophie positive, vol. 1: Leçons 1 à 45, M. Serres, F. Dagognet and A. Sinaceur, editors. Paris: Hermann, 1998. Engl. transl. in Andreski, editor, The Essential Comte, Routledge, London, 2014.
Courcier, J. (2006). "La scolastique et le thomisme", la réception du livre de Louis Rougier en 1925 par le milieu scolastique. Philosophia Scientiæ, 10(2):92–156.
Couturat, L. (1899). La logique mathématique de M. Peano. Revue de Métaphysique et de Morale, 7(5):616–646.
Couturat, L. (1900). L'algèbre universelle de M. Whitehead. Revue de Métaphysique et de Morale, 8 (3):323–362.
Couturat, L. (1904). Les principes des mathématiques. Revue de Metaphysique et de Morale, 12:19–50, 211–40, 664–698, 810–844.
Couturat, L. (1905). Les Principes des Mathématiques: avec un appendice sur la philosophie des mathématiques de Kant. Alcan, Paris.
de Libera, A. (2006). La scolastique: une faillite? Louis Rougier, historien in partibus. Philosophia Scientiæ, 10(2):177–206.
Enriques, F. (1906). Problemi della scienza. Zanichelli, Bologna. Engl. transl. Problems of science. Open Court Publishing Company, Chicago, 1914.
Enriques, F. (1917). Sur quelques questions soulevées par l'infini mathématique. Revue de Métaphysique et de Morale, 24(2):149–164.
Enriques, F. (1936a). Allocution de M. F. Enriques. In Rougier 1936e, vol. 1, page 12.
Enriques, F. (1936b). Philosophie scientifique. In Rougier 1936e, vol. 1, pages 23–27.
Freudenthal, H. (1958). Logique mathématique appliquée. Gauthier-Villars, Paris.
Gilson, E. (1926). Compte rendu de Louis Rougier, la scholastique et le thomisme. Revue d'histoire franciscaine, 3:313–318.
Goblot, E. (1907). La démonstration mathématique. L'année psychologique, 14(1):264–283.
Goblot, E. (1918). Traité de logique. Colin, Paris.
Granger, G.-G. (1956). Le traité de la connaissance de M. Louis Rougier. Les Études philosophiques, 11(1):pp. 15–19.
Grattan-Guinness, I. (2000). The Search for Mathematical Roots, 1870–1940: Logics, Set Theories and the Foundations of Mathematics from Cantor Through Russell to Gödel. Princeton University Press, Princeton (NJ).
Guillaume, M. (2009). La logique mathématique en France entre les deux guerres mondiales. Quelques repères. Revue d'histoire des sciences, 62(1):177–219.
Hallie, P. P. (1962). Review. La métaphysique et le langage by Louis Rougier. Philosophy and Phenomenological Research, 22(3):433–435.
Halsted, G. B. (1904). Rational Geometry: A Textbook for the Science of Space; Based on Hilbert's Foundations. J. Wiley & sons, New York.
Heinzmann, G., editor (1991). Poincaré, Russell, Zermelo et Peano. Textes de la discussion (1906–1912) sur les fondements des mathématiques: des antinomies à la prédicativité. Blanchard, Paris.
Heinzmann, G. (2010). Conventions in geometry and pragmatic reconstruction in Poincaré: a problematic reception in logical empiricism. Workshop: Logic and rational Reconstruction of Science from a Pragmatic point of view, 25–26 mars 2010, MSH Lorraine, Nancy Université. <halshs-00529135>.
Hyder, D. (2006). L'argument anti-empiriste de Rougier et sa relation à l'anti-réalisme de Putnam. Philosophia Scientiæ, 10(2):251–265.
Jacob, A. (1960). La métaphysique et le langage by Louis Rougier. Les Études philosophiques, 15(4):559.

Kennedy, H. C. (2002). Twelve Articles on Giuseppe Peano. Peremptory Publications, San Francisco.

Kennedy, H. C. (2006). Peano. Life and Works of Giuseppe Peano. Peremptory Publications, Concord, CA.

Kergomard, J. and Salzi, P. (1937). Edmond Goblot, 1858–1935, la vie et l'œuvre. Alcan, Paris.

Lacki, J. (2006). Rougier et la physique. Philosophia Scientiæ, 10(2):267–308.

Laforest, G. (1981). Du paradis à l'utopie by Louis Rougier. Canadian Journal of Political Science/Revue canadienne de science politique, 14(4):845–846.

Lalande, A. (1907). Le mouvement logique: Analyse et critique de quelques ouvrages récents. Revue Philosophique de la France et de l'étranger, 63:256–288.

Lalande, A. (1919). La logique de M. Goblot. Revue philosophique de la France et de l'étranger, 87:131–154.

Lenzen, V. F. (1956). Review. Traité de la connaissance by Louis Rougier. Philosophy and Phenomenological Research, 17(1):125–127.

Lévy-Bruhl, L. (1910). Les fonctions mentales dans les sociétés primitives. Alcan, Paris.

Lindenbaum, A. (1936). Sur la simplicité formelle des notions. In Rougier 1936e, vol 7, pages 29–38. Hermann, Paris.

Lolli, G. (2012). Federigo Enriques as a philosopher of science. In Coen, S., editor, Mathematicians in Bologna 1861–1960, pages 333–342. Springer, New York.

Luciano, E. (2012). Peano and his school between Leibniz and Couturat. The influence in mathematics and in international language. In Krömer, R. and Chin-Drian, Y., editors, New Essays on Leibniz Reception, Publications des Archives Henri Poincaré, Springer Basel, pages 41–64.

Malherbe, J.-F. (1973). Louis Rougier, la métaphysique et le langage. Essai. Revue Philosophique de Louvain, 71(12):801.

Marchisotto, E. A. and Smith, J. T. (2007). The legacy of Mario Pieri in geometry and arithmetic. Birkhäuser, Basel, Boston and Berlin.

Marion, M. (2004). Investigating Rougier. Cahiers d'épistémologie (Université du Québec à Montréal), 314(2):1–61.

Marion, M. (2005). Louis Rougier, the Vienna Circle and the unity of science. In Nemeth, E. and Roudet, N., editors, Paris–Wien (Veröffentlichungen des Instituts Wiener Kreis), Springer, Vienna, pages 151–177.

Marion, M. (2007). Une philosophie politique pour l'empirisme logique? Philosophia Scientiæ, Cahier Spécial 7:181–216.

Mazliak, L. and Sage, M. (2014). Au-delà des réels: Émile Borel et l'approche probabiliste de la réalité. In Bourdeau and Brenner, 2014, pages 331–357.

Minnaja, C. and Paccagnella, L. G. (2010). La scuola di Peano nella visione critica di B. Migliorini. In Roero, C. S., editor, Peano e la sua Scuola fra matematica, logica e interlingua. Atti del Congresso internazionale di Studi (Torino 6–7 ott. 2008), Università degli Studi di Torino, Torino, pages 575–604.

Müller, M. F. (1891). The science of language: founded on lectures delivered at the Royal Institution in 1861 and 1863. Longmans and Green, London.

Nabonnand, P. (2006). Rougier et l'histoire des sciences. Philosophia Scientiæ, 10(2): 208–222.

Padoa, A. (1900). Essai d'un théorie algébrique des nombres entiers précédé d'une introduction logique à une théorie déductive quelconque. In Bibliothèque du congrès international de philosophie, vol. 3, Colin, Paris, pages 309–365.

Padoa, A. (1909). Introduzione alla teoria delle frazioni. Bollettino della "Mathesis", 1(7-8-9):66–81.

Padoa, A. (1911a). D'où convient-il de commencer l'arithmétique. Revue de Métaphysique et de Morale, 19(6):549–55.

Padoa, A. (1911b). La logique déductive: dans sa dernière phase de développement. Revue de Métaphysique et de Morale, 19(6): 828–883.

Padoa, A. (1911c). Sur le principe de l'induction mathématique. Revue de Métaphysique et de Morale, 19(2–3):395, 246.
Padoa, A. (1912a). La logique deductive. Gauthier-Villars, Paris.
Padoa, A. (1912b). La logique déductive dans sa dernière phase de développement (suite et fin). Revue de Métaphysique et de Morale, 20(1):48–67, 20(2): 207–231.
Padoa, A. (1917). Des conséquences d'un changement d'idées primitives dans une théorie déductive quelconque. Revue de Métaphysique et de Morale, 24(3):315–325.
Padoa, A. (1936a). Ce que la logique doit à Peano. In Rougier 1936e, vol. 8, pages 31–37.
Padoa, A. (1936b). Classes et pseudoclasses. In Rougier 1936e, vol. 3. pages 26–28.
Padoa, A. (1936c). Les extensions successifs des ensembles des nombres au point de vue déductif. In Rougier 1936e, vol. 7, pages 52–59.
Padoa, A. (1938). La loi de la causalité. Revue de Métaphysique et de Morale, 2:203–212.
Padovani, F. (2006a). La correspondance Reichenbach-Rougier des années trente: une "collaboration amicale", entre empirisme logique et exil. Philosophia Scientiæ, 10(2):223–250.
Padovani, F. (2006b). Références bibliographiques. Philosophia Scientiæ, 10(2):321–380.
Padovani, F. (2007). Références bibliographiques. Philosophia Scientiæ, Cahier Spécial 7:217–276.
Pap, A. (1956). Logical empiricism and rationalism. Dialectica, 10(2):148–166.
Peano, G. (1889). I principii di geometria logicamente esposti. Fratelli Bocca.
Peano, G. (1891a). Sul concetto di numero. Nota I. Rivista di Matematica, 1:87–102.
Peano, G. (1891b). Sul concetto di numero. Nota II. Rivista di Matematica, 1:257–267.
Peano, G. (1894). Notations de Logique Mathématique (Introduction au Formulaire de Mathématiques). Guadagnini, Turin.
Peano, G. (1900). Les définitions mathématiques. In Bibliothèque du Congrès International de Philosophie, vol. 3, Armand Colin, Paris, pages 279–288.
Peano, G. (1901). Formulaire des mathématiques. Bocca, Torino. 3rd edition.
Peano, G. (1903). I. De latino sine flexione. II. Principio de permanentia. Extract. Revista de Mathematica, 8:1–14.
Peano, G. (1904). Il latino, quale lingua ausiliare internazionale. Atti della Reale Accademia delle Scienze di Torino. Adunanza del 3 Gennaio 1904, 39:273–283.
Peano, G. (1906). Formulario Mathematico. Editio V. Bocca, Torino.
Peano, G. (1911). Le definizioni in matematica. Arxivs de l'Institut de ciencies, 1:49–70.
Peano, G. (1915). Le definizioni per astrazione. Bollettino della Mathesis. Società Italiana di Matematica, pages 106–120.
Peano, G. (1918). Eguale. Il Bollettino di matematica, 15:195–198.
Peano, G. (1921). Le definizioni in matematica. Periodico di Matematiche, 4(1):175–189.
Peano, G. (2002). L'archivio Giuseppe Peano. Roero, C. S., Nervo, N., and Armano, T., editors Dipartimento di Matematica, Torino. CD-ROM.
Peirce, B. (1881). Linear associative algebra. American Journal of Mathematics, 4(1):97–229.
Pieri, M. (1906). Sur la compatibilité des axiomes de l'arithmétique. Revue de Métaphysique et de Morale, 14(2): 196–207.
Poincaré, H. (1902). La science et l'hypothèse. Paris: Flammarion.
Poincaré, H. (1912). The latest efforts of the logisticians. The Monist, 22(4):524–539.
Pont, J.-C. and Padovani, F., editors (2006). Louis Rougier: vie et œuvre d'un philosophe engagé, Philosophia Scientiæ, 10(2).
Pont, J.-C. and Padovani, F., editors (2007). Louis Rougier: vie et œuvre d'un philosophe engagé. Témoignages - écrits politiques. Philosophia Scientiae, Cahier Spécial 7.
Pont, J.-C. (2006). Louis Rougier, philosophe de l'axiomatique. Philosophia Scientiæ, 10(2): 309–319.
Pont, J.-C. (2007). Coup d'œil sur l'œuvre de Louis Rougier. Philosophia Scientiae, Cahier Spécial 7:9–18.
Renan, E. (1855). Histoire générale et système comparé des langues sémitiques, volume 1: Histoire générale des langues sémitiques. Imprimerie impériale, Paris.

Renan, E. (1876). Dialogues et fragments philosophiques. Calmann Lévy, Paris.
Rodriguez-Consuegra, F. A. (1991). The mathematical philosophy of Bertrand Russell: origins and development. Birkhäuser, Basel.
Rougier, L. (1913). Henri Poincaré et la mort des vérités nécessaires. La Phalange, 20 Juillet: 1–20.
Rougier, L. (1916). La démonstration géométrique et le raisonnement déductif. Revue de Métaphysique et de Morale, 23(6):809–858.
Rougier, L. (1917). De la nécessité d'une réforme dans l'enseignement de la logique. Revue de Métaphysique et de Morale, 24(5):569–594.
Rougier, L. (1919). A propos de la démonstration géométrique: Réponse a M. Goblot. Revue de Métaphysique et de Morale, 26(4):517–521.
Rougier, L. (1920a). La Philosophie géométrique de Henri Poincaré. Alcan, Paris.
Rougier, L. (1920b). Les paralogismes du rationalisme. Alcan, Paris.
Rougier, L. (1921a). La Structure des théories déductives. Théorie nouvelle de la deduction. Alcan, Paris.
Rougier, L. (1921b). La matière et l'énergie selon la théorie de la relativité et la théorie des quanta. Gauthier-Villars, Paris. Engl. transl. Philosophy and the new physics, an essay on the relativity theory and the theory of quanta. Routledge, London, 1921.
Rougier, L. (1925). La scolastique et le thomisme. Gauthier-Villars, Paris.
Rougier, L. (1931). La philosophie scientifique. Son développement depuis le début du XXe siècle. Larousse Mensuel illustré, 8(293):752–755.
Rougier, L. (1935). La scolastique et la logique. Erkenntnis, 5:100–111.
Rougier, L. (1936a). Allocution d'ouverture du congrès. In Rougier 1936e, vol. 1, pages 7–11.
Rougier, L. (1936b). Allocution finale. In Rougier 1936e, vol. 8, pages 88–91.
Rougier, L. (1936c). Avant-propos. In Rougier 1936e, vol. 1, pages 3–6.
Rougier, L. (1936d). Pseudo-problèmes résolus et soulevés par la logique d'Aristote. In Rougier 1936e, vol. 3, pages 35–40.
Rougier, L., editor (1936e). Actes du Congrès international de philosophie scientifique, Sorbonne, Paris 1935, 8 voll., Hermann, Paris.
Rougier, L. (1937/1938). Le langage de la physique est-il universel et autonome? Erkenntnis, 1461 7:189–194.
Rougier, L. (1939). La relativité de la logique. The Journal of Unified Science (Erkenntnis), 8 (4):193–217.
Rougier, L. (1941). The relativity of logic. Philosophy and Phenomenological Research, 2(2): 137–158.
Rougier, L. (1955). Traité de la connaissance. Gauthier-Villars, Paris.
Rougier, L. (1960). La Métaphysique et le Langage. Flammarion, Paris.
Rougier, L. (1961). Itinéraire philosophique. La revue libérale, 33:6–80.
Rougier, L. (1963). Itinéraire philosophique. In Deledalle, G. and Huisman, D., editors, Les philosophes français d'aujourd'hui par eux-mêmes, Centre de Documentation Universitaire, Paris, pages 117–128.
Rougier, L. (1979). Du paradis à l'utopie. Copernic, Paris.
Russell, B. (1901). Recent work on the principles of mathematics. The international Monthly, 4 (1):83–101. Repr. as "Mathematics and the Metaphysicians" in Bertrand Russell, Mysticism and Logic and Other Essays, New York, London: Longmans, Green & Co., 1918, 74–96.
Russell, B. (1903). The principles of mathematics, volume 1. Cambridge University Press, Cambridge.
Russell, B. (1911). Le réalisme analytique. Bulletin de la Société Française de Philosophie, 11:53. Engl. transl. in Logical and Philosophical Papers. 1909–13, ed.by J. G. Slater and B. Frohmann. Vol. 6. Psychology Press, 1992, pages 133–146.
Russell, B. (1936). The congress of scientific philosophy. In Rougier 1936e, vol. 1, pages 10–11.
Scholz, H. (1936). Die klassische deutsche Philosophie und die neue Logik. In Rougier 1936e, vol. 8, pages 1–8.

Stadler, F. (2002). The Vienna Circle and Logical Empiricism. Re-Evaluation and Future Perspectives. Vienna Circle Institute Year Series. Kluwer, Dordrecht.

Stumpf, D. (1991). Poincaré's Thesis of the Translatability of Euclidean and Non-Euclidean Geometries. Noûs, 25(5):639–657.

Tardieu, M. (1900). Le congrès international de philosophie. Revue de Métaphysique et de Morale, 8:503–698.

Vacca, G. (1911a). Sur la logique de la théorie des nombres (Réponse à M. Padoa et à M. Wickersheimer). Revue de Métaphysique et de Morale, 19(1–2): 252–257.

Vacca, G. (1911b). Sur le principe d'induction mathématique. Revue de Métaphysique et de Morale, 19(3):393–394.

Vailati, G. and Calderoni (1909). Le origini e l'idea fondamentale del pragmatismo. Rivista di psicologia applicata, 5(1). Repr. in Vailati (1911), pages 920–932. Engl. transl. in Arrighi et al. (2010), pages 233–247.

Vailati, G. (1898). La méthode deductive: Comme instrument de recherche. Revue de Métaphysique et de Morale, 6(6):667–703. French transl. of Il Metodo Deduttivo come Strumento di Ricerca. Prolusione ad un corso libero di Storia della Meccanica, 1897–1898. Torino: Roux-Frassati. Repr. in Vailati (1911), pages 118–148. Engl. transl. in Arrighi et al. (2010), pages 23–62.

Vailati, G. (1899a). Alcune osservazioni sulle Questioni di Parole nella Storia della Scienza e della Cultura. Prolusione ad un corso libero di Storia della Meccanica, 1898–98. Bocca, Torino. Repr. in Vailati (1911), pages 203–228. Engl. transl. "Some Observations on the Questions of Words in the History of Science and Culture," in Arrighi et al. (2010), pp. 63–92.

Vailati, G. (1899b). Recensione a C. Laisant. La Mathématique: philosophie, enseignement. Paris: Carré et Naud 1898. Il Nuovo Risorgimento, 9(8). Repr. in Vailati (1911), pages 258–259.

Vailati, G. (1900). Recensione a M. Begey Del lavoro manuale educativo. Rivista di biologia generale, 1–2. Repr. in Vailati (1911), pages 343–345.

Vailati, G. (1901). Sulla portata logica della classificazione dei fatti mentali proposta dal Prof. Franz Brentano (Comunicazione presentata al III Congresso Internazionale di psicologia di Parigi, agosto 1900). Rivista Filosofica, 2(1). Repr. in Vailati (1911), pages 336–340. Engl. transl. "On the Logical Import of the Classification of Mental Facts Proposed by Franz Brentano," in Arrighi et al. (2010), pages 107–112.

Vailati, G. (1903a). Di un'opera dimenticata del P. Gerolamo Saccheri ("Logica Demonstrativa" 1697). Rivista Filosofica, 4. Repr. in Vailati (1911), pages 477–484.

Vailati, G. (1903b). La teoria aristotelica della definizione. Rivista di Filosofia e scienze affini, 2 (5). Repr. in Vailati (1911), pages 485–496. Engl. transl. "The Aristotelian Theory of Definition," in Arrighi et al. (2010), pages 121–136.

Vailati, G. (1904a). La più recente definizione della matematica. Leonardo, pages 7–10. Repr. in Vailati (1911), pages 528–533. Engl. transl. in Arrighi et al. (2010), pages 137–143.

Vailati, G. (1904b). Sur une classe remarquable de raisonnements par réduction à l'absurde. Revue de Métaphysique et de Morale, 12(5):799–809. French Transl. of the original A proposito di un passo del Teeteto e di una dimostrazione di Euclide, Rivista di Filosofia e scienze affini, 6(1): 1904. Repr. in Vailati (1911), pages 536–537.

Vailati, G. (1904c). Intorno al significato della differenza tra gli assiomi ed i postulati nella geometria greca. In Verhandlungen des III Internationalen Mathematiker Kongresses in Heidelberg 8–13 Aug. 1903, Leipzig. Teubner, 1905. Repr. in Vailati (1911), pages 547–552. Engl. transl. "On the Meaning of the Difference Between Axioms and Postulates in Greek Geometry," in Arrighi et al. (2010), pages 145–152.

Vailati, G. (1905). Sull'arte di interrogare. Rivista di Psicologia, 1(2). Repr. in Vailati (1911), pages 572–576. Engl. transl. "The art of asking questions," in Arrighi et al. (2010), pages 153–157.

Vailati, G. (1906). Pragmatismo e logica matematica. Leonardo, 4(1): 16–25. Repr. in Vailati (1911), pages 689–694. Engl. transl. by H. D. Austin "Pragmatism and Mathematical Logic," Monist, 16:481–491, 1906. Repr. in Arrighi et al. (2010), pages 163–171.

Vailati, G. (1908). La grammatica dell'algebra. Rivista di Psicologia Applicata, 4. Repr. in Vailati (1911), pages 871–889. Engl. transl. "The Grammar of Algebra," in Arrighi et al. (2010), pages 205–226.

Vailati, G. (1911). Scritti di G. Vailati, 1863–1909. M. Calderoni, U. Ricci, and G. Vacca, editors. Barth-Seeber, Leipzig-Firenze.

van Haecht, L. (1967). Louis Rougier, La métaphysique et le langage. Revue Philosophique de Louvain, 65(86):255–255.

van Heijenoort, J., editor (1967). From Frege to Gödel. Harvard University Press, Cambridge, Mass.

van Heijenoort, J. (1968). Preface. In van Heijenoort, J., editor, Écrits logiques. Presses Universitaires de France, Paris.

Whitehead, A.-N. (1907). Introduction logique à la géométrie. Revue de Métaphysique et de Morale, 15(1): 34–39.

Whitehead, A.-N. (1916). La théorie relationniste de l'espace. Revue de Métaphysique et de Morale, 23(3): 423–454.

The Eyes of French Mathematicians on Tullio Levi-Civita—the Case of Hydrodynamics (1900–1930)

Rossana Tazzioli

1 Introduction

Tullio Levi-Civita (1871–1943), professor at the University of Rome from 1918 onwards, was a prominent Italian mathematician of the first part of the 20th century. He gave remarkable contributions to various mathematical fields, such as general relativity, the three-body problem, differential geometry, and hydrodynamics. French scholars generally appreciated Levi-Civita's work—for instance, his work on differential geometry influenced Élie Cartan and his school, and his papers on the regularization of the three-body were essential for Jean Chazy's research concerning the long-term behaviour of the solutions of the three-body problem. Henri Villat and other French mathematicians assimilated Levi-Civita's works on hydrodynamics and took inspiration from them, especially from those concerning the wake hypothesis and wave theory. Villat's thesis focused on the so-called Levi-Civita method for deducing the general integral of any plane motion with wake under some conditions on the shape of the body (Levi-Civita 1907a). In 1929 Villat, who was professor at the Sorbonne, was appointed director of the Institute of fluid mechanics (*Institut de mécanique des fluides*) of Paris, created by the Ministry of aeronautics. In the twenties and thirties, Levi-Civita in Italy and Villat in France were the two reference points for hydrodynamics in their respective countries, although Villat had an institutional role much more important than that played by Levi-Civita in Italy.

In this article we aim at giving an image of Levi-Civita's work on hydrodynamics and of its reception in France, starting from his personal and scientific relationship with Villat. In particular, we focus on the following questions: How did

R. Tazzioli (✉)
Laboratoire Paul Painlevé, Université de Lille 1, Sciences et Technologies, Villeneuve-d'Ascq, France
e-mail: rossana.tazzioli@math.univ-lille1.fr

F. Brechenmacher et al. (eds.), *Images of Italian Mathematics in France*, Trends in the History of Science, DOI 10.1007/978-3-319-40082-2_9

French scholars assess Levi-Civita's work? How did he influence French studies on hydrodynamics, and particularly those of Villat and his students? We point out that Villat supervised a large number of theses during the years between the two world wars and some of his students spent a period of study with Levi-Civita at the University of Rome.[1]

In this paper we argue that Levi-Civita's aim was to found hydrodynamics on rigorous mathematical bases by avoiding the use of physical evidence often invoked by his time scholars. We shall show that his approach to hydrodynamics was transmitted to the French school headed by Villat.

Our analysis makes use of letters and unpublished documents mainly contained in Levi-Civita's Archive (Library of the Accademia dei Lincei, Rome), Villat's Archive (Archives of the French Academy of Sciences, Paris), and Boussinesq's Archive (Institute de France, Paris). Such documents allow us to provide a fresh picture of Levi-Civita's relations with France. In particular, we observe that Villat was undoubtedly his French privileged correspondent; nevertheless, concerning hydrodynamics, Levi-Civita also exchanged letters with Jacques Hadamard, Marcel Brillouin, Marie-Louise Jacotin-Dubreil, Robert Mazet, and others.

We mention that Levi-Civita's scientific and personal relationships with his French colleagues were not affected by political opinions. In the twenties, the French generally shared feelings and policy of ostracism against scholars from Central Powers, to which Levi-Civita opposed his internationalist ideals. Levi-Civita was against imperialism and wars, as he often claimed in his correspondence. Just after the First World War, in a letter dated December 9, 1920, he wrote to his German colleague Arnold Sommerfeld: "I have always been, and not only in science, a convinced internationalist and, in consequence of this ideal, I consider au dessus de la mêlée all nationalisms indistinctly. I have the same opinion about the nationalisms preceding and following the horrible war, which upset Europe in a so ruinous way."[2] In 1922, in opposition to the decision to exclude Axis scientists from international meetings strongly supported by France, Levi-Civita organized, together with the director of the Laboratory of Aeronautics in Aachen, Theodor von Kármán, the first International Congress of Applied Mechanics opened to scholars from all countries.[3]

Section 2 gives a brief—and not exhaustive—overview on Levi-Civita's life and research. The following sections will focus on his contribution to hydrodynamics and how French scholars assimilated his ideas. Section 3 illustrates Levi-Civita's contributions to plane motions with wake, especially related to the works by Marcel Brillouin and Villat. Section 4 focuses on Levi-Civita's role as a master at the

---

[1]In France Villat and Henri Cartan supervised the largest number of thesis in the interwar period, see (Leloup 2009).

[2]The letter is contained in the Archives of the Deutsches Museum, and published in (Nastasi and Tazzioli 2005; 224).

[3]On Levi-Civita's attitude towards German and Austrian scholars see (Nastasi and Tazzioli 2013; 2014). For the organisation of the first International Congress of Applied Mechanics see (Battimelli 1988).

University of Rome; it describes the work of Levi-Civita on wave theory and jets, which attracted many students to Rome—Rockefeller fellows, post-doc students, and students at a distance who just exchanged letters with Levi-Civita without meeting him personally. Section 5 contains concluding remarks.

2 A Biographical Sketch of Tullio Levi-Civita

Levi-Civita graduated from the University of Padua in 1892 with a thesis on the theory of invariants supervised by Gregorio Ricci Curbastro (1853–1925). Levi-Civita's professional biography is divided into two great periods—the period at the University of Padua, where he was appointed professor of rational mechanics in 1897, and the period at the University of Rome, from 1918 onwards, where he contributed to create a mathematical school at an international level. Nevertheless, even during the time spent in Padua, Levi-Civita supervised some theses, especially in the field of hydrodynamics (see Sects. 3, 4).

Levi-Civita's main research fields are: the theory of relativity, the three-body problem, adiabatic invariants, analytic mechanics, hydrodynamics, differential geometry, and tensor calculus.[4] His contributions are essential for the development of some branches of these theories. Each of these subjects attracted most interest of French mathematicians, who sometimes strictly followed Levi-Civita's research.

On the eve of World War I, Levi-Civita had already published relevant papers, especially in the fields of differential geometry, hydrodynamics and the three-body problem, and was considered as one of the most prominent Italian mathematicians. For these reasons, he was asked to move to the Italian capital to join the other Roman mathematicians and to improve the scientific level of the University of Rome. In 1909 Vito Volterra (1860–1940), who had a great esteem of Levi-Civita from the early years of his scientific career, together with his colleague Guido Castelnuovo (1865–1952), offered him the opportunity to get a chair at the University of Rome.[5] At the time Levi-Civita declined—maybe his family ties prevented him from accepting the offer. Immediately after the war, however, in 1918, various personal reasons induced Levi-Civita to leave Padua and join the Roman group. In 1919, he was in fact appointed professor of higher analysis *per chiara fama*.

In the twenties, also the two algebraic geometers, Federigo Enriques (1871–1946) and Francesco Severi (1879–1961), joined Levi-Civita, Guido Castelnuovo, and Vito Volterra at the University of Rome—Castelnuovo and Volterra being in Rome since 1891 and 1900 respectively. The *International Education Board* (IEB), created by the Rockefeller Foundation in 1923 to encourage scientific studies, judged Rome

[4]For a more detailed analysis of Levi-Civita's work see (Nastasi and Tazzioli 2005).

[5]The letter by Castelnuovo to Levi-Civita is dated August 27, 1909 and published in (Nastasi and Tazzioli 2000; 260–261).

University as an international centre of excellence in mathematics.[6] George D. Birkhoff (1884–1944), a distinguished American mathematicians of the first half of the 20th century and a close friend of Levi-Civita, addressed to the director of the Paris office of the IEB, Augustus Trowbridge (1870–1934), his Report on European mathematical centres based on his trip through Europe of 1926: "The numerical strength of the mathematical group and the power of tradition at Paris, Göttingen and Rome far transcend those at the other centers named." (Siegmund-Schultze, 2001; 268) Already in 1923, Birkhoff suggested Levi-Civita as one of the advisors for mathematics in Europe—he also mentioned Borel (France), Brouwer (Netherlands), Hardy (Great Britain), Mittag-Leffler (Sweden), Nörlund (Denmark), de la Vallée Poussin (Belgium) (see Sigmund-Schultze, 2001; 37). There were several reasons to mention Levi-Civita as one of the main references, together with Borel, Mittag-Leffler and de la Vallée Poussin who were pillars of European mathematics. First, since the early years of the twentieth century, Birkhoff and Levi-Civita shared common scientific interests, in particular both contributed to the three-body problem; secondly, the scientific interests of Levi-Civita were at the forefront of mathematics of his time, in particular his studies on the theory of general relativity and his related work on differential geometry and tensor calculus; finally, his well-known scientific internationalism was another aspect surely taken into account.

In the twenties and thirties, Levi-Civita and his Roman colleagues welcomed many foreign students, often Rockefeller fellows, from several countries including several French (see Sect. 4), and made mathematical studies in Rome really international. In addition to his position of advisor for IEB, Levi-Civita was the reference point in Italy of Ernest Vessiot (1865–1952), the director of the *Ecole Normale Supérieure* (ENS) of Paris. Vessiot often recommended to Levi-Civita his students of the ENS. In a letter to Levi-Civita, dated August 8th, 1925, Vessiot wrote that "he was pleased" to have obtained grants for Michel Guérard des Lauriers (1898–1988), Robert Mazet (see Sect. 4.2) and André Weil (1906–1998), three ENS students, who were to spend a period of study at the University of Rome. Vessiot asked Levi-Civita to supervise his students' research and to introduce them to the other colleagues at the University of Rome.[7] He remarked that André Weil was an "exceptional" student.[8]

---

[6]On the role of the IEB in the internationalization of mathematics between the two world wars see (Sigmund-Schultze 2001).

[7]For instance, on September 24th, 1937, Vessiot recommended to Levi-Civita his student Christian Pauc: "M. Pauc travaille dans le champ de la théorie des ensembles, et les éléments de sa thèse sont soumis actuellement à votre collègue M. Fréchet; de sorte qu'il est, dans une certaine nature, libre d'étudier d'autres sujets dans lesquels la théorie des ensembles pourrait servir d'instrument utile. Il pense qu'il serait intéressant pour lui d'être mis, à cet effet, en relation avec votre collègue M. Severi. Voudriez-vous être assez aimable pour mettre ces jeunes gens en rapport avec lui, et pour présenter, à cette occasion, à Monsieur Severi mes meilleurs souvenirs?".

[8]On August 8th, 1925, Vessiot wrote to Levi-Civita: "Notre troisième boursier est M. *Weil*; il sort cette année de l'Ecole Normale; c'est un esprit d'une précocité et d'une faculté d'assimilation exceptionnelles, car il n'a pas encore 20 ans. Il travaille sans doute sur la direction de M. Volterra, à qui il a été présenté dernièrement. Mais je serai très heureux que vous vous vouliez bien vous intéresser aussi à lui." The letters by Vessiot to Levi-Civita are contained in Fondo Levi-Civita, Biblioteca dell'Accademia dei Lincei, Rome.

Levi-Civita's scientific reputation in Italy and abroad as a mathematician explains, at least partially, his success as a *Master* at the University of Rome and his wide network of mathematicians, physicists and more generally intellectuals, who were in touch or corresponded with him for personal and scientific reasons. In 1922 he was awarded the Sylvester Medal by the Royal Society of London, and in 1930 elected a foreign member of the Royal Society. He was often invited to give lectures in all Europe—including in the Soviet scientific institutes (for instance, the Institute of Tensorial Calculus, the Institute of Aerodynamics, and the Institute of Theoretical Astronomy)—US and South America. In France, he became a member of the Academy of Sciences in 1911, and was awarded the title of Doctor honoris causa from the University of Toulouse in 1929[9] and from the University of Paris (Sorbonne) in 1933.[10] From 1929 onwards, Henri Villat invited him several times to the Institute of fluid mechanics. In 1932 and 1934 the celebrated *Séminaire Hadamard* devoted a series of conferences to Levi-Civita's (and his students') works on adiabatic invariants.

In 1931, despite his hostility to the fascist regime, Levi-Civita decided to swear a loyalty oath to Fascism. He made that difficult decision to save the Roman school of mathematics. Various sources support this argument. In some letters, Guido Fubini (1879–1943) and Volterra showed him their utmost gratitude for his sacrifice—the oath to Fascism—in the name of the *school*.[11] Moreover, according the witness of Gino Arrighi (1906–2001), Levi-Civita swore also by following Volterra's explicit advice—Volterra and Levi-Civita shared the responsibility of the Roman school and did not want to betray their students, but to give them the necessary support in such a difficult period.[12]

In September 1938 the Italian government promulgated the Laws for the defence of the race, according to which Jews were dismissed from public positions. In particular, Jewish university professors were excluded from universities, academies and cultural institutions. Levi-Civita was forced to retire and prevented from participating in scientific meetings. Moreover, he was expelled from the editorial board

[9]See the report by Adolph Buhl in *Enseignement Mathématique*, vol. 28, 1929, p. 129–132.

[10]Villat was charged to write the "Report", which the Dean of the Faculty of Sciences of Paris read during the celebration. See the letter by Villat to Levi-Civita on March 3rd, 1933 (in Nastasi and Tazzioli 2003; 397–398).

[11]See for instance the letter by Fubini to Levi-Civita dated December 1, 1931 (in Nastasi and Tazzioli 2003; 124) and the letter by Volterra to Levi-Civita dated February 23, 1932 (in Nastasi and Tazzioli 2000; 151).

[12]The testimony by Gino Arrighi is reported in Lettera Matematica Pristem, vol. 7, February 1993. Arrighi wrote: "L'argomento sostenuto da Volterra—che ebbe valore essenziale e determinante—era che quando Levi-Civita avesse lasciato l'Università, come comportava il rifiuto a giurare, si sarebbe creato un vuoto incolmabile nella scuola matematica di Roma, mentre gli scolari che lo avevano seguito sino ad allora non dovevano assolutamente essere abbandonati. Queste considerazioni sulla scuola—mi assicurava Levi-Civita—furono quelle che decisero il successivo suo comportamento dinanzi all'odiosa prescrizione."

of the *Zentralblatt füt Mathematik*—he was the only Italian!—and Francesco Severi and Enrico Bompiani (1889–1975) took his place. That replacement of Levi-Civita was a scandal and produced a strong international reaction. Some members of the editorial staff of the journal resigned, and among them the director Otto Neugebauer (1899–1990). The project of abstracting and reviewing mathematical journals moved to US where *Mathematical Reviews* was founded.[13]

In the thirties, in spite of his difficult personal and scientific situation, Levi-Civita remained faithful to the ideal of scientific internationalism and helped colleagues and students victims of anti-Semitism. Thanks to him, several of them found positions in the US or South America.[14] His death occurred in December, 1941 and was ignored by the Italian mathematical community that did not officially commemorate Levi-Civita.

3 The D'Alembert Paradox and the Wake Hypothesis. Villat and Levi-Civita.

In this section we analyse some aspects of the scientific relationship between Levi-Civita and his French colleagues with regard to hydrodynamics and, in particular, to wake theory.[15] We focus on Villat's PhD thesis (Villat 1911) directly inspired by Levi-Civita's ideas, in order to assess the French reception of Levi-Civita's work on hydrodynamics. We shall notice that Boussinesq had certainly encouraged Villat to do his thesis on hydrodynamic issues, but it was M. Brillouin who suggested to his young student the subject and closely followed his work. We argue that Levi-Civita and his French colleagues shared a similar approach to hydrodynamics based on mathematical and rigorous methods.

3.1 Levi-Civita's Works on Wake Theory

"The group of experimental facts, which for longest disconcerted mathematicians, concerns the drag opposed by a fluid to the advancement of a solid body fully immersed in it"—wrote Villat in his survey on recent advances of hydrodynamics published in 1918 (Villat 1918; 44). In particular, D'Alembert had proved that in a perfect fluid the drag force is zero on a body moving with constant velocity relative to the fluid. That statement, which is in contradiction to the observation of non-zero

[13]This episode is reported in detail in (Sigmund-Schultze 1994).

[14]For instance Guido Fubini, Alessandro Terracini, Berud Steinlerger, Enrico Volterra. On that subject see the epistolary exchanges with Levi-Civita published in (Nastasi and Tazzioli 2000).

[15]For a general overview on the history of hydrodynamics see (Darrigol 2005).

drag on bodies moving relative to fluids, is called the D'Alembert paradox. Besides his own papers and those published by his *patron* (supervisor) Marcel Brillouin (1854–1948), Pierre Duhem (1861–1916) and Jacques Hadamard (1865–1963), in his work Villat mainly cited Italian mathematicians, and Levi-Civita in particular.

In 1901 Levi-Civita published his first note on hydrodynamics aiming at reassessing the D'Alembert paradox and succeeded in explaining it by introducing the "wake hypothesis", according to which there are two different regions in the fluid—inside and outside the wake—separated by a surface of discontinuity.[16] He deduced "theoretically"—as he was proud to claim—"Newton's law on incompressible fluids (drag is proportional to the square of velocity)" (Levi-Civita 1901; 130).

The relevance of that result is corroborated by an exchange of letters between Levi-Civita and Hadamard, who since 1901 was editing his lectures on hydrodynamics given in the academic years 1898–99 and 1899–1900 and then published in 1903 as *Leçons sur la propagation des ondes et les équations de l'hydrodynamique* (Hadamard, 1903). Hadamard moved a critical objection to Levi-Civita's wake hypothesis (19 April, 1902): "there is one case in which I cannot agree with you: that of liquids. It seems to me that the true theory of the phenomenon cannot be found (for liquids) in discontinuities of the kind you introduce. Indeed, these discontinuities should propagate being affected at different moments by different molecules (unless your surface [...] has the shape of a cylinder with generators parallel to the motion), which is impossible in the case of liquids. On the contrary, I believe like you that in the case of gases one must introduce the discontinuities." (Nastasi and Tazzioli, 2006; 87–88) We do not have Levi-Civita's answer, but in a following undated letter Hadamard recognised that Levi-Civita was right, and that his "objection" to the hypothesis on the existence of a discontinuity in the case of the liquids was "a simple inadvertence".

In another letter from February 19, 1903, Hadamard asked Levi-Civita some questions about "phenomena presented by fluids and which rational mechanics fails to explain" concerning friction and (above all) "the mixing of layers, local turbulence or *swirls*".[17] In 1907 Levi-Civita published a paper in the *Rendiconti del Circolo matematico di Palermo* that partially answers these questions (Levi-Civita 1907a). By starting from the wake hypothesis introduced in his 1901 paper, Levi-Civita deduced the general integral of any plane motion involving a wake if the shape of the body moving in the fluid (supposed as to be incompressible) is polygonal or curvilinear with a unique angular point O called bow. In his proof, he used an adequate conformal transformation of complex analysis by starting from

[16]For a detailed analysis of Levi-Civita's papers on the wake hypothesis and motions with wake see (Nastasi and Tazzioli 2006).

[17]For the complete text of the letters see (Nastasi and Tazzioli 2006; 88–90).

some works by von Helmholtz, Kirchhoff, Rayleigh and Joukowski.[18] Levi-Civita dealt with the wake problem in a "purely mathematical way", but he remarked that "actual applications to ship movements" are very important. The latter question—he pointed out—has not been based on "scientific foundations" yet. Levi-Civita used the wake hypothesis already described in his 1901 paper, though actually "the nature of discontinuity is more complicated than the one supposed theoretically". In this regard, he quoted the photographs of the wake Étienne Marey had taken and published some years before (Marey 1900), and added in a footnote some remarks on the actual surfaces of discontinuity that recall Prandtl's boundary-layer concept (Prandtl 1905).[19] In particular, Levi-Civita pointed out, the wake is not infinitely long and not even rigidly attached to the body.

Levi-Civita's ideas in hydrodynamics—especially his wake hypothesis—influenced the work of many students, and scholars in general, who read and followed his research. At the University of Padua Levi-Civita had given rise to an Italian school of hydrodynamics, mainly formed by mathematicians. Some of them generalized his methods and contributed to obtain new results. Umberto Cisotti (1882–1946) extended the validity of D'Alembert's paradox to bodies with various shapes and motions (Cisotti 1905–1906); Tommaso Boggio (1877–1963) and Emilio Almansi (1869–1948) published some papers in which they improved and simplified Levi-Civita's 1907 method (Boggio 1910a, 1910b) (Almansi 1909–1910).

In his lectures held at the University of Pavia in the academic year 1918–19 and published in 1922, Cisotti gives an interesting overview of the knowledge on plane hydrodynamics in that period. He significantly writes in the Preface to his book: "The development of plane hydrodynamics in recent years—due for the most part to Italian contributions, concerning both the development of methods and the large number of results—made me feel that time is appropriate not only to bring the echo of the school, but also to spread the knowledge to a wider circle of scientists who are outside the field of mathematics, more particularly among the specialists of hydraulic disciplines." (Cisotti 1921–1922, I; V) Besides Boggio, Almansi and Cisotti himself, we also mention Gustavo Colonnetti (1886–1968), Bruto Caldonazzo (1886–1960), Attilio Palatini (1889–1949), and Giuseppe Picciati (1868–1908), who generalized Levi-Civita's results on plane motions with wake, and whose contributions are collected and organized in Cisotti's book *Idromeccanica piana* (Cisotti 1921–1922, I). They all belonged to Levi-Civita's school of hydrodynamics.

[18]On the polemics about drag measurements that poorly agreed with late nineteenth-century hydrodynamic theories by Helmholtz, Kirchhoff and Rayleigh, see (Darrigol 2008). For a detailed analysis of Levi-Civita's 1907 paper (Levi-Civita 1907a) see (Nastasi and Tazzioli 2006).

[19]On Prandtl's theory see for instance the Part III of (Anderson 1997). In his footnote, Levi-Civita (1907a; 522) wrote: "New vortex rings must continually come out from the first part of the surface of discontinuity, in contact with the body, descending along the surface, in substitution for those which separate at the end.".

3.2 Levi-Civita's Influence in France. Villat's Thesis

In France, Levi-Civita's 1907 paper inspired in particular Brillouin and Villat, who tried to improve his method and to adapt it to actual cases. Henri Villat (1879–1972), who entered the ENS in 1899, defended his PhD thesis in 1911. Brillouin and Boussinesq were among his advisors. From 1902 to 1911, Villat taught in Caen, his native town—at first at the Lycée and then at the University as *chargé de conférences*. During the years he spent in Caen, Villat prepared his thesis and exchanged letters with Brillouin and Boussinesq. His correspondence allows us to reconstruct the genesis of Villat's thesis, the choice of the subject, the difficulties of the task, and the achievement of the final result. Moreover, it enlightens the point of view of French mathematicians on Levi-Civita's approach to hydrodynamics.

Marcel Brillouin (1854–1948), who was professor of theoretical physics at the *Collège de France* from 1900 to 1931, held a series of lectures on general properties of fluids and the kinetic theory of gas when Villat was a student of ENS.[20] Brillouin lectured on the theory of "Jets" in the academic year 1908–1909, and on "Viscous fluids" the year later. Villat explicitly mentioned Brillouin's lectures of 1909 in his thesis (Villat 1911; 205). As it appears by reading the notes of his course, Brillouin deeply analysed Levi-Civita's 1907 paper in March 1909, and other papers by Italian mathematicians in the following months, such as (Boggio 1910a, 1910b) and (Picciati 1907) concerning the motion of a sphere in a viscous fluid.[21]

Another French *master* of hydrodynamics, Joseph Boussinesq (1842–1929), highly appreciated the Italian works on that subject. Boussinesq got the chair of calculus of probability and mathematical physics at the Sorbonne in 1896 as successor of Poincaré, and was one of the reference points for hydrodynamics in France. Villat and Boussinesq exchanged letters in spring 1910 in which Boussinesq "provided Villat with a precious reference to Italian literature which he had not read due to the language in which they were written" (Aubin 2010; 6). Boussinesq mentioned "the Italians"[22] and explicitly referred to the works by Levi-Civita and Almansi (very probably Levi-Civita 1907a and Almansi 1909–1910). In his five notes, Almansi deals with the action of a fluid on a body plunged in it; he considers the very general problem, but he is forced to introduce special assumptions in order to obtain some interesting results. Actually, Villat was already aware of the Italian papers cited in Boussinesq's letters: "Besides the papers in Rendiconti you have suggested, I have already studied during the last months the beautiful papers you

[20]The manuscript of Brillouin's course (1898–1900) by Henri Bénard is contained at the Library of the Institut de France (Code: MS 5592).

[21]The manuscript of Brillouin's course at Collège de France of the academic year 1908–1909 entitled "Jets" is kept at the Library of the Institut de France (Code: MS 5601). In March 1909 Brillouin developed a detailed analysis of Levi-Civita's 1907 paper on hydrodynamics (Levi-Civita 1907a). The year later (1909–1910), Brillouin lectured on the subject "Viscosité" (Code: MS5602).

[22]Boussinesq to Villat, letter dated May 11th, 1910 (Fonds Villat, Archives de l'Académie des Sciences, Paris).

mentioned", he wrote to Boussinesq on May 1910.[23] In the same letter, Villat illustrates his work project with the utmost detail. Though Villat had already read the Italian works cited by Bousinnesq, he was certainly comforted by the fact that a recognized expert in hydrodynamics showed him the same way he had begun to follow some time before.

Both Boussinesq[24] and Brillouin highlighted that the Italian authors had restricted their attention to perfect fluids when it was obvious that other physical aspects—such as small speed, vortices, and so on—played an essential role. Nevertheless, they were aware that "the Italians" dealt with perfect fluids in the hope of finding a generalization to concrete fluids later on—their choice of working with perfect fluids being imposed by the mathematical difficulty of the general problem.

In his lectures of 1909, Brillouin criticized some points of Levi-Civita's article and his letters to Villat[25] pointed out his "physical" approach in contrast with Levi-Civita's point of view. In a letter dated 27 September, 1910, Brillouin remarked: "Your plan is very interesting and will take you very far, even without adding anything to Levi-Civita."[26] And he pointed out that he was trying to improve Levi-Civita's work. In fact, Levi-Civita's method was too general and applicable to cases that were not *physically* possible. In his 1911 note (Brillouin 1911), Brillouin analysed Levi-Civita's wake hypothesis by trying to define its limits of validity. He proved that, if the usual assumptions on wake are valid, the discontinuity surfaces must be infinite, otherwise some points of the fluid have a negative pressure. Duhem proposed to denote that statement as "the Brillouin paradox" and extended it from two to three dimensions (Duhem 1914).

In another letter to Villat dated 10 June, 1910, Brillouin suggested analysing Levi-Civita's memoir in details but avoiding "the Italian fashion of publication"; he mainly referred to the lack of examples.[27] Some remarks are required to put in perspective the contribution of Levi-Civita to hydrodynamics and his concerns on the status of the discipline. Levi-Civita's approach to hydrodynamics is indeed the same he shows towards other fields of mathematical physics—he aims to found physical principles on rigorous mathematical theory. Levi-Civita elaborates the rigorous mathematical theory by starting from ideal fluids. He is aware that it is

[23]"A part les articles des Rendiconti que vous m'indiquez, j'ai déjà étudié pendant ces derniers mois les beaux mémoires dont vous me parlez." The letters by Villat to Boussinesq are contained in Papiers Boussinesq, MS4229, Bibliothèque de l'Institut de France, Paris.

[24]On Boussinesq and his relationship with Villat see (Aubin 2010).

[25]The letters by Marcel Brillouin and Levi-Civita to Villat are kept in Dossier Villat, Archives de l'Académie des Sciences, Paris.

[26]"Votre plan est très intéressant, et vous amènera fort loin, même sans rien ajouter à Levi-Civita".

[27]Brillouin wrote to Villat (June 10th, 1910): "toutefois je vous conseille de ne pas trop imiter le mode de publication des Italiens, qui modifient infiniment peu l'exposition déjà connue, la généralisent un peu, et finalement ne traitent aucun exemple ou groupe d'exemples nouveaux.".

only the first stage—even if an important stage—towards attaining the final goal. In his paper on the wake hypothesis, Levi-Civita remarks that, "because of the difficulty of the question, it is convenient to study the problem in two dimensions" (Levi-Civita, 1907a; 521); in that special case, he can apply complex analysis, which is the key method for solving the general integral of plane motions with wake. However, he adds that "in spite of such a simplification, we do not miss the practical interest", and gives the example of a ship whose motion can be seen on horizontal sections. (Levi-Civita 1907a; 521) He is then actually concerned with the applications of his results. Furthermore, we shall see that in some cases he takes inspiration from experimental facts and tries to deduce them from his new theoretical principles. (See Sect. 4.2)

His French colleagues appreciated Levi-Civita's work on hydrodynamics, and tried to improve it by making it applicable to concrete cases. After all, this is the usual way to proceed in mathematical physics today; at the beginning of the 20th century, however, scientists—and especially specialists of hydraulics and engineers—often preferred to deal with hydrodynamic theories empirically and often rejected mathematical rigor.

In a letter on June 20th, Brillouin persuaded his student Villat to focus on some special topics of Levi-Civita's 1907 paper, which "deserve" the most attention. In Brillouin's opinion, in his thesis Villat had to re-deal with, develop and generalize such crucial points of (Levi-Civita 1907a). Brillouin added that a mathematician specialist in the theory of complex functions, as Villat was, could be able to make important advances in "that sense"—in improving Levi-Civita's work.

Villat followed Brillouin's advice and actually generalized some Levi-Civita's results in his dissertation. In the first part of his work, he established the general integral of a permanent plane motion in a fluid bounded by an infinite wall, and where a given body is immersed. Villat's solution is expressed in terms of elliptic functions, and becomes Levi-Civita's formula when the distance from obstacle and wall tends to infinity. The second part of Villat's thesis is devoted to solving the following question: "To find the motion and all its elements, if the obstacle shape is given." (Villat 1911a; 206). It is the inverse of Levi-Civita's problem—for every regular analytical function in a circle that assumes real values on the diameter, an irrotational motion with wake exists (the shape of the obstacle is deduced a posteriori). Villat considered the case of an indefinite fluid, and found a new arbitrary function that allows us to find the general solution of the problem in a simple way, and whose expression is linked to the obstacle shape. In Roy's opinion, that is "one of the fundamental merits" of Villat's memoir (Roy 1972). In the conclusive section, Villat showed some remarkable examples. Even though he tried to extend Levi-Civita's results by making applications and showing actual examples, Villat's mathematical methods are directly inspired by Levi-Civita's ones.

Villat sent his memoir to Levi-Civita, who answered with his usual modesty (letter dated June 19th, 1911): "I am very pleased that the perfection of the Helmholtz-Kirchhoff method that I have indicated has given rise to many scientific studies". (In Nastasi and Tazzioli 2003; 373) The rich epistolary exchanged between Villat and Levi-Civita began in 1911 and went on until 1939.[28]

3.3 Henri Villat

Villat is the French colleague with whom Levi-Civita had a privileged relationship, both personal and scientific. In the years 1911–1912 their letters mainly concern some points of Levi-Civita's 1907 method and Villat's thesis, and especially focus on a special functional relation, due to Ulisse Dini (1845–1918), and its application to hydrodynamics. This formula connects the values of a function f on the circumference of a circle with the values of its normal derivative on the same circumference, if f is assumed to be harmonic in the circle.[29] We note that their discussion is also connected to both the theory of integral equations and the use of the Green function for solving the Dirichlet problem. Villat published several notes (Villat 1911b, 1911c, 1912) concerning the Dirichlet principle in circle, annulus and other special figures, and exchanged some letters with Émile Picard (1856–1941) on that subject. On October 25th, 1911 Picard wrote to Villat: "Dear Sir, yesterday I communicated your note to our Academy [Académie des Sciences]. Your elegant results strongly interested me [he refers to (Villat 1911d)]. Taking the question in its most general form, for my part I get a lot less simple analysis [...]"[30] In the years 1906–1920, Boggio, Marcolongo, Orlando, Lauricella and other Italian mathematicians often applied the theory of integral equations in order to solve questions of mathematical physics, including heat theory, theory of elasticity, and hydrodynamics.[31] In particular, Boggio clearly highlighted the fundamental links between hydrodynamic problems and the so-called Fredholm method for solving some special classes of integral equations.[32]

After defending his PhD thesis, Villat got a position as assistant professor (*maître de conférence*) at the University of Montpellier. Villat played a very important role in applied hydrodynamics and aeronautics in France after World War

[28]The letters by Villat to Levi-Civita are kept in Fondo Levi-Civita; the letters by Villat to Levi-Civita are contained in Fonds Villat, Archives de l'Académie des Sciences, Paris. Their correspondence (43 letters) is published in (Nastasi and Tazzioli 2003; 371–410).

[29]For a detailed analysis of the correspondence between Villat and Levi-Civita in the couple of years 1911–1912 and its connection to Dini's formula, see (Nastasi and Tazzioli 2006; 97–108). See (Nastasi and Tazzioli 2004) for a historical analysis of the Dirichlet problem in connection to problems of mathematical physics.

[30]"Cher Monsieur, j'ai communiqué hier votre note à l'Académie. Vos résultats élégants m'ont vivement intéressé. En prenant la question sous sa forme la plus générale, j'ai de mon côté une analyse beaucoup moins simple...". The letters by Emile Picard to Villat are contained in Fonds Villat, Archives de l'Académie des Sciences, Paris.

[31]See (Archibald, Tazzioli 2014) for a detailed analysis.

[32]See (Nastasi and Tazzioli 2006; 106–108).

I—in 1920, just after World War I, he moved to Strasbourg, when the University of Strasbourg became the symbol of the French victory against the Germans,[33] obtained the chair of mechanics at the Sorbonne as successor of Paul Painlevé (1863–1933) in 1927, was appointed director of the Institute of fluid mechanics in 1929, and became professor of aerodynamics at the School of Aeronautics of Paris in 1930.[34]

In the twenties and thirties Villat became more and more interested in questions of applied hydrodynamics both for intellectual reasons and for his institutional position as the director of the Institute of fluid mechanics. In spite of that, Villat often mobilized mathematical tools in his works on fluid dynamics, such as Dini's formula or the theory of integral equations. In particular, he investigated some questions connected to the first section of his dissertation and deduced the explicit solution for an obstacle of any shape, discussed the uniqueness of the solution, and established some necessary conditions for avoiding solutions that are physically impossible (Villat 1921). In another paper he succeeded in applying integral equations to fluids with vortices (Villat 1923). His taste for abstract mathematical methods, especially related to functional analysis and the theory of complex functions, explains his admiration for the work of Levi-Civita and his school. However, while Levi-Civita showed some attention to applications especially concerning hydraulic experiments, Villat's works did not explicitly refer to experimental facts.

Moreover, some of Villat's students analysed topics connected to Levi-Civita's hydrodynamic research. For instance the thesis by René Thiry (1886–1968) was directly inspired by "an important Memoir of Levi-Civita (Levi-Civita 1907a)" and its "important developments" made by Villat and Brillouin (Thiry 1921; 1). He started from Villat's statement that hydrodynamic equations can lead to more than one solution—in particular, if the obstacle is made by two concurrent segments with concavity towards the current, then two solutions are possible and neither of the two solutions is better than the other a priori (Villat 1914). Thiry proved that, under certain conditions, the solutions constitute a continuous succession between the two solutions deduced by Villat. Thiry's research is clearly inspired by Levi-Civita's search of rigour apparently shared by his *patron* Villat.

We mention that, still in the 1940s when he was already an authority in the field of aerodynamics, Villat seems more interested in mathematical rigour rather than in experimental data. His lectures on the theory of viscous fluids published in 1943 (Villat 1943) put in evidence the "extreme" difficulties of research involving viscosity that make the approach to these issues "very dangerous". Villat pointed out that there were several "rudimentary" books about viscous fluids that are not

[33]On the University of Strasbourg and its strategic importance in France see (Crawford, Olff-Nathan 2005).

[34]A wide image of French mathematics and Villat's role in the interwar period is in (Gispert and Leloup 2009). An analysis of Villat's role as a specialist of fluid-dynamics and the director of the Institute of fluid mechanics is in (Aubin 2010). For a survey on the history of aeronautics in France see (Weber 2008).

rigorous at all from the mathematical point of view; here "the experimenter" generally misses the crucial points. On the contrary, the aim of Villat's book is to specify some "essential points" of the theory, in order "to facilitate theoretical research without which any practical application will always devoid of real meaning" (Villat 1943; V). Anyway, he added he did not intend to replace Brillouin's (1907) treatise on viscosity (Brillouin 1907), whose approach was "rather experimental", with his 1943 lectures. Once more, Villat's point of view on *mathematical* hydrodynamics seems to be close to that of Levi-Civita.

4 The Wave Theory. Levi-Civita as a Master at the University of Rome

In Sect. 4.1 we illustrate Levi-Civita's original works on wave theory, which mainly concern progressive permanent waves and the existence proof of the irrotational wave in a canal with infinite depth. We focus on some works of his French students, which he often supervised in collaboration with other French colleagues—such as Villat, M. Brillouin, Hadamard, and Painlevé. Section 4.2 is devoted to Levi-Civita's work on the theory of jets, often neglected by historians of science, and its influence in France.

4.1 Levi-Civita's Contribution to Wave Theory

Levi-Civita's interest in waves starts very early, when he was professor at the University of Padua, and concerns the very definition of "wave". In a letter to Volterra (dated April 26th, 1903), Levi-Civita wrote: "Before leaving [for the Easter holidays] I started and now go on studying wave theory [...]—it is the subject I shall develop in my course of higher mechanics. I met, however, a preliminary difficulty [...] concerning the lack of a definition of wave, both general and rigorous." (Nastasi and Tazzioli 2000; 75–76) Levi-Civita's interest in wave theory remains central in his research throughout his scientific career. Again in 1935, in a fundamental article written in collaboration with Ugo Amaldi (1875–1957) for the *Enciclopedia Italiana*, the authors observe how it is difficult to give a definition of wave motion, which on the one hand embraces all cases where waves are physically detectable, and on the other hand is mathematically rigorous (Amaldi and Levi-Civita, 1935). We point out that, in his research on hydrodynamics—both on the wake hypothesis and wave theory—Levi-Civita often contributed by founding some intuitive concepts on rigorous mathematical formalism.

In 1907, Levi-Civita published a paper on progressive permanent waves in a canal with horizontal bed—"progressive" means that the motion appears as stationary to an observer moving with the apparent translation of the fluid (Levi-Civita 1907b). In this research Levi-Civita considered vertical motions and included gravity as an actual force. So, he considered the velocity potential φ with $d\varphi =$

$udx + vdy$ (u, v the velocity components of the point (x, y)) in a suitable region, and the stream function ψ such that $d\psi = -vdx + udy$. He introduced the complex variable z, and the functions f and w:

$$z = x + iy;$$
$$w = u - iv;$$
$$f = \varphi + i\psi$$

such that $w = \frac{df}{dz}$. By following step by step his method illustrated in (Levi-Civita 1907a), he still applied complex analysis and proved that f is a conformal map, in order to deduce the rigorous solution of wave motion in a finite depth canal. Finally, he reduced his original problem to the solution of a mixed equation (differential and finite difference equation) related to the holomorphic function $w(f)$.

Levi-Civita's results on various mathematical fields, including hydrodynamics, allowed him to be awarded the prestigious *Premio Reale* from the Academy of Lincei in 1907 (*ex aequo* with Federigo Enriques). His contributions to wave theory inspired several scholars. In particular his paper (Levi-Civita 1907b), together with some contributions published by his students and colleagues on the same topics, influenced the French René Risser (1870–1958). A *polytechnicien*, Risser soon turned his interests into actuarial science. In 1898 he founded and directed, together with Lucien Marchis, the *Statistique Générale de la France*. In 1907 he was employed in the Ministry of Labour, where he became the director of the actuarial service in 1912. In parallel to his career as an actuary, Risser obtained his doctorate with a thesis on hydrodynamics (Risser 1925) inspired by Levi-Civita's ideas on progressive permanent waves. Risser's work provides a historical overview on wave theory, and especially refers to the works of Boussinesq, and Levi-Civita and his students. Moreover, his dissertation deals with actual applications by considering some special examples, such as waves produced by an emerging cylinder whose generatrixes are perpendicular to the axis of the canal and as large as the canal width. Risser is proud to send his work to Levi-Civita, whom he had met in Italy during World War I.[35] In fact during the war Risser had served as an artillery officer near Padua and received the War Cross for his military actions. He also contributed to war work by improving ballistic tables for several artillery pieces.

After defending his PhD thesis, Risser took part in the International Congress of Mathematicians in Bologna (1928), where he met again Levi-Civita and his wife; just after the Congress he wrote him (letter dated December 1st): "I regret not having been your student, because you are not only incomparable as a master, but

[35]Risser wrote to Levi-Civita on May 20th, 1925: "Cher Maître, Dans les premiers jours de Janvier 1918, j'ai eu l'honneur de vous être présenté à Padoue par un ami de ma famille, alors que je venais d'être détaché en mission à l'armée d'Italie par Monsieur le Général Bourgeois aujourd'hui membre de l'Institut, et à cette époque grand Directeur de l'Artillerie." The letters by Risser to Levi-Civita are contained in Fondo Levi-Civita.

also a friend to all those who work with you."[36] Risser went on working on hydrodynamics and in 1935 published a paper, still inspired "by Hadamard, Levi-Civita and their students", in the book for celebrating the Jubilee of Marcel Brillouin (Risser 1935). In the same years, Risser contributed to Borel's *Traité du calcul des probabiblités et de ses applications* by publishing in 1936 (together with Claude-Emile Traynard) the 4th issue on the principles of mathematical statistics.[37]

Levi-Civita's approach focuses on the use of functional and complex analysis as prominent methods for dealing with hydrodynamic problems in general, and permanent waves in particular. As already mentioned, Dini's formula in functional analysis has a central tool in some contributions by Levi-Civita and Villat (see Sect. 3.3). Levi-Civita devoted a note, communicated to the Academy of Lincei, to Dini's formula and its deep connections to an analytic method due to Lord Rayleigh and related to the solitary wave (Levi-Civita 1911). A solitary wave is a wave that propagates without any temporal evolution in shape or size. It consists of a single intumescence and was observed and experimentally studied by Scott Russell and theoretically (in an approximating way) by Boussinesq and Lord Rayleigh. Alexander Weinstein (1897–1979), a Rockefeller student of Levi-Civita at the University of Rome, obtained further approximations for solitary waves by improving Boussinesq-Lord Rayleigh ones (Weinstein 1926).

In (Levi-Civita 1912) Levi-Civita analysed permanent waves in a canal submitted to certain physical conditions. He extended Lord Rayleigh's well-known theorem—according to which the global transport of fluid increases infinitely with time—to permanent waves in a canal with finite or infinite depth. Nevertheless, his main contribution to wave theory concerns irrotational waves with finite amplitude (Levi-Civita 1925)—in particular, it deals with periodic waves that propagate without changing their shape. A periodic wave that attracted Levi-Civita's interest is given by trochoidal waves found by Gerstner (1802) and then by Rankine (1863) independently.[38] The wave profile is that of a trochoid and the fluid particles describe little circles decreasing with the distance from water surface.[39] As in Fig. 1 the *limit* wave is a cycloid. Gerstner's waves have interesting applications to hydraulics and nautical science.

The fluid particles have, however, a turbulent (or rotational) character, while in a perfect fluid—where only conservative forces act—only irrotational motions are admitted; a mathematical inconvenience that "detracts somewhat from the physical interest of the result" (Lamb 1932; 421), as also pointed out by Amaldi and Levi-Civita in their 1935 paper published in the Italian Encyclopaedia (Amaldi and Levi-Civita 1935).

---

[36]"Je regrette de n'avoir point été votre élève, car vous êtes non seulement le maitre incomparable, mais aussi l'ami de tous ceux qui travaillent avec vous.".

[37]On Risser's contribution to statistics see (Bustamante et al. 2015).

[38]For a survey of Gerstner's waves see (Darrigol 2005; 72–75).

[39]A trochoid is the curve described by a fixed point on the radius of a circle when the circle rolls without sliding along a straight line.

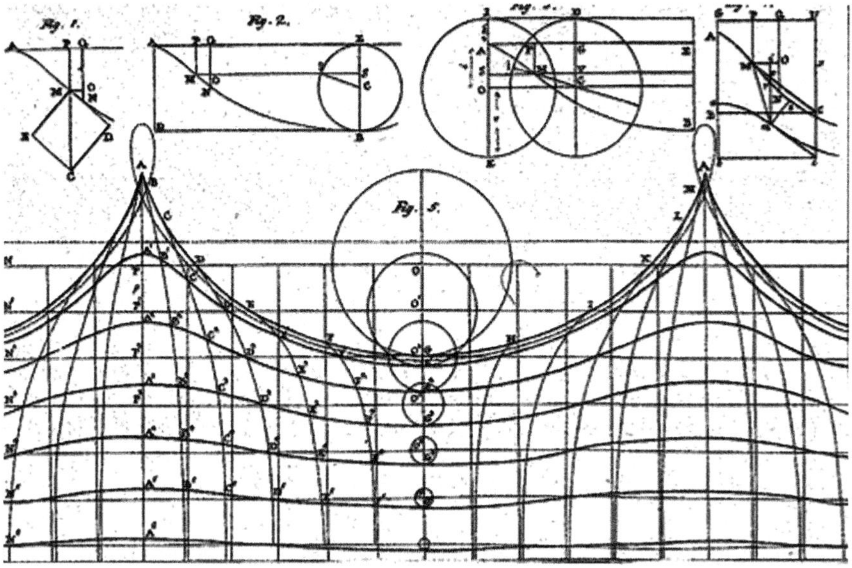

Fig. 1 The lines represent possible wave profiles the limit of which is a cycloid. The circles represent the orbits of fluid particles. [F.J. Gerstner, Theorie der Wellen, Annalen der Physik, vol. 32 (1809), p. 402–445]

In 1847 Stokes had elaborated an approximating method for calculating periodic permanent irrotational waves by using a series development. Lord Rayleigh obtained more precise approximated solutions. However, their methods did not prove the convergence of the approximating series. In his 1925 paper, Levi-Civita solved the problem with the utmost rigour, by means of a new approximating expression that he called the "stokian" (in honour of Stokes). In Lamb's words, Levi-Civita closed "an historic controversy" (Lamb 1932; 420) by proving the existence of the irrotational wave in a canal with infinite depth and deducing simple mathematical formulae linking together height, length, transport, and velocity of propagation of waves. An important stage towards that result is represented by an article, published in *Mathematische Annalen* in 1922, that concerns the functional equation representing periodic waves in a canal of infinite depth (Levi-Civita 1922a).

Of course, the extension of his 1925 existence proof to more general cases was central for Levi-Civita's research, and some of his students dealt with it. His Rockefeller student Dirk J. Struik (1894–2000) was able to generalize his result to canals with finite depth (Struik 1926), and Marie-Louise Dubreil-Jacotin (1905–1972) proved the existence of infinite rotational waves (Dubreil-Jacotin 1934). Dubreil-Jacotin, graduated from the ENS and followed Villat's course on fluid mechanics in Paris, got a grant that allowed her to collaborate with the meteorologist Vilhelm Bjerknes (1862–1951) in Oslo. In 1930 she married the mathematician Paul Dubreil (1904–1994) and the same year the couple travelled in

Germany—where they met, among others, Emmy Noether (1882–1935) and David Hilbert (1862–1943)—and spent a period in Rome, where Paul Dubreil was a Rockefeller fellow and collaborated with the group of algebraic geometry. In his souvenirs, Paul Dubreil wrote about his wife's research: "Having studied the work (Levi-Civita 1925) my wife arrived in Rome with a curious remark: one of the conditions introduced by Levi-Civita, *the absence of mass transport in the deeper layers*, was not satisfied by the wave of Gerstner, also called cycloid wave, which is an exact solution as well, but not irrotational.[40] [...] That remark on *the deeper layers* was indeed the starting point of my wife's thesis—by assuming a hypothesis valid for both the wave of Gerstner and that of Levi-Civita, she rigorously establishes the existence of an infinite number of waves including the two mentioned above" (Dubreil 1983; 69).

Marie-Louise Dubreil proved that statement for the two cases of finite and infinite depth of the canal; for both cases she reduced the problem to a system of integral equations. She defended her PhD thesis at the University of Paris in 1934—Vessiot, Julia, and Villat were the members of the PhD committee—and published her memoir in the *Journal des mathématiques pures et appliquées* directed by Villat. Immediately after obtaining her doctorate, she was put in charge of the *Cours Peccot* at the prestigious *Collège de France* by giving lectures on wave theory. In this regard, she wrote on October 31st, 1935 to Levi-Civita: "Having been asked to give the course *Peccot* at the *Collège de France* this winter, I have to hold a series of lectures on waves. In this regard, I plan to talk a little also of the solitary wave." At this point, she asked Levi-Civita for some references on that subject. In other letters, she sent him some notes he communicated to the Academy of Lincei. Throughout her letters to Levi-Civita (Nastasi and Tazzioli 2003; 159–172) it is possible to follow some aspects of her scientific career and also the difficulties linked to the fact that she got positions at provincial universities and came back to Paris only in 1956.[41]

In conclusion of the section, we mention Levi-Civita's treatises on wave theory. Since Levi-Civita's activity at the University of Rome focused on the creation and development of an international mathematical school, it is not surprising that he published most of his university treatises during this period.

In 1921 Levi-Civita held a series of lectures in Barcelona on various questions of classic and relativistic mechanics, including a lecture on waves in canals.[42] Such lectures were collected in a book and immediately published by the Institute of Catalan Studies (Levi-Civita 1922b); two years later, they appeared in Italian and in German translation. Though some French scholars wished a French translation as

[40]We point out that Paul Dubreil denotes as "Gerstner wave" the limit of the trochoidal waves.

[41]We mention that some years after her doctorate, she turned her interests to algebra by following Emmy Noether's ideas.

[42]In 1921 Esteve Terradas invited Levi-Civita for giving a series of lectures in the "Cursos Monogràfics d'Alts Estudis i d'Intercanvi" in Barcelona. The Catalan Institute organized the "Cursos Monogràfics" from 1915 to 1925 and invited, among the others, Jacques Hadamard, Hermann Weyl, Arnold Sommerfeld, and Albert Einstein. On the subject see (Sallent Del Colombo and Roca i Rosell 2006).

well, it never appeared. In this connection, Risser wrote to Levi-Civita (30 May 1925): "I am convinced that your book entitled «Questions concerning mechanics and relativity» (Levi-Civita 1922b) would be a very great success here; for my part, I deeply regret that my very imperfect knowledge of the Italian language does not allow me at all an immediate reading, and I admit that I would like to have a translation. Such a translation would have the advantage that all French teachers could profit of your remarkable work on hydrodynamics and mechanics of relativity..."[43]

Another book on theoretical hydrodynamics was, however, translated in French. This treatise is a course held by Levi-Civita at the Mathematical Seminar of the University of Rome on the theory of characteristics during the academic year 1930–31 (Levi-Civita 1931). These lectures, drafted by one of Levi-Civita's students Giovanni Lampariello (1903–1964), provide an account of the theory of differential equations, and focus on the problem of Cauchy and its exceptional cases naturally leading to the idea of characteristics. Levi-Civita analyses characteristics in relation to waves of discontinuity, and elucidates some compatibility conditions at the wave front. In special cases, Levi-Civita proves the existence theorems and then introduces the concept of "characteristic manifold" by following Hadamard's theory and language.[44] The theory is well developed and illustrated by examples from hydrodynamics, optics and from the theory of de Broglie waves.

It was Marcel Brelot (1903–1987), *normalien* and then Rockefeller student of Vito Volterra in Rome (1929–30), who proposed the French translation of his treatise on the theory of characteristics to Levi-Civita. In a letter, dated Berlin, December 26th 1931, Brelot informed Levi-Civita that a scientific commission recommended his book (Levi-Civita 1931) for a French translation.[45] In his letter he also mentioned that Mussolini was aware of the decision and that a similar Italian scientific commission had been created. At the end of his letter, Brelot wrote about "a strange decision of the Italian government, which of course leaves all astonished abroad [...]" Apparently he referred to the loyalty oath to the Fascist regime,

[43]"Je suis convaincu que votre traité intitulé «Questions touchant à la mécanique et à la relativité» aurait ici le plus grand succès; pour ma part je regrette profondément que ma connaissance très imparfaite de la langue italienne ne m'en permette point une lecture immédiate et j'avoue que je serais ravi d'en avoir une traduction. Une semblable traduction aurait l'avantage de faire bénéficier tous les professeurs français de votre connaissance si remarquable de l'Hydrodynamique et de la mécanique de la Relativité...".

[44]For details on Hadamard's theory see the chapter "Lines on the horizon. Hadamard and Fréchet, readers of Volterra" by Guerraggio, Jaëck, Mazliak in this book.

[45]More in detail, in his letter Brelot wrote: "Hanno creato, poco fa, una commissione di scienziati in Francia per scegliere le opere francesi e straniere degne di essere tradotte in diverse lingue; il signor Paul Gaultier direttore della "Revue Scientifique" si occupa dell'impressione. E' andato poco fa a Roma, ha parlato col Mussolini e una commissione italiana è stata anche scelta nello stesso scopo; forse che Lei lo sa? Il signor Gaultier sta questi giorni a Berlino per occuparsi della stessa cosa in Germania, e riunire scienziati di tutti i campi. E' venuto a vedermi questa mattina perché le commissioni italiana e francese desiderano far tradurre la sua opera "Caratteristiche dei sistemi differenziali e propagazione ondosa", e per questa traduzione in francese, il Prof. Picard mi ha proposto." The letter is contained in Fondo Levi-Civita.

instituted in that period and that all university professors were required to take. We remark that Volterra had to retire as a consequence of his refusal.

4.2 Jets

Levi-Civita published just one contribution not related to wave theory during the long period he spent at the University of Rome—it is a lecture held at the mathematical and physical Seminar of Milan in 1931, entitled "On liquid jets". He gave the same lecture at the Institute of fluid mechanics of Paris and published it in 1932 in French translation in the *Journal de mathématiques pures et appliquées*, directed by Villat, and in Polish translation in 1933. We shall analyse the content of that memoir (Levi-Civita 1932) in order to discuss, once more, Levi-Civita's approach to hydrodynamics. We shall see that Levi-Civita was very interested in applying his results to hydraulics, and in deducing concrete results from a theory founded on solid mathematical bases. However, actual issues in hydrodynamics generally lead to very complicated mathematical problems and, therefore, simplified assumptions are required in order to find mathematical solutions. Of course, such assumptions do not generally describe concrete problems—that is quite natural in mathematical physics!

In his 1932 paper Levi-Civita writes: "Eminent specialists in hydraulics and physicists have been doing research on experimental study of liquid jets for a long time", and quotes the works by Borda, Bidone, Bazin, Plateau, Tyndall, Rayleigh, and others. However, a mathematical theory of this phenomenon has not been elaborated yet and that is the aim of the paper. (Levi-Civita 1932; 37) In the conclusive section, Levi-Civita discusses his theoretical results and relates them to experiments done in the hydraulic laboratory of the engineering school of Rome (Levi-Civita 1932; 55–56).

In his paper, Levi-Civita deals with the theory of jets in 3 dimensions, and deduces results in agreement with experimental data. Once more, Levi-Civita tries to found hydrodynamic principles on rigorous mathematical bases. Before him, fluid particles were considered as independent one of the other, and fluid pressure was supposed constant during the motion (it is called *free regime*). Nevertheless, such assumptions are true only for stationary motion. In the case of thin jets, the internal pressure is much higher than atmospheric pressure (rather than equal as in free regime). That is the *linear regime* analysed in Levi-Civita's paper by using an analogy between hydrodynamics and mechanics—the liquid jet is indeed interpreted as a flexible and inextensible wire; the pressure of the jet as the pressure of the wire; the continuity equation as an equation between the derivatives of the wire tensions, and so on. Therefore, he deduces the hydrodynamic equations of the jet from the corresponding differential equations of classical mechanics. These equations, in the case of a stationary regime, lead to the following statement: "For usual heavy jets [...], jets take the form of a catenary with the concavity downwards (while a heavy wire in equilibrium turns its concavity upward)" (Levi-Civita 1932; 193).

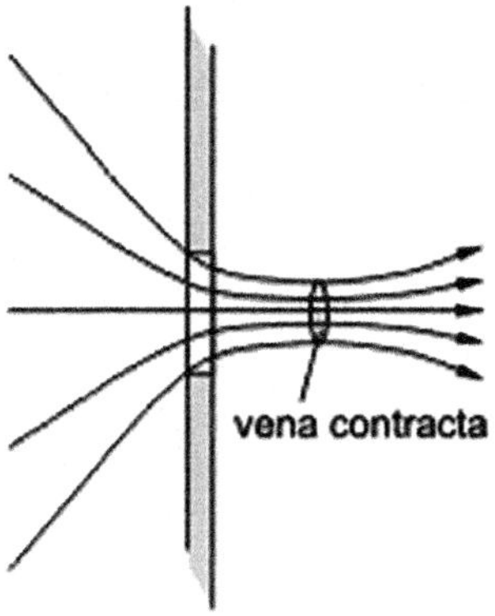

Fig. 2 The external conoid in the figure is called the «vena contracta» and its least section the section of the vena contracta

Already in a paper published in 1905, Levi-Civita (Levi-Civita 1905) dealt with two-dimensional jets, called "vena contracta". As Weinstein wrote in his commemorative memoir on Levi-Civita's contribution to hydrodynamics: "More than a century ago and only a few years before T. Levi-Civita was born, H. von Helmholtz gave the first mathematical example of a two-dimensional *jet*, the so-called vena contracta." And he added: "However, it was only in 1905–1908 that fundamental progress was made by Levi-Civita". (Weinstein 1973; 269) Progress concerns mathematical rigour especially.

In modern terms, "vena contracta" denotes the reduction in the area/diameter of a fluid jet after it emerges from an orifice in a tank (see Fig. 2). The maximum contraction takes place at a section slightly downstream of the orifice, where the jet is more or less horizontal.

In his note, Levi-Civita remarked: "The phenomenon of vena contracta has been the subject of experimental studies that, at least for the needs of hydraulics, can be said to be exhaustive. Not so advanced is the theoretical study". He just quoted the classic papers by Helmholtz and Stokes, which gave the complete theory of two-dimensional jets issuing from "long enough" nozzles. But what happens if the nozzle has a closed border with shape, for example, circular or square?

Levi-Civita considered the coefficient of contraction, defined as the ratio between the area of the jet at the vena contracta to the area of the orifice. In equation:

$$C_c = \text{Area at vena contracta}/\text{Area of orifice}$$

Until then, in fact, physicists had only recognised that the coefficient of contraction for ordinary nozzles (independent of what happens downstream) is always > ½, and becomes very close to ½ for orifices with a cylindrical mouth (Borda's tube). It is possible, and it is the purpose of Levi-Civita's work, to deduce a general and rigorous formula leading to these statements and to prove that the coefficient of contraction is less than ½ if the mouth of the internal orifice is divergent (Fig. 3).

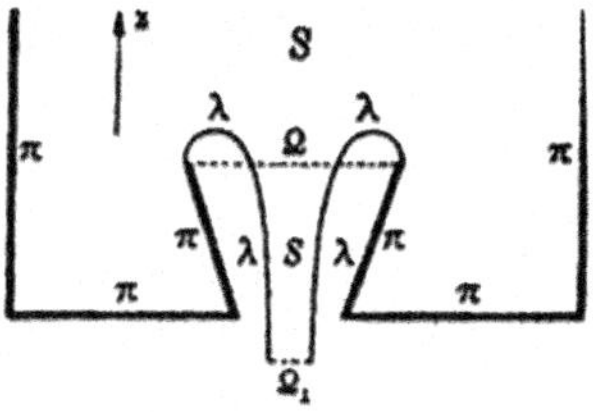

Fig. 3 In the figure π is the plane boundary, Ω is the orifice, Ω_1 is the *contracta* section, λ is the vein surface between the orifice Ω and Ω_1, S the whole space filled by the fluid that has as boundary $\Omega_1 + \lambda + \pi$ (in Levi-Civita 1905; 461)

Nevertheless, Levi-Civita was forced to introduce some hypotheses: in the tank S the fluid motion is irrotational and permanent (then the velocity potential is submitted to some conditions) and no external forces act. Moreover, for experimental confirmations he refers to some specialists of hydraulics at the University of Rome and Padua, in particular to his colleague Giacinto Turazza (1853–1925),[46] the director of the *Gabinetto di Idraulica* in Padua.

Some years later, one of Levi-Civita's students, Cisotti, improved the theory of two-dimensional jets by using the method of conformal representation developed by Levi-Civita in his article on the wake (Levi-Civita 1907a), briefly described in Sect. 3.1. In his book on two-dimensional hydrodynamics (Cisotti, 1921–1922, II), Cisotti applies Levi-Civita's method in order to deduce the general solution of irrotational two-dimensional motions of a fluid passing through an orifice. Cisotti highlights that any analytic function regular in a circle, which is real on the real axis and remains finite and continuous on the circumference, corresponds to a motion analytically possible (the shape of the walls is deduced a posteriori); however, further conditions are necessary if the motion is required to be *physically* possible. The duality between mathematical and physical solution often arises in works of this period. Even in his 1907 paper (Levi-Civita 1907a) Levi-Civita found solutions that are mathematically possible but that, by a closer examination, are not admissible from the physical point of view. M. Brillouin highlighted this aspect (Brillouin 1911), as we pointed out in Sect. 3.2.

Still devoted to the principles of hydrodynamics is the article on Torricelli's theorem that Levi-Civita published on the *Comptes Rendus* of the Academy of Sciences of Paris in 1913. The paper aims at extending Torricelli's theorem to more

[46]His father was Domenico Turazza (1813–1892), professor of geodesy and hydrometry at the University of Padua from 1842. Domenico Turazza founded the *Scuola di Applicazione per gli Ingegneri* in Padua that later became the Faculty of Engineering. Author of a famous treatise on hydraulics, D. Turazza wrote in the Preface to his book that hydraulics is "a pure experimental science", and pointed out that his book is not "a mathematical work" but it concerns "applicable experimental hydraulics" addressed "to engineers" (Turazza 1867; VI–VII). On the attitude of Italian engineers and mathematicians towards rational hydraulics in the 19th century see (Garibaldi 1994).

general case: "The speed of the jet outflowing through a small nozzle is expressed, by using Torricelli's theorem, in the form

$$(*)\, v^2 = gh$$

h being the level of the orifice below the free surface of the liquid." (Levi-Civita 1913; 481) As Levi-Civita claimed, the classic proof concerns the stationary regime only. "I am not aware that the validity of (*) has been pointed out when the motion starts, that is at the moment in which the fluid begins to outflow through a nozzle suddenly opened in the wall of a tank containing a liquid at rest." (Levi-Civita 1913; 481) By applying the energy theorem only, Levi-Civita is able to prove that formula (*) defines "in a rigorous way" the initial velocity for each element dW, where W is the width of the orifice.

Levi-Civita's ideas on jet theory influenced several scholars in Italy and abroad. Here, we shall consider the case of the French Phd-student, Robert Mazet.

Robert Mazet (1903–1991) graduated from the ENS of Paris and was a student of Villat and Painlevé. He spent two years at the University of Rome (from 1925 to 1927) with Levi-Civita and Volterra.

Both Painlevé and Vessiot recommended Mazet to Levi-Civita. Painlevé wrote to Levi-Civita that "his student" Mazet knew very well "such questions" of fluid mechanics, since he had attended his lectures on fluid resistance at the Sorbonne and drafted them for publication (Painlevé 1930).[47] Vessiot, the director of the ENS, recommended Mazet as one of his *normaliens*. Vessiot remarked in his letter to Levi-Civita (August 8, 1925) that though Mazet's subject for his thesis was not yet "delimitated", he had attended Painlevé's lectures on fluid dynamics and read the memoirs of Villat and Thiry.[48] Finally, in 1925 Mazet left for Rome with a David-Weill fellowship, and the year later he was happy to come back to Rome as a Rockefeller fellow.[49] Levi-Civita supervised Mazet's research. From 1926 to 1928, Mazet published many notes on issues related to the theory of two-dimensional jets

[47]On October 17th, 1925, Painlevé wrote to Levi-Civita: "Je recommande tout particulièrement à votre bon accueil mon élève, Monsieur MAZET, qui se rend en Italie pour préparer une thèse sur la mécanique des fluides. J'ai chargé Monsieur Mazet de rédiger et publier le cours que j'ai fait l'année dernière sur la mécanique des fluides. Il connaît très bien ces questions, et sera à même de comprendre et d'appliquer les conseils que vous pourriez lui donner." The letter is in Fondo Levi-Civita.

[48]In particular, Vessiot wrote to Levi-Civita: "M. *Mazet* n'a pas encore un sujet aussi délimité; mais il a déjà bien étudié la dynamique des fluides et surtout le problème du sillage; il a lu, en particulier les mémoires de M. Villat et de M. Thiry. Il vient, d'autre part de rédiger les leçons d'aérodynamique que M. Painlevé a professé cette année à la Sorbonne." The letter is in Fondo Levi-Civita.

[49]In his letter to Levi-Civita (dated August 10, 1926), Mazet wrote: "Je vais soumettre l'ensemble de mon travail à M. Villat pour savoir si ce que j'ai fait est suffisant pour ma thèse. L'International Education Board m'a avisé récemment qu'une bourse Rockefeller m'était accordée. J'aurai donc le plaisir de retourner à Rome au mois de novembre". The letter is in Fondo Levi-Civita. For more details on French students at the University of Rome see (Mazliak 2015).

in the *Comptes Rendus* of the French Academy of Sciences and in the *Rendiconti* of the Academy of Lincei, communicated by Levi-Civita.

Mazet defended his PhD thesis at the Sorbonne in 1929. In his thesis (Mazet 1929) he collected and developed the content of his published notes. Mazet mainly investigated two-dimensional flows of a liquid through an orifice, oscillations of a fluid in communicating vessels, and motions of a permanent liquid under the action of gravity. He aimed to extend some results obtained by Levi-Civita and his students to more general cases. In particular, in the second part of his dissertation, he intended to focus on the following "unsolved" problem Levi-Civita suggested to him—to determine the two-dimensional motion of a fluid in vacumm, submitted to gravity, variable, and reaching stationary state in some time (Mazet 1929; 3–4). However, as Mazet confessed, such a problem presents mathematical difficulties he was not able to overcome. As he explained in his PhD thesis, some discussions with Brillouin, Vergne and Villat convinced him to simplify the original problem and to introduce a further hypothesis, by considering a horizontal circular nozzle opening progressively in time. In his work, Mazet applied Levi-Civita's results on two-dimensional jets, and used his "fruitful" method of conformal transformation as developed by Cisotti in his treatise (1921–1922).

As it used to be in the case of France at that time, Mazet had to print his PhD thesis before defending it. Nevertheless, the impression was generally very expensive. Young PhD students tried to publish their works on scientific journals that, however, did not meet all demands and hardly accepted very long memoirs for publication. A good solution was to publish in the *Rendiconti del Circolo matematico di Palermo*, an international and well-reputed journal. By following, once more, the suggestion of Vessiot and Villat, Mazet asked Levi-Civita for publishing his memoir in the *Rendiconti* of Palermo. Thanks to the intercession of Levi-Civita, his request was accepted and his work was actually published in volume 53 of the journal.[50]

Finally, Mazet defended his thesis in 1929—Painlevé, Vessiot, and Villat (*rapporteur*) belonged to his PhD committee. Anyway, as already remarked, Mazet worked on a subject proposed by Levi-Civita and then developed by his students, especially by Cisotti. Mazet went on researching on hydrodynamics, in particular on fluid friction, during his scientific career. He soon involved himself in the creation of new institutions related to practical mechanics and applied hydrodynamics. Mazet spent most of his career at the University of Lille, where he became assistant professor in 1932 and professor of mechanics in 1936; the same year he was appointed "director of studies" at the *Institut Industriel du Nord*. Very involved in local scientific life, in 1934 Mazet founded a *Laboratoire de Mécanique Expérimentale* related to the University of Lille, the first in France, aiming at

[50]Mazet wrote to Levi-Civita on December 6th, 1927: "... Je viens recevoir le permis d'imprimer [ma thèse] et je pourrai soutenir ma thèse dès que l'impression sera terminée. Comme celle-ci entraîne toujours à des gros frais, M. Vessiot et M. Villat m'ont suggéré de m'adresser à une publication scientifique susceptible d'accueillir mon manuscrit. [...] J'ai pensé que je serais peut-être plus favorisé en m'adressant à un recueil italien tel que les Rendiconti del Circolo di Palermo qui ont déjà publié plusieurs thèse françaises". The letter is in Fondo Levi-Civita.

practising students in experimental mechanics and making them able to put concrete problems into equations. This Laboratory was highly appreciated[51] and had vocation to become a great institution of applied mechanics involving local industries.[52] Since 1947 Mazet was director of the ONERA (*Office National d'Etudes et de Recherches Aérospatiales*) of Lille, succeeding to Joseph Kampé de Fériet (1893–1982).

Levi-Civita was then attentive to applications and was proud to show that his theoretical results were in agreement with practical problems of hydraulics. We mentioned the case of Mazet, who was forced to give up the original and more general subject of his thesis, proposed by Levi-Civita, and finally reduced it to a case easier to solve mathematically. It was Villat himself who suggested the new version of the problem to him. Villat, Levi-Civita and the other *mathematicians* working on hydrodynamics in Italy and France shared a similar approach to hydrodynamics—to simplify the physical problem in order to find solutions without renouncing mathematical rigour.

5 Concluding Remarks

In this section we focus on some crucial aspects, which explain why Levi-Civita became a reference point in the field of hydrodynamics for French students and colleagues. Finally, we put in evidence the historiographical elements highlighted by our analysis.

5.1 Levi-Civita as a *Master*

We have shown that Levi-Civita's ideas influenced Villat's thesis and the work of many French scholars. He was a Master for young French students, eager to improve their mathematical education in Rome. We add that Levi-Civita played an important role for the publication strategy of French students. In fact, several PhD students published (in French) their thesis in *Rendiconti* of the *Circolo Matematico di Palermo* also thanks to his intercession with the *Circolo*. In addition, during their stay in Italy and even later, young French mathematicians regularly published brief notes in *Rendiconti* of the Academy of the Lincei, often communicated by Levi-Civita.

[51]For instance Paul Montel, in a letter addressed to the Dean of the Faculty of Sciences of Lille University (dated November 8, 1935), put in evidence the importance of Mazet's Laboratoire de Mécanique Expérimentale in France. In particular he wrote: "Enfin j'ai proposé aussi pour une promotion Mazet. […] Il a realise ici une organisation unique en France de travaux pratiques de Mécanique Rationnelle qui fait d'ailleurs l'objet d'un ouvrage en cours de publication." The letter is in Dossier Mazet, ASA, Université de Lille 1.

[52]See the "Rapport de la commission chargée d'examiner les conséquences pour la Faculté de la nomination à Paris de M. Fleury", in Dossier Mazet, ASA, Université de Lille 1.

Moreover, our transnational approach allows us to observe that ENS students could count not only on their supervisor (*patron*)—currently Villat for hydrodynamics—but also on other scholars (internal or external to ENS) with whom they regularly discussed or exchanged letters, as in the case of Levi-Civita. Let us give some examples. Jacotin-Dubreil's supervisor was Villat; nevertheless, she decided by herself the subject of her thesis by reading Levi-Civita's paper on wave theory. Levi-Civita suggested the subject of the thesis to Mazet, though Villat was his supervisor. Mazet mentioned Volterra, Levi-Civita, but also Villat, Painlevé, and Vessiot as supervisors. Moreover, he referred to his essential discussions with Vergne and Brillouin, though none of them belonged to his PhD scientific committee (*jury*). Mazet, and other PhD-students of ENS, accomplished their work by attending lectures, discussing and corresponding with many researchers.

Finally, we point out that Levi-Civita's activity as a Master is not limited to an institutional engagement. In fact, he not only supervised several PhD theses or Rockefeller research, but he also had exchanges with students and colleagues, who had never studied with him and to whom he gave his help. For instance, we mention the case of Dimitri Pavlovitch Riabouchinski (1882–1962), a Russian student of Jukowski, who had founded in 1904 in Kotchino (Russia) an aerodynamics laboratory of which he was director. In 1919 he moved to Paris where he met Painlevé, Villat, and other experts of aerodynamics. Riabouchinski asked Levi-Civita to support him in order to obtain a Rockefeller fellowship. Also thanks to Levi-Civita's help, he got the grant and went to Oxford with Augustus Edward Hough Love (1863–1940).[53] The following year, he asked Levi-Civita for another testimonial in order to obtain the renewal of his Rockefeller fellowship.[54]

5.2 Levi-Civita's Activity in France

Levi-Civita strongly participated in scientific activities in France—he published French translations of his treatises on hydrodynamics, accepted invitations to publish papers in French journals or books, and gave lectures at the Institute of fluid mechanics of Villat and in other French institutions. Corresponding member of the Academy of Sciences of Paris since 1911, Levi-Civita published in the *Comptes Rendus* of the French Academy, but also in the *Journal des Mathématiques pures et*

[53]On March 6th, 1926 Riabouchinski wrote to Levi-Civita: "Permettez moi de vous entretenir d'une question personnelle. Le Prof. Paul Painlevé a bien voulu me présenter comme candidat à une bourse de l'International Education Board fondé par Rockefeller. Cette bourse me permettrait de continuer pendant un an mes recherches théoriques et expérimentales et, peut-être, d'obtenir ensuite un Laboratoire. La proposition de M. Painlevé est secondé par le Prof. Love de Oxford." The letters by Riabouchinski to Levi-Civita are contained in Fondo Levi-Civita.

[54]Riabouchinski wrote to Levi-Civita on August 10th, 1927: "Ne m'en voulez pas trop si je prends la liberté de vous prier aussi à me faire l'honneur de me donner encore une fois l'appui de votre grand nom, dont le poids, comme j'ai eu encore maintes fois m'en convaincre en Angleterre, est très grand, en insistant une fois de plus que vous considérez si je puis être utile comme directeur d'un Laboratoire d'Aérodynamique ou comme research-professor.".

appliquées, accepting the invitations of its director Villat. For instance, the paper on liquid jets, which summarized a lecture held at the Institute of fluid mechanics, was published in *Journal des Mathématiques* in 1932.

Just after his appointment as director of the Institute of Fluid Mechanics on November 22nd, 1930, Villat asked Levi-Civita to give lectures on "a topic of your choice concerning your most recent research" (Nastasi and Tazzioli 2003; 393). In the following years, Villat would renew his request. In Villat's Institute several students and colleagues could meet the Italian Master and discuss with him. Levi-Civita was pleased to accept Villat's invitations, although in the thirties the fascist regime required a special authorization from the government upon submission of documentation. A reference to this fact is for example in an undated letter—but very probably drafted about 1930—where Levi-Civita asked Villat for a "diplomatic invitation" addressed to the Italian Ministry of Foreign Affairs. However, Levi-Civita added, permission was usually granted even to those who are not "in the good graces of the regime" (in Nastasi and Tazzioli 2003; 394); the reference to himself is implicit.

Villat got Levi-Civita's support for all his most important enterprises. In particular, Villat asked Levi-Civita for his collaboration in his *Mémorial des Sciences mathématiques*. On June 5th, 1925, he wrote to Levi-Civita: "Thank you very much for your kind letter and your cordial wishes you sent me for the Mémorial des Sciences Mathématiques. I am particularly pleased with your high approval, which evidence to me our current initiative responded well to a need, and will render service to mathematical public. Your authoritative opinion reinforces my conviction." (Nastasi and Tazzioli 2003; 392) In 1934 Levi-Civita accepted to write a volume of the *Mémorial* on the relativistic two-body problem, after preparing the result "in the paper devoted to Brillouin" and detailing it in "following lectures". For the moment, he preferred not to precise when the volume should be ready.[55] His volume, actually, was to be published posthumously, in 1950 (Levi-Civita 1950), because of criticism moved by American scholars—especially by Howard Percy Robertson (1903–1961)—against crucial points of his theory.[56] The above-mentioned letter refers to the book for Brillouin's Jubilee to which Levi-Civita contributed with a paper on the relativistic two-body problem (Levi-Civita 1935). Grateful, Villat wrote him on January 31st, 1935: "Your manuscript will be one of the jewels of the book to honour Mr. Marcel Brillouin, and I read it with intense pleasure." (Nastasi and Tazzioli 2003; 403).

5.3 Hydrodynamics in the Inter-War Period in France

Historiography has often claimed the preponderance of pure mathematics (especially analysis) over applied mathematics during the inter-war period. As Gispert

[55]See the letter by Levi-Civita to Villat on November 23, 1934, in (Nastasi and Tazzioli 2003; 403).

[56]For details about this history see (Nastasi and Tazzioli 2006; 207–210).

and Leloup argue, oral and written testimonies on mathematics in the period between the two world wars concern almost exclusively members of the Bourbakist movement.[57] That can explain, at least partially, the historiographical distortion of French mathematics. In fact, such an image should be nuanced; several mathematicians—among whom some specialists of pure mathematics such as Maurice Fréchet—began to work on the theory of probability from the early years of the 20th century.[58] As we have seen, in the twenties and thirties, hydrodynamics also seems a very dynamic research field thanks to the work by M. Brillouin, Villat, Painlevé and their students.

The strong scientific and institutional interests of Paul Painlevé in aviation certainly contributed to the development of hydrodynamic studies in France. In 1895 Painlevé gave a course on friction, a topic related to aerodynamics (Painlevé 1895), and, moreover, was actively involved in the first research on mechanical flight and on the propulsion of aeroplanes.[59] A professor at the Sorbonne, Painlevé played an important role in supporting mathematicians, such as Villat, leading major research centres of hydrodynamics and aerodynamics, and in giving impetus to mathematical research on such topics. Villat, who was Painlevé's successor in the chair of fluid mechanics, supervised 21 theses on fluid mechanics defended at the Sorbonne between 1914 and 1945,[60] and was one of the most influential scholars in France in the inter-war period.

5.4 Mathematical Methods in Hydrodynamics

The fact that mathematical methods are the only prerequisite for concrete applications underlies the contribution of Levi-Civita and Villat to hydrodynamics. We argue that the work of Levi-Civita's school was relevant for the development of mathematical hydrodynamics in France, i.e. hydrodynamics based on mathematical principles.

Concerning the judgement on "the Italians" expressed by Brillouin and Boussinesq and discussed in Sect. 3, we think that their criticism should be downsized. In fact, both appreciated the Italian contribution to hydrodynamics, as evidenced by the fact that in 1909 and 1910 Brillouin focused his courses at the *Collège de France* on the memoirs by Levi-Civita, Picciati and Boggio, while Boussinesq recommended Villat to read carefully the results on fluid dynamics published by Levi-Civita and Almansi.

Brillouin and Boussinesq feared that a too theoretical point of view could undermine the discipline and warned Villat of that danger. On September 9th, 1919,

[57]For bibliography on Bourbakist historiography see (Leloup 2009; 21), (Gispert and Leloup 2009; 41).

[58]See for instance (Mazliak 2010).

[59]See the biography of Paul Painlevé (Anizan 2012). For Painlevé's engagement in aviation see (Fontanon 2005).

[60]See (Gispert and Leloup 2009; 92).

when Villat was appointed professor at the University of Strasburg, Brillouin congratulated Villat, who had been able "to forge […] the mathematical instrument needed to approach the physical problem of hydrodynamics little by little."[61] Villat's courses at the *École Superieure d'Aéronautique* had the same purpose, i.e. to strengthen the theory in view of applications, and similarly the works of Villat's students, as in the case of Thiry, Mazet, Jacotin-Dubreil.

In his course on fluid mechanics held at the same school of aeronautics (Villat 1930), Villat focused more on theorems than on applications—he dealt with Levi-Civita's method of conformal representation for wake theory, the theorem of Kutta-Jukowski and its generalizations, and the theory of Prandtl. This course is the same as the one held at the Sorbonne. By using the words of the reviewer, Adolphe Buhl: "Nothing is more natural […] that Mr. Villat does not speak at the *École d'Aéronautique* a language essentially different from that spoken at the Sorbonne" And he added that "the author certainly wanted to put [in his lectures] only things that are aesthetically pleasing and relatively elementary" (Buhl 1930; 361).

The second edition of Villat's treatise appeared in 1938. Buhl's criticism is even harsher: "it must be remembered that the author, who has an eminently elegant spirit, has always presented his topics in aesthetic forms, which sometimes expressed regret concerning their non complete agreement with experimental verifications" (Buhl 1938; 227). Buhl concludes his review by mentioning *inter alia* the *École de Toulouse* of physics, headed by Charles Camichel, whose experiences have made great contributions to hydrodynamics; then, he hopes that is the path that we will follow in future.

Similarly to Villat, in his lectures at the *École Supérieur d'Aéronautique* held in the years 1925–27, Painlevé underlines that "we tried to push as far as possible the theoretical solution of this great problem." (Painlevé 1930; 2) Apparently, theoretical hydrodynamics represents for Painlevé the solid foundation for any actual applications; that is why he mainly deals with the "rigorous" analysis of the motion of a non-deformable body in a perfect fluid.

Therefore, the polemics between mathematical hydrodynamics and experimental hydrodynamics does not involve "the Italians" on one side and French scholars on the other; on the contrary, Levi-Civita, Villat, Painlevé and their students agree on the essential points. The controversy seems, rather, much more extensive. On the one hand, there were hydraulic engineers and experimental physicists who were proponents of a highly experimental hydrodynamics (in Villat's words, "rudimentary hydrodynamics"), and, on the other hand, the supporters of a "mathematical" or "rational hydrodynamics" founded on rigorous mathematical bases but not always in agreement with experimental data.[62] In 1929 Paolo Straneo, an Italian mathematician interested in aerodynamics, wrote that "the flight by aircraft has lead

[61]"chercher […] à forger l'instrument mathématique nécessaire pour approcher peu à peu le problème physique de l'hydrodynamique".

[62]On the difficulty between mathematical hydrodynamics and practical problems of fluid mechanics see (Darrigol 2008).

several specialists of fluid dynamics to revise their discipline in order to better understand the intolerable gaps between reality [and theory] [...] Therefore, many older theories accepted should be amended." (Straneo 1929; 298) We point out that Levi-Civita did not actively work in a laboratory, but he was proud to show that his theoretical results are in agreement with experimental data found in some hydraulic laboratories (see Sect. 4). Villat did not mention any physical experiments explicitly—even when he dealt with "applied hydrodynamics", he did not go out of the field of mathematical physics.

We conclude by quoting a letter by Theodore von Kármán (1881–1963) to Jerome Clarke Hunsaker (1886–1984),[63] an organiser of the International Congress of Applied Mechanics in Cambridge (Mass.) in 1938: "Concerning Levi-Civita's recommendation of French representatives of Applied Mechanics, I agree that Villat and Pérès are excellent men; however, Villat is far beyond the line of what we would call the frontier of useful or applied mechanics. I really believe that the man we could use best for a general lecture is Kampé de Fériet, director of the Institute for fluid mechanics, in Lille. In the last years, he published two reviews on recent progress concerning waves and turbulence. Both reports were excellent and just on the limit between the practical and theoretical viewpoint as we like it. Besides that, he follows the experimental research, whereas Villat, in spite of the fact that he is director of an experimental institution, has no idea of experimental questions".

This letter shows once more the sharp contrast between Villat's theoretical research and his institutional position as director of the Institute of Fluid Mechanics; a contradiction that did not escape his physicist and engineer colleagues even in France. Indeed, Villat's name does not appear in the *Proceedings* of the Fifth International Congress of Applied Mechanics, while Pérès gave one of the three general lectures on "the methods of analogies in applied mechanics" (Pérès 1939). In the same Congress, Kampé de Fériet lectured on his recent research about turbulence (Kampé de Fériet 1939). Furthermore, except for Jules Drach (1871–1949) and Henri Beghin (1876–1969) who were academic-mathematicians, the other French lecturers at the Cambridge Congress were "engineers" working in technical laboratories and teaching in various Engineering Schools.[64] In the forties, various research institutions in aerodynamics appeared even in France, where mathematicians, physicists and engineers worked together.[65]

[63]The letter dated 2 March 1937 is contained in the von Karman Papers, box 47, folder 3, California Institute of Technology. I thank Giovanni Battimelli for giving me the content of this letter.

[64]We mention for instance: Georges Darrieus (1888–1979), an electric engineer working in the Compagnie Électro-Mécanique; Henri de Leiris (1903–1991), a General of the Marine Engineering; Albert Métral (1902–1962), polytechnicien, both teacher and engineer working in private industries; Louis Bergeron (1876–1948), a hydraulic engineer and professor at the École Centrale, École Supérieure d'Électricité, and École de Physique et Chimie.

[65]See (Weber 2008).

References

AAVV (1975) Tullio Levi-Civita. Convegno internazionale celebrativo del centenario della nascita (Roma, 17–19 dicembre 1973) (ed. by Accademia Nazionale dei Lincei). Accademia Nazionale dei Lincei, Roma.

Almansi E (1909–1910) Azione esercitata da una massa liquida in moto sopra un corpo fisso (5 Notes). Rendiconti della R. Accademia dei Lincei (5) 18: 587–594; 19: 56–63; 19: 116–118; 19: 244–260; 19: 437–443.

Amaldi U, Levi-Civita T (1935) Teoria matematica dei fenomeni ondosi. Enciclopedia Italiana 25: 356–362; in Levi-Civita's Opere V: 493–518.

Anderson J D, Jr, A History of Aerodynamics, Cambridge University Press, 1997.

Anizan AL (2012) Paul Painlevé. Political Science et de la Belle Epoque aux années trente. PUR, Rennes.

Archibal T, Tazzioli R (2014) Integral equations between theory and practice: the cases of Italy and France to 1920. Archive for history of exact sciences 68: 547–597.

Aubin D (2010) "Audacity or Precision": The Paradoxes of Henri Villat's Fluid Mechanics in Interwar France. Proceedings of the Workshop on the History of Fluid Mechanics, Rauischholzhauschen, 15–18 October 2006. [preprint; on hal-upmc]

Battimelli G (1988) The Early International Congresses of Applied Mechanics, in IUTAM: A Short History, ed. by S. Juhasz, Springer, New York, pp. 9–13.

Boggio T (1910a) Sul moto permanente di un solido in un fluido indefinito. Atti del R. Istituto Veneto 69: 883–891.

Boggio T (1910b) Sul moto stazionario di una sfera in un liquido viscoso. Rendiconti del Circolo matematico di Palermo 50: 65–81.

Brillouin M (1907) Leçons sur la viscosité des liquides et des gaz. Gauthier-Villars, Paris.

Brillouin M (1911) Les surfaces de glissement d'Helmholtz et la résistance des fluides. Annales de Chimie et Physique (8) 23: 145–230.

Buhl A (1930) Report on «Mécanique de Fluides» by H. Villat (Gauthie-Villar, Paris, 1930). L'Enseignement Mathématique, 29: 360–361.

Buhl A (1938) Review: Henri Villat. Mécanique des Fluides. Cours de l'Ecole nationale supérieure d'Aéronautique. Deuxième edition revue et augmentée. L'Enseignement Mathématique 37: 227–228.

Bustamante, M C, Cléry, M, Mazliak, L (2015), Le Traité de calcul des probabilités et de ses applications. Étendue et limites d'un projet borélien de grande envergure. North-Western European Journal of Mathematics 1: 85–123.

Cisotti U (1905–1906) Sul paradosso di D'Alembert. Atti del R. Istituto Veneto 63: 423–426; 65: 1291–1295.

Cisotti U (1921–1922) Idromeccanica piana, 2 vols. Libreria Editrice Politecnica, Milano.

Crawford E, Olff-Nathan J (2005) (ed. by) La Science sous influence. L'Université de Strasbourg, enjeu des conflits franco-allemands (1872–1945), Université Louis Pasteur et La Nuée Bleue.

Darrigol O (2005) Worlds of Flow. A History of Hydrodynamics from the Bernoullis to Prandtl. Oxford University Press.

Darrigol O (2008) Empirical Challenges and Concept Formation in the History of Hydrodynamics. Centaurus 50: 214–232.

Dubreil P (1983) Souvenirs d'un boursier Rockefeller 1929–1931. Cahiers du séminaire d'histoire des mathématiques 4: 61–73.

Dubreil-Jacotin M-L (1934) Sur la détermination rigoureuse des ondes permanentes périodiques d'ampleur finie. Journal de Mathématiques pures et appliquées (9) 13: 217–291.

Duhem P (1914) Sur le paradoxe hydrodynamique de M. Brillouin. Comptes Rendus hebdomadaires des séances de l'Académie des Sciences 159: 790–792.

Fontanon C, Frank R (2005) (by) Paul Painlevé (1863–1933). Un savant en politique. PUR, Rennes.

Fontanon C (2005) Paul Painlevé et l'aviation; in (Fontanon, Franck, 2005; 41–56).

Garibaldi A C, I contributi di Giusto Bellavitis e di Domenico Turazza alla discussion e sull'idraulica razionale in Italia nella prima metà del secolo XIX; in Le scienze matematiche nel Veneto dell'Ottocento. Istituto Veneto di Scienze, Lettere ed Arti, Venezia, 1994, pp. 163–197.

Gispert H, Leloup J (2009) Des patrons des mathématiques en France dans l'entre-deux-guerres. Revue d'histoire des sciences 62: 39–117.

Hadamard J (1903) Leçons sur la propagation des ondes et les équations de l'hydrodynamique. Hermann, Paris.

Kampé de Fériet J (1939) Some Recent Researches on Turbolence. In Proceedings of the Fifth International Congress for Applied Mechanics (ed. by J.P. den Hartog, H. Peters. John Wiley, New York and Chapman & Hall, London, 1939), 352–355.

Lamb H (1932) Hydrodynamics. Cambridge University Press, Cambridge (6th edition).

Leloup J (2009) L'entre-deux-guerre mathématique à travers les thèses soutenues en France. Thèse. Institut Mathématique de Jussieu, Paris.

Levi-Civita T (1901) Sulla resistenza dei mezzi fluidi. Rendiconti dell'Accademia dei Lincei (5) 10: 3–9; in (Levi-Civita 1954–73 II; 129–135).

Levi-Civita T (1905) Sulla contrazione delle vene liquide. Atti del R. Istituto Veneto di Scienze Lettere e Arti 64: 1465–1472; in (Levi-Civita 1954–73 II; 459–464).

Levi-Civita T (1907a) Scie e leggi di resistenza. Rendiconti del Circolo matematico di Palermo 23: 1–37; in (Levi-Civita 1954–73 II; 519–562).

Levi-Civita (1907b) Sulle onde progressive di tipo permanente. Rendiconti della R. Accademia dei Lincei (5) 15: 777–799; in (Levi-Civita 1954–73 II; 615–629).

Levi-Civita T (1911) Trasformazione di una relazione funzionale dovuta al Dini. Nota Ie II. Rendiconti della R. Accademia dei Lincei (5) 21: 285–296; 381–391; in (Levi-Civita 1954–73 III; 163–185).

Levi-Civita T (1912) Sulle onde di canale. Rendiconti della R. Accademia dei Lincei (5) 21: 3–14; in (Levi-Civita 1954–73 III; 261–273).

Levi-Civita T (1913) Théorème de Torricelli et début de l'écoulement. Comptes Rendus hebdomadaires des séances de l'Académie des Sciences 157: 481–484; in (Levi-Civita 1954–73 III; 385–388).

Levi-Civita T (1922a) Risoluzione dell'equazione funzionale che caratterizza le onde periodiche in un canale molto profondo. Mathematische Annalen 85: 256–279; in (Levi-Civita 1954–73 IV; 287–314).

Levi-Civita T (1922b) Qüestions de Mecànica clàssica i relativista. Institut d'Estudios Catalans, Barcelona; Italian translation: Zanichelli, Bologna, 1924; German translation, Springer, 1924.

Levi-Civita T (1925a) Détermination rigoureuse des ondes permanentes d'ampleur finie. Mathematische Annalen 93: 264–314; in (Levi-Civita 1954–73 IV; 199–260).

Levi-Civita T (1931) Caratteristiche dei sistemi differenziali e propagazione ondosa (ed. by G. Lampariello). Zanichelli, Bologna.

Levi-Civita T (1932) Sur les jets liquides. Journal de mathématiques pures et appliquées (9) 11: 37–56.

Levi-Civita T (1935) Points matériels et corps célestes en relativité générale. Étude préliminaire; in (Villat, 1935; 203–212).

Levi-Civita T (1950) Le problème de *n* corps en relativité générale. Mémorial des sciences mathématiques, vol. 116. Gauthier-Villar, Paris.

Levi-Civita T (1954–73) Opere matematiche (ed. by Accademia Nazionale dei Lincei), 6 vols. Bologna, Zanichelli.

Marey É (1900) Des mouvemens de l'air lorsqu'il rencontre des surfaces des différentes formes. Comptes Rendus hebdomadaires de séances de l'Académie des Sciences 131: 160–163.

Mazet R (1929) Sur deux mouvements permanents d'un liquide sous l'action de la pesanteur. Rendiconti del Circolo matematico di Palermo 53: 1–78.

Mazliak L (2010) Borel, Fréchet, Darmois. La découverte des statistique par les probabilistes français. Journal électronique d'histoire des probabilités et de la statistique 6, n. 2. http://www.jehps.net/Decembre2010/Mazliak.pdf

Mazliak L (2015) Volterra and the journeys of French students to Italy in the 1910s. Revista Brasileira de História de Matemática 14: 1–30.

Nastasi P, Tazzioli R (2000) Aspetti scientifici e umani nella corrispondenza di Tullio Levi-Civita (1873–1941). Quaderni Pristem, N. 12, Palermo.

Nastasi P, Tazzioli R (2003) Aspetti di meccanica e di meccanica applicata nella corrispondenza di Tullio Levi-Civita (1873–1941). Quaderni Pristem, N. 14, Palermo.

Nastasi P, Tazzioli R (2004) Sulla determinazione della m.ma funzione di Green e questioni connesse, in Supplemento ai Rendiconti del Circolo Matematico di Palermo 74: 71–101.

Nastasi P, Tazzioli R (2005) Towards a scientific and personal biography of Tullio Levi-Civita (1873–1941). Historia Mathematica 32: 203–236.

Nastasi P, Tazzioli R (2006) Problems of method in Levi-Civita's contributions to hydrodynamics. Revue d'histoire des mathématiques 12: 81–118.

Nastasi P, Tazzioli R (2013) I matematici italiani e l'internazionalismo scientifico. La matematica nella Società e nella Cultura. Rivista dell'Unione Matematica Italiana (1), 6: 355–405.

Nastasi P, Tazzioli R (2014) Italian Mathematicians and the First World War: Intellectual Debates and Institutional Innovations; in The Wars of Guns and Mathematics (ed. by D. Aubin & C. Goldstein), AMS/LMS, 2014.

Painlevé P (1895) Leçons sur les frottements. Hermann, Paris.

Painlevé P (1930) Leçons sur la Résistance des Fluides non visqueux, rédigées par A. Métral et R. Mazet. Gauthier-Villars, Paris.

Pérès J (1939) Les méthodes d'analogie en mécanique appliquée. In Proceedings of the Fifth International Congress for Applied Mechanics (ed. by J.P. den Hartog, H. Peters. John Wiley, New York and Chapman & Hall, London, 1939), 9–19.

Picciati G (1907) Integrazione dell' equazione funzionale che regge la caduta di una sfera in un liquido viscoso. Rendiconti della R. Accademia dei Lincei (5) 16: 45–50.

Prandtl L (1905), Ueber Flüssigkeitsbexegung bei sehr kleiner Reibung. Verhandlungen des III Internationalen Mathematiker-Kongress, Heidelberg, 1904. Leipzig, Teubner, 484–491.

Risser R (1925) 1925, Essai sur la théorie des ondes par émersion. Thèse. Paris.

Risser R (1935) Essai sur la théorie des ondes par émersion; in (Villat, 1935; 310–320).

Roy M (1972) Henri Villat. Comptes Rendus hebdomadaires des séances l'Académie des Sciences 274: 127–132.

Sallent Del Colombo, E, Roca i Rosell, A (2006) La visita de Tullio Levi-Civita a Barcelona el gener de 1921. In Actes de la VIII Trobada d'Història de la Ciència i de la Tècnica (ed. by J. Batlló-Ortiz, J. Ferran Boleda, M. Piqueras, SCHCT, IEC, Barcelona), 431–439.

Straneo P. (1929) Dall'aerodinamica classica all'aerodinamica dell'aeronautica. L'Aerotecnica 9: 287–314.

Siegmund-Schultze R (1994) "Scientific control" in mathematical reviewing and German-U.S. American relations between the two world wars. Historia Mathematica 21: 306–329.

Siegmund-Schultze R (2001) Rockefeller and the Internationalization of Mathematics Between the Two World Wars. Birkhäuser.

Struik D (1926) Détermination rigoureuse des ondes irrotationelles périodiques dans un canal à profondeur finie. Mathematische Annalen 95: 595–634.

Thiry R (1921) Sur les solutions multiples des problèmes d'hydrodynamiques relatifs aux mouvements glissants. Thèse, Gauthier-Villars, Paris.

Turazza D (1867) Trattato di Idrometria o di Idraulica pratica. Tipografia Editrice Sacchetto, Padova.

Villat H (1911a) Sur la résistance des fluides. Annales scientifiques de l'Ecole Normale Supérieure (3) 28: 203–311.

Villat H (1911b) Sur le problème de Dirichlet relatif à une couronne circulaire. Comptes Rendus hebdomadaires des séances de l'Académie des Sciences 152: 680–682.

Villat H (1911c) Sur le problème de Dirichlet relatif au cercle. Bulletin de la Société mathématique de France 39: 443–456.

Villat H (1911d) Sur certaines équations intégrales d'un type nouveau, et sur quelques problèmes qui s'y rattachent. Comptes Rendus hebdomadaires des séances de l'Académie des Sciences 153: 758–761.

Villat H (1912) Le problème de Dirichlet dans une aire anulaire. Rendiconti del Circolo Matematico di Palermo 33: 134–175.

Villat H (1914) Sur la détermination des problèmes d'Hydrodynamique relatifs à la résistance des fluides. Annales scientifiques de l'Ecole Normale Supérieure (3) 31: 455–493.

Villat H (1918) Quelques récents progress des theories hydrodynamiques". Bulletin des sciences mathématiques (2) 42: 43–92.

Villat H (1921) Sur certaines équations intégrales possédant une infinité de solutions avec un nombre illimité de paramètres arbitraires. Comptes Rendus hebdomadaires des séances de l'Académie des Sciences 173: 816–818.

Villat H (1923) Sur une équation intégrale singulière et sur un problème de la théorie des tourbillons. Comptes Rendus hebdomadaires des séances de l'Académie des Sciences 177: 11–15.

Villat H (1930) Mécanique des fluides. Gauthier-Villars, Paris; 2nd edition, Gauthier-Villars, Paris, 1938.

Villat H (ed. by) (1935) Jubilé de Marcel Brillouin. Gauthier-Villars, Paris.

Villat H (1943) Leçons sur les fluides visqueux (ed. by J. Kravtchenko). Gauthier-Villars, Paris.

Weber J-M (2008), Un demi-siècle d'aéronautique en France. Études et recherches. Comité pour l'Histoire de l'Aéronautique. Centre des hautes études de l'armement, Paris.

Weinstein A (1926) Sur la vitesse de propagation de l'onde solitaire. Rendiconti della R. Accademia dei Lincei (6) 3: 463–468.

Weinstein A (1973) Jets and Wakes. In (AAVV, 1975; 269–296).

French Mathematicians at the Bologna Congress (1928). Between Participation and Boycott

Annalisa Capristo

The International Congress of Mathematicians, held in Bologna from September 3 to 10, 1928 (with the closing session in Florence at the Palazzo Vecchio), was a significant scientific event: for the first time since World War I, mathematicians from many nations, that had opposed one another as members of the warring factions, the Triple Entente and the Central Powers, came together in an international conference. Attendance was very high and the Congress proved a great success (Fehr 1929; 28; Lehto 1998; 47).

Despite this, however, no book has been dedicated to this congress, in which its contributions to scientific advancement and its broader political meaning are analyzed. To date, the Congress has been studied almost exclusively as an episode in the history of the International Mathematical Union (IMU) and its internal conflicts.

This chapter aims to highlight two aspects which are of great interest and deserve to be explored further.

The first aspect is the political and institutional background of the Congress. Of particular interest are the relations of the Congress with both the Fascist Government and the Italian National Research Council (Consiglio Nazionale delle Ricerche, CNR). The fact that big scientific conferences were used by the regime as a propaganda display was clear enough during the interwar period (Poupault 2014; 142–146). The government sponsored the Congress of Mathematicians, organized by professor Salvatore Pincherle (1853–1936) (President of the International Mathematical Union and of the Italian Mathematical Union (Amaldi [1937] 1954; Bortolotti 1937, 1938; Tonelli 1937; Vivanti 1936)) under the auspices of the University of Bologna and in close collaboration with the Prefecture of Bologna. The CNR was the Italian organization dealing with scientific research, which also cooperated with foreign scientific institutions. The Bologna Congress was held at a rather difficult

A. Capristo (✉)
Centro Studi Americani, Rome, Italy
e-mail: a.capristo@centrostudiamericani.org

F. Brechenmacher et al. (eds.), *Images of Italian Mathematics in France*,
Trends in the History of Science, DOI 10.1007/978-3-319-40082-2_10

time, when relations between the Italian Mathematical Union (Unione Matematica Italiana, UMI) and the Mathematical Committee of the CNR were being overhauled and a shake-up was under way in the top echelons of the Italian mathematical milieu (Nastasi 1998; Nastasi 2001; 243–261; Guerraggio and Nastasi, 2005; 183–194).

The second aspect of interest this chapter focuses on is the participation in the Congress of the French mathematicians, who were divided between those who continued to uphold an intransigent attitude towards the German scientists, as in the immediate post-war period, and those who were in favor of a more conciliatory approach. The former included the President of the International Research Council (IRC), Émile Picard, and the Secretary of the International Mathematical Union, Gabriel Koenigs. Their position is relatively well-known because it is documented in the introduction to the *Proceedings* of the Congress, although, as we shall see further on, documents have emerged from the Italian archives that shed light on the correspondence of both with Pincherle.

The position of the latter group, which included, among others, Élie Cartan, whose opinion Pincherle repeatedly requested before the Congress, is less well-known and this aspect will be considered further on.

To begin with, it seems necessary to provide an overview of the main international and Italian studies, produced to date, that touch on the Bologna Congress. The main account of the Congress can be found in Olli Lehto's history of the International Mathematical Union (Lehto 1998) and is largely based on secondary sources, first and foremost the *Proceedings* of the Congress, published in 6 volumes between 1929 and 1932 by Zanichelli. The author also uses private and institutional correspondence, albeit not from Italian archives, except for a letter by Pincherle to Vito Volterra, dated September 1, 1925. This letter was published by Giovanni Paoloni in 1990, in the catalogue of the historical and documentary exhibition on Volterra (Paoloni 1990; 156–157). Yet the unpublished correspondence between the two mathematicians includes other interesting letters on the Bologna Congress, and there are also other important Italian documentary sources.

Lehto dedicates a chapter of his book (about 6 pages long) to the Bologna Congress. The title is "The IMU Separates from the Congresses," and, in fact, it highlights the "critical turning point" represented by Bologna 1928 in the history of the IMU. The crisis had begun some time before, but it culminated at Bologna with the break, within the board of the Union, between the hardliners who opposed the participation of mathematicians from the countries defeated in the war, and those in favor of healing the rift between the nations and who opposed any exclusions. The early resignation of Pincherle from the position of President of the IMU, on September 9, 1928, during the Bologna Congress, was a tangible sign of the crisis under way.

According to Lehto, the Bologna Congress was an important milestone in the history of the IMU:

> The exclusion policy was broken at the ICM in Bologna, Italy, in 1928 […] The Italian policy of openness was widely applauded […] Pincherle and the other Italian organizers of the Bologna Congress very likely rendered an invaluable service to international mathematical cooperation (Lehto 1998; 44, 45, 49).

Even Pincherle's violation of the formal rules governing admission to the Congress was, according to the author, a necessary step for breaking the scientific ostracism of the mathematicians from the former Central Power countries, which was by then considered by many to be unacceptable (Dauben 1980; Rasmussen 2007; Siegmund-Schultze 2011; Nastasi and Tazzioli 2013). Lehto's interpretation has been substantially incorporated in all subsequent studies dealing with the Congress.

More recently, the issue has been addressed by the Spanish mathematician Guillermo Curbera in his book on international mathematical congresses (Curbera 2009).[1] Curbera highlights an aspect disregarded by Lehto, namely, the political implications of the Bologna Congress. In fact, he points out that "the Bologna congress was a great international exhibition for the Fascist state of Benito Mussolini" (Curbera 2009; 88). The author also stresses the great amount of public funds granted to Pincherle and he adds additional considerations. First, he notes the presence of the religious Catholic authorities at the opening ceremony:

> Even the presence of the Archbishop of Bologna at the opening ceremony was an indication of political influence (that had not occurred in the 1908 congress in Rome); Mussolini was close to signing the first official agreement between the Catholic Church and the Italian State (Curbera 2009; 88).

Then, Curbera mentions the choice of official languages. In this respect, the main innovation was the readmission of German, after years of exclusion (Schroeder-Gudehus 1973; 99). Curbera's observation regarding the political significance of the inclusion of Spanish is also significant:

> The inclusion of Spanish as an official language could only be interpreted as a political gesture in support of the political regime in Spain at the time (an authoritarian military dictatorship) (Curbera 2009; 88–89).

Recently, Elaine McKinon Riehm and Frances Hoffman, in their biography of John C. Fields, have dedicated a chapter to the Bologna Congress, highlighting the fact that "the Bologna Congress was in many ways a showcase for Italian Fascism" (McKinon Riehm and Hoffman 2011; 172).

Finally, some other contributions deserve to be mentioned briefly here, as they too help to build a larger picture. First, the biography of the Dutch topologist L.E.J. Brouwer (Van Dalen 2005; 587–599), which focuses on his firm opposition to the IRC politics. Brouwer sided with the German hardliners, who protested against any initiative sponsored by the international scientific organisations hostile to Germany (Segal 2002, 2003; 349–355). On September 27, 1928 Brouwer received a detailed report on the Bologna Congress by Hasso Härlen (Van Dalen 2011; 334–338). Another contribution is Reinhard Siegmund-Schultze's recent work on Hilbert's speech at Bologna 1928,

[1]Curbera also highlights the fact that "the first photographs that we have from a social event of an international congress are from the Bologna 1928 ICM" (Curbera 2009; 239). His book includes the photos of "Jacques Hadamard on the beach during the excursion to Ravenna" and that of "Gaston Julia and Charles de la Vallée Poussin at the lunch in the Littoriale sports complex" (Curbera 2009; 233, 239), from the George Pólya picture album in the possession of Gerald L. Alexanderson (Alexanderson 1987).

which sheds new light on the participation of the leading German mathematician to the Congress (Siegmund-Schultze 2014, 2016). And finally, the recent book by Christophe Poupault on French travellers in Fascist Italy includes many interesting considerations, particularly on the participation of French personalities to international congresses and events, such as the celebrations in honor of Francesco Petrarca in 1928 (Poupault 2014; 142–150 and 583–589, see also Poupault 2009; 71–72).[2]

With regard to the Italian sources that treat the congress, they, too, are mainly based on the conference proceedings and on Lehto's book. Let us mention here the book by Angelo Guerraggio and Pietro Nastasi on the congress of mathematicians held in Rome in 1908, the closing chapter of which is dedicated to the Bologna Congress of 1928 (Guerraggio and Nastasi 2008; 150–160). Of course, the two Italian scholars stress the important role played by the Fascist government in the event.

Italian primary sources are included in some other essays. The first is an article by Carlo Pucci on the *Unione Matematica Italiana* (Pucci 1986). Although rather dated, it deserves to be mentioned here because the author based his work on some documents from the archives of the UMI, of which he was the Honorary President until his death (Talenti 2005). Pucci (who was a nephew of the antifascist Ernesto Rossi, and was himself a Resistance fighter during World War II) (Figà Talamanca and Pepe 2014) also underlines the political and financial support received by Pincherle from the Fascist regime.

A second essay is the chapter by Pietro Nastasi in the first volume of the history of the Italian National Research Council, edited by Raffaella Simili and Giovanni Paoloni (Nastasi 2001). This chapter is mainly based on the records found among the papers of Enrico Bompiani, kept in the archive of the National Academy of Sciences, also known as the Academy of the Forty (Accademia Nazionale delle Scienze detta dei XL). Although Nastasi's contribution contains only some references to the Bologna Congress, it is useful for contextualizing the organization of the event in the midst of the overhauling process of the Mathematical Committee of the CNR.[3]

A reference to the 1928 Congress is also included in Simona Salustri's book on the University of Bologna under Fascism, which is based on the University archival documents and the local press (Salustri 2010; 95–100). Finally, in her recent book (which I was not able to utilize because my own essay was already being published) Monica Cioli dedicated a chapter to the Congress of Bologna, based on information from the documents of the Presidency of the Council of Ministers held in Rome's Central State Archives (Cioli 2016; 104–129).

Let us now examine the organization of the Congress and Pincherle's efforts to involve his colleagues, in particular the French mathematicians. We shall present and comment on several useful documents for understanding his strategy.

Before and after the Bologna Congress, Pincherle appealed several times to Mussolini for financial and political support. From the beginning, the scientist asked the head of the government to accept the honorary chairmanship of the event.

[2]However, the author does not mention that Borel did not actually take part in the Bologna congress (Poupault, 2014; 143).

[3]See also the paragraph "The Dualism U.M.I-C.N.R." in (Guerraggio and Nastasi, 2005; 183–194).

The first time that Pincherle wrote to the *Duce* was on February 16, 1926.

> Before I officially propose to my colleagues in the bureau of the International Mathematical Union which city ought to host the Congress, I believe that it is necessary to make sure that the proposed Italian city meets the favour and support of the National Government; and there can be no doubts about this, *because this Government supports any initiative that can benefit and bring honour to our Country* [the italics are mine].[4]

The Government's support was necessary in order to adequately organize the Congress and make sure that the foreign scientists were welcomed in a manner worthy of the occasion. Therefore, Pincherle wanted to be sure of Mussolini's support before proposing to the IMU Bologna as a candidate city. As he wrote on this occasion, it was "a proposal in keeping with the vision of national greatness that He [Mussolini himself] has signified to us as the Italian people's reason for being."[5]

This first letter by Pincherle was delivered to the *Duce* by Balbino Giuliano, who was at the time a member of the Chamber of Deputies for the National Fascist Party and future Minister of National Education; it was under his administration that, in 1931, all university professors were ordered to take an oath of allegiance to the Fascist regime (Pertici 2001).

After the official announcement of Bologna as a candidate for hosting the 1928 Congress in November 1926 (Lehto 1998; 44), Pincherle wrote to Mussolini once again, requesting concrete financial support. On December 7, 1926, the scientist was received in audience by the *Duce*, together with the Mayor of Bologna Umberto Puppini (Supino 1946). The latter was professor of Hydraulics at the Bologna School of Engineering, of which he became director in 1927, in addition to being a founding member of the UMI, a future Undersecretary of State for Finance (1932–34) and Minister of Telecommunications (1934–35).[6]

The meeting of Pincherle with Mussolini was also reported by the official publication of the Italian Mathematical Union (*Bollettino dell'Unione Matematica Italiana*):

> The Prime Minister, who appeared convinced of the importance of the upcoming scientific congress, has benevolently assured his moral and material support and warmly agreed to chair the Honorary Committee.[7]

---

[4]"Prima di proporre ufficialmente ai colleghi dell'Ufficio di Presidenza dell'Unione Matematica Internazionale quale debba essere la Sede del futuro Congresso, il sottoscritto crede necessario di essere assicurato che proponendosi come Sede una città italiana, la proposta incontrerebbe il favore e l'appoggio del Governo Nazionale; né di ciò egli può dubitare, *poiché questo Governo fa sempre proprie quelle iniziative che alla nostra Patria possono recare vantaggio ed onore" [the italics are mine]* ACS, PCM (1928–30), b. 1328, f. 14.3.3303, Bologna. Congresso Internazionale dei Matematici (settembre 1928-VI).

[5]"Una proposta che si collega essenzialmente con quei fini di grandezza nazionale, che Egli ha indicati, come prima ragione di vita, al popolo italiano" (ACS, PCM (1928–30), b. 1328, f. 14.3.3303, Bologna. Congresso Internazionale dei Matematici (settembre 1928-VI).

[6]On September 3, 1928, at the Bologna Conference, Puppini gave a keynote speech on "Land Reclamation in Italy" highlighting the "impressive financial and technical effort" of the Fascist state (ICM Proceedings 1929; vol. 1: 166).

[7]"Il Primo Ministro, che si è mostrato convinto dell'alta importanza della prossima riunione scientifica, ha benevolmente promesso al Congresso il Suo appoggio morale e materiale, ed ha accettato di buon grado la presidenza del costituendo Comitato d'onore" (*Bollettino dell'Unione Matematica Italiana*, 1927; 41; Pucci, 1986; 195).

Initially, the Ministry of Education allocated the sum of 70,000 lire as a State contribution towards the cost of the Congress. On December 10, 1926, in a letter following up on Mussolini's undertaking 3 days earlier, Pincherle pointed out that the estimated budget for the organisation of the Congress was 400,000 lire, a quarter of which would be provided by local bodies (among which the Municipality and Province of Bologna) and scientific institutions (such as the University of Bologna). Pincherle therefore appealed to the *Duce* for the remaining 300,000 lire. Mussolini acceded to the request; a note on the memorandum sent to his attention reads: "24/XII/926: Approved by H.E. the Head of the Government, who confirms a grant by the Government of 250,000 lire (including the 70,000 lire)". Then the grant was reduced to 200,000 lire, which was nevertheless a good amount of money for the time.[8]

About three weeks after his meeting with Mussolini, on December 31, 1926, Pincherle joined the National Fascist Party.[9] The previous year, in April 1925, his name had featured in the list of intellectuals who had signed the Manifesto of Fascist Intellectuals *(Manifesto degli intellettuali fascisti*), drafted by Giovanni Gentile (Papa 1958; 47; Gennaro 2013; 88).[10]

On February 3, 1927, Pincherle once again wrote to Mussolini, to inform him that

> the General Assembly of the Members of the Italian Mathematical Union, at its latest meeting, resolved by acclamation to pay tribute to your Excellency and express its gratitude for the support granted to the preparation of the forthcoming 1928 International Congress of Mathematicians in Bologna.[11]

[8]On 15 May 1929, Pincherle sent to the Prefecture of Bologna a detailed account of the expenses incurred in relation to the organization of the Congress; this report was transmitted to the PCM on 25 May (ACS, PCM (1928–30), b. 1328, f. 14.3.3303, Bologna. Congresso Internazionale dei Matematici (settembre 1928-VI).

[9]The information is drawn from the personal files of Pincherle's second wife, Natalia Gentilomo, and his daughter Edvige (born from Pincherle's first wife, Emma Morpurgo), which are preserved in the archive of the General Administration for Demography and Race (shortened as Demorazza), the Ministry of the Interior department set up in 1938 to enforce the Race Laws. Both the widow and the daughter of Pincherle attached his National Fascist Party certificate to their request for being partially exempted from the persecution. In his report, dated February 23, 1939, the Prefect of Bologna added that Pincherle was "a loyal member of the Party until his death" and that all his university assistants are "of Aryan race" (ACS, Demorazza, personal files, b. 107, f. BEN 7465, Pincherle Edvige and b. 118, f. BEN 7883, Gentilomo in Pincherle Natalia). Salvatore Pincherle professed Judaism; in his will (1931) he wanted to be buried in the Jewish cemetery in Bologna, together with "the French Jewish Tephillah given me by my father on my thirteen birthday." He also wished that his nephews would have avoided mixed marriages. I wish to thank Maurizio Pincherle, great-granson of Salvatore Pincherle, for having sent me a copy of his will. Pincherle's profile was included in the *Jüdisches Lexikon* (Zoller 1930).

[10]The Honorary Committee included the philosopher Giovanni Gentile, senator, former Minister of Education in Benito Mussolini's first cabinet, then President of the Supreme Council of Public Education and member of the Fascist Grand Council. Pincherle's letter of invitation is kept in the Gentile's archive (AFG, Corrispondenza, Lettere inviate a Gentile, f. Pincherle Salvatore).

[11]"L'Assemblea Generale dei Soci dell'Unione Matematica Italiana nella sua ultima seduta ha deliberato, per acclamazione, di rivolgere alla E.V. un voto di reverente omaggio e di viva gratitudine per il generoso appoggio concesso dalla E.V. alla preparazione del Congresso Internazionale dei Matematici che avrà luogo in Bologna nel 1928" (ACS, PCM (1928–30), b. 1328, f. 14.3.3303, Bologna. Congresso Internazionale dei Matematici (settembre 1928-VI).

Then, on March 9, 1927, the mathematician again addressed Mussolini to request authorization to officially announce Mussolini's acceptance of the honorary presidency of the Congress. The *Duce* replied personally in the affirmative,[12] and the news was even reported abroad (for instance, by the journal *L'Enseignement mathématique* (t. 26, 1927; 142). However, in September 1928, the head of the government did not attend the opening ceremony of the Congress, sending the Minister of Education, Giuseppe Belluzzo (Minesso 2012), to represent him.

Pincherle then repeatedly expressed his appreciation for the support received from Mussolini, in both his public speeches and institutional correspondence. In the inaugural address of the Congress, which he delivered in French,[13] he defined Mussolini as "the exceptional man who has arisen, thanks to the good fortunes of Italy, to guide its destiny," and stressed that it was only thanks to the support received from the Government and the Municipality of Bologna that "the executive commission of the Congress has been able to achieve its aim, which was anything but simple."[14]

Pincherle emphasized the importance of the Congress, from both a political and a scientific perspective. In the report addressed to the *Duce*, dated September 11, 1928, the day after the conclusion of the Congress, he included among the "successful results" achieved "the explicit recognition, on all sides, of the well-being and regular operation of all services under the Fascist Government and regime founded by Your Excellency".[15]

On September 14, in his cover letter accompanying Pincherle's report, the Prefect of Bologna, Giuseppe Guadagnini (Urso 2003), highlighted the propaganda success of the Congress and added a personal consideration:

> The foreign scientists have unanimously expressed on numerous occasions, to the members of the organising Committee, to myself and to the Podestà [Mayor] their unconditional

[12]Handwritten note on the memo: "Sì" (ACS, PCM (1928–30), b. 1328, f. 14.3.3303, Bologna. Congresso Internazionale dei Matematici (settembre 1928-VI).

[13]From the city of his birth, Trieste, which, at the time, was still part of the Austro-Hungarian Empire, Pincherle, still a child, had moved with his family to the French port city of Marseilles, where he completed his basic schooling. He returned to Italy in 1869, to study mathematics at the University of Pisa. He was, therefore, very proficient in French.

[14]"L'Homme exceptionnel que la fortune de l'Italie a fait surgir pour qu'il en dirige les destinées a approuvé notre ligne de conduite; le Congrès a eu Son appui, comme celui du représentant du gouvernement et du premier magistrat de la ville; grâce à cet appui, la Commission exécutive de Congrès a pu remplir sa tâche, qui était loin de se présenter comme facile (ICM Proceedings 1929; vol. 1: 73).

[15]"Il riconoscimento più esplicito, avuto da ogni parte, del benessere, del regolare funzionamento di tutti i servizi sotto il Regime fascista, sotto il Governo dell'E.V. che ne è l'instauratore" (ACS, PCM (1928–30), b. 1328, f. 14.3.3303, Bologna. Congresso Internazionale dei Matematici (settembre 1928-VI). See also: Annuario della Regia Università di Bologna per l'anno accademico 1928–1929: 30–32; ASUBo, Pos. 78, b. 14, f. 7: Congresso Matematico: Prot. 1421, Sept. 12, 1928 (I wish to thank Daniela Negrini for her help). On October 19, 1928 Pincherle was received in audience by Mussolini and reported on the appreciation expressed by all the foreign scientists attending the Congress (ASBo, Prefettura di Bologna, Gabinetto, b. 1485, f. Congresso Internazionale dei Matematici 1927–1928).

> admiration for the hospitality received in our Country, and for the order, discipline and industriousness that reign here.
>
> I believe that this is the best form of patriotic propaganda and the most effective way of denying the slander spread abroad by the enemies of the regime.
>
> *The petty attempt to discourage foreign participation in the Congress, made last year in France, has thus entirely failed* [the italics are mine]: attendance was huge at the Congress, by French scientists as well, indeed France has participated with its most prominent mathematicians, such as Hadamard, Fréchet, Cartan, Julia, Denjoy.[16]

In the final passage of his letter, the Prefect of Bologna was referring to information that had alarmed the organisers the previous year: on October 21, 1927 the Prefect himself had written to the Prime Minister's Office about a letter Pincherle had received from "a member of the Belgian Chamber of Deputies, Count De la Vallie [i.e. De la Vallée]" containing "mention of a political campaign spreading abroad and directed at boycotting the Congress".[17]

The Prefect attached an excerpt from the letter received by Pincherle, in which his correspondent referred that the stories had been invented by people who "hated Fascism." According to these rumours, the honorary chairmanship assigned to Mussolini would characterize the Congress as a Fascist manifestation, and this would have inevitable political implications, such as the exclusion of any persons declared "unwelcome." Hence, the French-inspired "propaganda against the Congress."[18]

After receiving this information, the Prime Minister's Office requested the Ministry for Foreign Affairs to order the Italian Embassy in Paris to look into the matter, but with no results.

On February 21, 1928 the Prefect of Bologna informed Rome that the alarm concerning the potential political boycott of the Congress by the French had subsided:

> The Professor [Pincherle] has told me that, based on the most recent letters received from France, he is under the impression that the attitude in certain intellectual circles, which opposes the International Congress of Mathematicians, has changed.

---

[16]"Gli scienziati stranieri, concordemente, ed in più occasioni, hanno espresso ai componenti del Comitato promotore, a me ed al Podestà la loro incondizionata ammirazione per l'ospitalità avuta nel nostro Paese, e per l'ordine, la disciplina, l'operosità che vi regnano. Ritengo che questa sia la forma migliore di propaganda patriottica, e la più efficace smentita alle calunnie che si diffondono, all'estero, dai nemici del regime. *La meschina manovra che, in certi ambienti francesi, si andava svolgendo lo scorso anno per scoraggiare la partecipazione straniera al Congresso è, così, completamente fallita* [the italics are mine]: la partecipazione al Congresso anche da parte degli scienziati francesi è stata imponente ed anche la Francia ha partecipato al Congresso con i suoi matematici più insigni, come Hadamard, Fréchet, Cartan, Julia, Denjoy" (ACS, PCM (1928–30), b. 1328, f. 14.3.3303, Bologna. Congresso Internazionale dei Matematici (settembre 1928-VI). For the drafts of this letter, with handwritten corrections see ASBo, Prefettura di Bologna, Gabinetto, b. 1485, f. Congresso Internazionale dei Matematici 1927–1928.

[17]ACS, PCM (1928–30), b. 1328, f. 14.3.3303, Bologna. Congresso Internazionale dei Matematici (settembre 1928-VI). Comparing this letter with the draft kept in ASBo, Prefettura di Bologna, Gabinetto, b. 1485, f. Congresso Internazionale dei Matematici 1927–1928 it is clear that the name of Pincherle's correspondent ("de la Vallée") had been misspelled ("de la Vallie").

[18]In the archive of the Prefecture of Bologna the file containing this correspondence was entitled "French campaign against the Congress" ("Congresso Matematici—Campagna francese contro il Congresso"). See ASBo, Prefettura di Bologna, Gabinetto, b. 1485, f. Congresso Internazionale dei Matematici 1927–1928.

> Indeed, in a letter received a few days ago, the Secretary of the Union of French Mathematicians has assured him that the French mathematicians are very much looking forward to attending the Congress at Bologna and, therefore, are already making their preparations. Therefore, Prof. Pincherle now assumes that the rumours reported by the Comte de la Valleè [sic], regarding the campaign against the Congress, were simply the result and impression of that frame of mind that had swept French public opinion at the end of last year, in consequence of the foreign policy developments.[19]

This reference by the Prefect to the "foreign policy developments" probably alluded to the tensions between France and Italy as a result of the two countries' respective agreements with Yugoslavia and Albania in November 1927 and to the related press campaigns. (Shorrock 1988; 48–49, 55; Guillen 1991; 323, 325, 1998; 151–153; see also Milza 1987; 148-150; Decleva and Milza 1996).

Based on our current knowledge, we have no way of verifying the truth of the information received by Pincherle in 1927, nor of identifying with certainty who his correspondent was.

The most likely assumption is that it was the mathematician Charles-Jean de La Vallée Poussin (1866–1962), professor of mathematical analysis at the Catholic University of Louvain, first President of the IMU, of which he then became Honorary President (Montel 1962; Bubkill 1964; Butzer and Mawhin 2000). De La Vallée Poussin was not a "member of the Chamber of Deputies," but had been delegated by the Belgian Government to represent his country at the Bologna Congress, nor was he a "count," although the King of Belgium granted him the title of baron.[20]

1 The Absence of Borel

In his essay *Souvenirs de Bologne* (Bru 2003), Bernard Bru's mostly focus on a mathematical development, the "emergence in Bologna and the recognition of one of the great probabilistic theories of the 20th century, the theory of Markov chains"

[19]"Detto Professore [Pincherle] mi riferisce che dalle ultime lettere ricevute dalla Francia egli ha riportata l'impressione che l'atteggiamento di quelle sfere intellettuali verso il Congresso Internazionale dei Matematici si sia ora modificato. In una lettera, anzi, pervenutagli in questi giorni, il Segretario dell'Unione Matematici Francesi afferma che i matematici di Francia si mostrano desiderosi di partecipare al Congresso di Bologna e, a tal fine, già fanno preparativi. E ciò induce a supporre il Prof. Pincherle che le voci giunte al Conte de la Valleè [sic], circa la campagna contro il Congresso siano state il prodotto e l'impressione di quello stato d'animo che aveva pervaso l'opinione pubblica Francese negli ultimi mesi dell'anno scorso in conseguenza degli avvenimenti di politica estera del tempo" (ACS, PCM (1928–30), b. 1328, f. 14.3.3303, Bologna. Congresso Internazionale dei Matematici (settembre 1928-VI).

[20]A celebration in honour of de La Vallée Poussin was held on May 13, 1928, after 35 years of office. "At this occasion, King Albert I congratulated VP for his great service to Louvain and Belgium and on September 6 1930 conferred on him the title of Baron" (Butzer and Mawhin 2000; 6). The list of foreign mathematicians attending the celebration includes the name of Pincherle (Butzer and Mawhin 2000; 6). Cioli claims that the informer was Etienne de la Vallée Poussin (Cioli 2016; 122). But this identification is not convincing; in fact, Etienne de la Vallée Poussin (1903-1996) became a member of Belgian Parliament (sénateur) after the second World War (Munster 1999).

(Bru 2003; 135; see also Mazliak 2010; Mazliak and Catellier 2012). Bru however also formulates a "political" conjecture about the absence of the French mathematician Émile Borel (Bru 2003; 138–139). He assumes that Borel, as "an independent Republican member of the French Parliament", may have decided to stay away from Bologna, in order not to have to take part in an event designed to celebrate Fascism. Émile Borel had indeed been invited to hold the first plenary conference at Bologna on September 4, 1928. That morning, however, the chairman of the session, the other prominent French mathematician Jacques Hadamard, announced that the mathematician was not present and that his paper, which was then published in the *Proceedings*, would be presented by Élie Cartan (ICM Proceedings 1929; vol. 1: 67). In all likelihood, Cartan was selected because both he and Borel had been officially designated by the University of Paris to attend the Congress.[21]

No mention is made in the *Proceedings* about the reasons for the French mathematician's absence. It is merely stated that he was "unable to attend" (ICM Proceedings 1929; vol. 1: 77). In other cases, such as the absence of Federigo Enriques, for example, it is specified that he was in Buenos Aires (ICM Proceedings 1929; vol. 1: 68, 102). Enriques had apologized well in advance for his absence, citing the extension of his conference tour in Latin America. The letter he sent from Buenos Aires on August 11, 1928, is still preserved in the archives of the Italian Mathematical Union.[22] On the contrary, no letters by Borel have been found so far, and his absence may just have been due to a sudden impediment.

According to Adolphe Buhl, professor at the University of Toulouse, who was present at Bologna, Borel's absence was due to bad health. He published his account of the Congress in the journal *L'Enseignement mathématique,* of which he was one of the two editors, stating:

> Unfortunately, it was not possible for M. Borel to hold his conference on the *Calculation of Probabilities and the Exact Sciences*, because the author's indisposition had prevented him from travelling to Bologna. It was presented by M. Cartan instead.[23]

The Swiss mathematician Henri Fehr, the other editor of the periodical, and Vice President of the IMU, also present at Bologna, advanced the same reason: "In the author's absence, due to an indisposition that prevented him from reaching Bologna, the paper was read by M. Cartan".[24]

---

[21]AScP, Fonds Élie Cartan 38 J. Paris, 13 June 1928. Letter by S. Charléty, Rector of the University of Paris to Elie Cartan, regarding the International Congress of Mathematicians, 6176/08. Borel was also the delegate of the French Society of Physics and the French Mathematical Society (ICM Proceedings 1929; vol. 1: 27). On Borel (1871-1956) see also (Broglie 1957; Marbo 1968; Guiraldenq 1999). On Cartan (1869-1951) see (Whitehead 1952).

[22]The rearrangement of the Unione Matematica Italiana's Historical Archive is currently underway. I wish to thank Salvatore Coen and Rossana Tazzioli for their help.

[23]"La Conférence de M. Borel, sur *Le Calcul des Probabilités et les sciences exactes* n'a malheureusement pas pu être faite par son auteur qu'une indisposition empêcha de venir à Bologne. Elle a été lue par M. Cartan." (Buhl 1928; 199).

[24]"En l'absence de l'auteur empêché par une indisposition de se rendre à Bologne, le manuscrit a été lu par M. Cartan" (Fehr 1929; 31).

Nothing about the matter seems to transpire in the correspondence between Borel and Vito Volterra, preserved at the Accademia dei Lincei. On September 8, 1928, while the Congress was still under way, Volterra wrote to his friend and colleague telling him that he had stayed in Bologna only 3 days, just the time to deliver his own paper and to listen to those he was most interested in.[25] Among these was Borel's. Volterra regretted not having been able to meet him and assured his friend that his contribution, read by Cartan, had met great success. Volterra also congratulated Borel for his speech at the Berlin assembly, which was reported from the press.[26] In his reply, dated September 20, Borel thanked Volterra for the information about the Congress, without however touching on the reasons for his absence.[27]

We can therefore only speculate, as Bru does, about a plausible and possible political motivation to explain why Borel decided not to come. As already mentioned, after the Great War, during which he was deeply involved in the scientific mobilization, Borel entered a political career as a Member of the French Parliament, elected by the constituency of St Affrique in the department of Aveyron where he was born. Close to Painlevé (Borel was even briefly Painlevé's minister of Navy in 1925), he belonged to the radical-socialist tendency which was the main leftist current in French politics during the 1920s. He was also an active member of the international commissions born in the aftermath of the Great War, dealing with intellectual exchanges or disarmament (Guieu 1998; 2008). For instance, immediately before Bologna Congress, Borel attended the 25th Interparliamentary Union assembly, which took place in Berlin between August 23 and 25, 1928; he delivered his speech on peace, Europe and disarmament on August 24 (Union interparlementaire 1928; 270–272).

Borel moreover had a profound friendship with Volterra. Their relation was born at the end of the 1890s when both men were among the most fervent supporters of the International Congresses of Mathematicians (see in particular (Mazliak 2015)). Among the French scientists, Borel was Volterra's main contact, or at least the one to whom he was the closest, as in proved by the collection of more than 400 letters kept at the Academy of Lincei. Moreover, during the war, Volterra was particularly interested in studying Borel's involvement in the scientific mobilization in order to

[25]Following his early departure, Volterra (who was one of the Honorary Presidents of the IMU) avoided to participate in the unofficial meeting held on 9 September 1928. Although disavowed by the Secretary General of the IMU Koenigs and the President of the IRC Picard, the meeting approved the policy of inclusion upheld by Pincherle in the organization of the Congress and publicly sanctioned the rift within the Union. Consistently with his opinions, Pincherle subsequently resigned from the position of President of the IMU.

[26]ANL, Archivio Vito Volterra, Ser. 1, Correspondence, f. 162: Borel Émile, c. 287. In his letter Volterra, just awarded the Légion d'Honneur, thanked Borel for his support and friendship. In March 1928 the French government asked the approval of the Italian government regarding the award to Volterra, as "former president of the Accademia dei Lincei". Mussolini gave his personal approval (ACS, PCM 1928–1930, f. 12.3.1647, Conferimento di una onorificenza francese al Senatore M. Vito Volterra) (Capristo 2015; 645).

[27]ANL, Archivio Vito Volterra, Ser. 1, Correspondence, f. 162: Borel Émile, c. 288 ("Je vous suis bien reconnaissant de m'avoir donné des nouvelles du Congrès de Bologne; si j'avais pu y aller, j'aurais eu un grand plaisir à vous y rencontrer […]").

implement such a mobilization in Italy (on this, see in particular (Mazliak and Tazzioli 2009).

To our best knowledge, Borel had never publicly expressed his feeling about the advent of fascism in Italy, at least in a written form. No doubt he was personally hostile to the idea of Mussolini's personal power, but as many members of the moderate left side, his attitude was probably above all that of reservation. The Italian situation seen from France was so chaotic in the aftermath of the war (Borel's letters to Volterra contain some allusions to that, Borel mentioning that Italy would really need some stability), that before 1924 and the Matteotti's murder, there was a kind of resignation (and often an enthusiastic acceptation, of course) to think that a strong populist government may have been a necessary evil.

We have an interesting testimony of a fact which may have implied some change in Borel's mind. In the Volterra archive are also kept two letters exchanged between him and Borel on October 16 and 17, 1924 (Capristo 2015; 641–642). Borel had invited Volterra, who was then in Paris, to join the radical-socialist congress which would be held in Boulogne-sur-Mer between October 16 and 20. Volterra declined the invitation for both personal and political reasons. After having explained that he was very tired due to the long journey to the French capital, he added that he considered not advisable for him to take a public stance in that moment, speaking abroad about the political situation in Italy: "It is not a moment for us to speak from abroad about what happens in Italy! I could only remain silent".[28] At least, after the consolidation of Mussolini's dictatorship, Borel's preventions against Fascism were probably firmly implanted on his mind. The gradual exclusion of his friend Volterra from Italian academic life (which would take a dramatic turn after 1931 and Volterra's refusal of the oath to Mussolini) certainly increased them. Thus, Bru's hypothesis, that he may have been reluctant to participate as a guest star to an event in which propaganda for the regime would clearly play an essential part, cannot be absolutely discarded. Besides, Borel, who traveled so often to Italy until 1922, appears to have never been in the country during the fascist era.

2 Picard's Refusal

Contrary to Borel's somewhat mysterious absence, Picard's (1856–1941) behavior was crystal clear. In the introduction to the *Proceedings*, Pincherle mentions the correspondence with Koenigs, Secretary General of the IMU, and Picard (1856–1941), President of the IRC (on him see Broglie 1942; Hadamard 1942),

[28]ANL, Archivio Volterra, Serie 1., f. 162: Borel Émile ("Ce n'est pas le moment pour nous de parler à l'étranger de ce qui arrive chez nous! Je ne pourrais que me taire"). For a report on the Boulogne congress by the Italian press see for instance "Herriot e la tradizione repubblicana al Congresso radicale francese." *La Stampa*, 19 October 1924, p. 4.

dating between the end of May and the beginning of June 1928, and regarding the efforts made to resolve the dispute between a literal interpretation of the rules and the inclusive policy he had decided to adopt for sending out invitations to the Congress. In particular, Pincherle published several excerpts from Koenigs's stern letter of May 29, as well as the full text of the letter he himself sent to Picard on June 8, 1928, in an extreme attempt to convince the President of the IRC to attend the conference. However, he chose not to publish Picard's answer to his letter, summarizing it thus: "the reply by the President of the C.I.d.R. confirmed, albeit in a less severe way, the declarations by the Secretary General" (ICM Proceedings 1929; vol. 1: 8–9).

Lehto repeats the contents of the *Proceedings* in his book, adding—from the information in a memo preserved among the papers of William Henry Young—that Young, Pincherle's successor as President of the IMU, "had heard that Picard's reply was two lines on a postcard asserting that the President of the Council could not attend a Congress in which Germans would be present" (Lehto 1998; 47, 337).

But now, thanks to the documents preserved at the Archivio Centrale dello Stato, we know the full text of the letter. On June 21, in fact, Pincherle himself sent a transcript of it to the Vice President of the Italian National Research Council, Amedeo Giannini. A trusted advisor of Mussolini at the Ministry for Foreign Affairs (Santoro 1999; Melis 2000), in 1928, Giannini was a fierce supporter of Vito Volterra's full exclusion from any international scientific role, because of his antifascist positions (Paoloni 1990; 173; Simili 2001; 117; Capristo 2015; 644–645). Based on the documents found in the archives of the CNR, it emerges that Pincherle remained in close contact with Giannini, after having been confirmed, on March 2, 1928, in his role in the organization of the Bologna Congress. This role had been in doubt during the previous February, due to the reorganization of the Italian Mathematical Committee of the CNR and the consequent downsizing of the UMI's international role.[29]

On June 22, a copy of Picard's letter was also transmitted by the Prefect of Bologna to the Prime Minister's Office. But let's begin with the correspondence between Pincherle and Giannini.

On June 14, 1928, the mathematician transmitted to the Vice President of the CNR the letter of disavowal received from Koenigs; a copy of the letter which Pincherle himself had already sent on June 8 to Picard—whose content he had checked with Giannini; and the detailed report that Ettore Bortolotti, secretary of the organizing committee of the Congress and trusted assistant to Pincherle, had sent on June 12 to the above mentioned Prof. Fehr of Geneva, one of the Vice Presidents of the IMU. These letters were relayed in order to document "the manner in which certain parties have attempted to sabotage the impartial and conciliatory

[29]The documents related to this matter are preserved at the Archives of the Unione Matematica Italiana in Bologna; among the Enrico Bompiani Papers at the Accademia delle Scienze dei XL, in Rome, and among the papers of the Consiglio Nazionale delle Ricerche at the Archivio Centrale dello Stato, also in Rome. For a reconstruction of this episode, see (Nastasi 2001; Guerraggio and Nastasi, 2005; 183–194).

efforts by the Italian Committee, efforts which are firmly set out in the directives issued by the Fascist Government."[30] This reference was clearly directed to Koenigs himself, who Pincherle considered unmindful and hostile.

On June 21, 1928, Pincherle wrote another letter to Giannini, which also contained the transcription of part of Picard's reply, although he omitted the date and skipped several words. Therefore, in presenting the text, I will use the copy that the Prefect of Bologna transmitted, on June 22, to the Prime Minister's Office. This copy dates Picard's letter to Pincherle as June 15.

> You are aware of my feelings of respect and friendship for you. Therefore, I do not wish to enter into a discussion about the Congress of Bologna, which would be useless. Those who wish to meet German scientists at Bologna undoubtedly have their good reasons. You certainly acknowledge that the President of the International Research Council has not less convincing reasons for not attending. This Council unanimously invited Germany to take part, and the member Unions as well. This occurred in Brussels, in June 1926. Since then, the Germans have failed to answer and we know that, after mature resolutions, they have refused. Given these conditions, it is easy to draw one's conclusions.[31]

Writing to Giannini, Pincherle commented with a certain annoyance:

> As Your Excellency may see, this does not reply to my letter, the purpose of which was to raise the matter above such petty stubbornness. *But I am comforted by the fact that French scientists such as Cartan, delegated to attend the Congress by the University of Paris, approve our line of conduct with the words "you could not have done otherwise"* [the italics are mine].[32]

Let's see how Cartan replied to Pincherle.

[30]"Il modo in cui, da qualche parte, si è cercato di sabotare l'opera obiettiva e di conciliazione del Comitato Italiano, opera che riteniamo fermamente nelle direttive del Governo fascista"(ACS, CNR, Posizione 9, Convegni e Congressi, Serie III, 9b, Sottoserie II Congressi internazionali, b. 22, f. 655: Congresso Internazionale dei Matematici 1928 settembre 3–10).

[31]"Vous savez tous mes sentiments d'estime et d'amitié pour vous. Aussi je ne veux pas engager une discussion, d'ailleurs bien inutile, au sujet du Congrès de Bologne. Ceux qui désirent rencontrer à Bologne des savants allemands ont sans doute de bonnes raisons. Vous reconnaîtrez que le Président du Conseil international de Recherches en a de non moins bonnes pour ne pas y aller. Ce Conseil a, à l'unanimité, invité l'Allemagne à se joindre à lui et aux Unions qui lui sont associées. Ceci s'est passé à Bruxelles en Juin 1926. Depuis lors, les allemands n'ont pas daigné répondre et nous savons que, après mûres délibérations, ils ont en fait refusé. Dans ces conditions, la conclusion me paraît facile à tirer." (ACS, PCM (1928–30), b. 1328, f. 14.3.3303, Bologna. Congresso Internazionale dei Matematici (settembre 1928-VI). See also ASBo, Prefettura di Bologna, Gabinetto, b. 1485, f. Congresso Internazionale dei Matematici 1927–1928. Also quoted in (Cioli 2016; 126).

[32]"Come vede V.E., questo non risponde per nulla alla mia lettera, che tentava di sollevare la questione al di sopra dei piccoli puntigli. Ma ho la soddisfazione che scienziati francesi come il Cartan, che sarà delegato al Congresso dall'Università di Parigi, approvano la nostra linea di condotta dicendo "vous ne pouvriez faire autrement" (ACS, CNR, Posizione 9, Convegni e Congressi, Serie III, 9b, Sottoserie II Congressi internazionali, b. 22, f. 655: Congresso Internazionale dei Matematici 1928 settembre 3–10).

On November 10, 1927, Pincherle had written to Cartan asking him to intercede with the French mathematicians to persuade them to attend the Congress. "We attach great value to their participation," Pincherle added.[33]

On June 11, 1928, the Italian mathematician transmitted to his French colleague a copy of the letter he had sent to Picard 3 days earlier, to justify the policy adopted for sending out the invitations to the Congress and to obtain his approval.[34] Having received no comments, on June 15 Pincherle explicitly asked Cartan to give his opinion on the letter.[35]

And Cartan replied, as can be inferred from both the exact quotation in Pincherle's letter to Giannini and from the letter that Pincherle sent to Cartan on June 21, to thank him for his stand on the matter. Below is the most important passage of the letter from Pincherle to Cartan, kept at the archive of the Academy of Sciences in Paris.

> My dear Colleague,
>
> [...] I would like to speak to you about my letter to M. Picard, who expresses the situation quite frankly; the solution we have adopted seems to us to be the only way of getting us out of this impasse. It seems to have the advantage of protecting the Union, and therefore the Research Council, from a likely veritable fiasco.
>
> Therefore, I am pleased that you do not disapprove and I trust that various other French colleagues also share the same view. If you think it appropriate you may show my letter to M. Picard, because I consider the letter an explanation. Unfortunately, M. Picard has replied to me (in a rather convoluted way, to be honest) that, the situation being as it is, the President of the Research Council was in no position to attend the Congress at Bologna. We regret this very much, but we hope that, before September, this eminent geometrician may view the matter from a higher position and reconsider his decision.
>
> It is with great pleasure that we have learned that the University of Paris has delegated M. Borel and yourself as representatives at the Congress [...].[36]

---

[33]AScP, Fonds Élie Cartan 38 J: Bologne, 10 novembre 1927. Carte de S. Pincherle à Élie Cartan, 6.176/06. I wish to thank Florence Greffe and Rossana Tazzioli for their help.

[34]AScP, Fonds Élie Cartan 38 J: Bologne, 11 juin 1928. Lettre de S. Pincherle à Élie Cartan, 6.176/07.

[35]AScP, Fonds Élie Cartan 38 J: Bologne, 15 juin 1928. Carte de S. Pincherle à Élie Cartan, 6.176/09.

[36]"Mon cher Collègue, [...] J'ai tenu à vous faire connaître ma lettre à M. Picard, qui expose la situation avec toute franchise; la solution à laquelle nous avons recours nous a semblé la seule qui pouvait nous tirer d'une véritable impasse. Cette solution me semble avoir l'avantage de ne pas exposer l'Union, et par suite le Conseil des recherches, à un fiasco inévitable. J'ai donc été heureux que vous ne nous désapprouviez pas, et je pense que plusieurs de vos collègues français seront de même avis. Si vous le jugez convenable, vous pouvez faire connaître ma lettre à M. Picard, lettre que je considère comme une pièce justificative. Malheureusement M. Picard m'a répondu (de la façon plus conforme à la verité) qu'en cet état de choses le Président du Conseil des recherches ne pouvait se rendre au Congrès de Bologne. C'est pour nous un motif de véritable chagrin: mais nous espérons que d'ici à Septembre l'illustre géomètre considérera les choses à un point de vue plus élevé, et reviendra sur sa decision. Nous avons appris avec le plus grand plaisir que l'Université de Paris a délégué M. Borel et vous-même comme ses représentants au Congrès [...]" (AScP, Fonds Élie Cartan 38 J: Bologne, 21 juin 1928. Lettre de S. Pincherle à Élie Cartan, 6.176/10).

Ultimately, Picard did not reconsider his decision, as Pincherle had hoped. Despite this, however, at the first plenary session of the Congress, on September 3, Pincherle proposed to the assembly to send Picard a telegram of greetings, as a conciliatory gesture. The assembly approved (ICM Proceedings 1929; vol. 1: 77).

Thus, there surely was a boycott of the Bologna Congress on the part of the most intransigent faction of the anti-German group of French mathematicians (as well as on the part of the most intransigent faction of the German mathematicians). There are also various indications that led us to believe that, between 1927 and 1928, other reservations of political nature also arose, but which subsequently were largely overcome. Though perhaps not entirely.

3 The Ambiguous Position of the IIIC

As said before, direct testimonies about the congress coming from France are rare. It seems therefore worth mentioning here a curious report written for the Institut de Coopération Intellectuelle of the League of Nations (the Archives of the IIIC are kept at the UNESCO in Paris).[37] "The International Institute of Intellectual Cooperation [was] set up in Paris in 1924 by, and at the expense of, the French Government. It began work in 1926" as "the executive organ of the International Committee on Intellectual Cooperation".[38] Though the Institute was therefore officially an international organ, the French succeeded in keeping it largely under their influence. It is in particular remarkable that, though it was placed under the authority of the plenary committee, the Institute was directed by a French member. The first director, Julien Luchaire (director from 1926 to 1930), was the main Italianist in France, founding the French Institute of Florence, and deeply involved in the cultural exchanges and propaganda between Italy and France during the war (Mazliak and Tazzioli 2009; 171–172) see also (Broche 2005; Jeannelle 2011). Hardly after the creation of the Institute, the British historian Alfred Zimmern, Deputy Director of the Institute for Intellectual Cooperation (Markwell 1986), communicated to Luchaire's other adjoint, the Dutch astronomer and (mostly) politician Jacob Evert de Vos van Steenwijk (1889–1978), head of the section of scientific relations, the circular letter of Edward V. Hustington, chairman of the American Section of the International Mathematical Union. Dated from January 1926, this letter mentioned that the American Mathematical Society required from the American members on the International Mathematical Union that they must resign unless the IMU obtains from the International Research Council the amendment of "its rules so that membership in the Union may be entirely

[37]UNESCO, Paris. Archives de l'Institut International de Coopération Intellectuelle, Dossier D IV, Congrès de Bologne. I wish to thank Laurent Mazliak for bringing these documents to my attention.

[38]http://biblio-archive.unog.ch/report.aspx?rpt=1&id=408. See also (Valderrama 1995; Renoliet 1999; Laqua 2011).

international". de Vos proposed immediately to communicate the letter to Luchaire. Zimmern added that he told his "friend Prof. [Willcox] that we are working actively to broaden the IRC but that I cannot speak about the technical details though I understand there [must be] every desire to ease them." Zimmern mentions also to be in contact with Lord Balfour to convince the "anti-German element in England". A note by Luchaire from March 1926 shows that he read the letter shortly after. In spring 1926, Pincherle, then president of the IMU, wrote to Luchaire to inform him that the American, British, Italian, Dutch and Scandinavian delegates of the IMU asked the IRC to cancel the exclusion of several nationalities from the IMU, before deciding of the location of the next Congress in 1928. In June 1928, Luchaire received from the aforementioned Amedeo Giannini, a letter informing him that he advised Pincherle to send to the IIIC an invitation to participate to the Congress in Bologna. Luchaire replied by a letter to the chancellor of the University of Bologna Giuseppe Albini, as President of the organizing committee of the Congress: he wrote that he would not be able to come to Bologna, but that his collaborator de Vos would represent the IIIC.

De Vos came indeed. It is remarkable that the proceedings of the Congress includes his name as delegate of the IIIC considered as a French institution (ICM Proceedings 1929; vol. 1: 27). de Vos wrote a report to Luchaire shortly after the Congress, on September 17th. In his text, de Vos sums up the situation which allowed the German participation at the Bologna conference despite the intricate institutional situation resulting from the absence of decision concerning the rules of the IMU. de Vos explains that the Germans received a personal invitation from the organizing committee and not from the IMU. The success of this manœuvre ("ce petit stratagème") was complete and the Congress may be considered as the real continuation of the pre-war meetings. Though he mentions some protests (and the absence of Picard), de Vos insists that these were not presented in an official way. It is not really strange that de Vos, being not a mathematician, had not been really interested by the mathematical sessions (it is in fact amazing that he seems to have been unpleasantly surprised that the conference was mostly filled by such sessions). The only session attracting his positive appreciation was that about mathematical education, presided by H. Fehr. He mentions in particular to have participated to one of the debates of this session where he explained the role of the IIIC and how it works on "intellectual statistics" ("statistique intellectuelle", that is to say the statistics concerning intellectual life and education in various countries) could be used to help implementing some of the suggestions made during the session. In a final section, de Vos mentions that he found the general organization of the Congress "mediocre" ("médiocre"), a slightly surprising comment when one considers how the Congress was used as a display of the successes of Italian fascism. A major problem, in de Vos' eyes, had been the absence of translation of the conferences, most of which were in Italian. The last lines may reveal the main frustration of the

IIIC delegate, explaining that "he had not been welcomed on the same foot as the delegates of the governments" and that "the organizers of the Congress had not been informed of his presence".[39]

4 Archival Sources

ACS = Archivio Centrale dello Stato (Central State Archives), Rome
- CNR = Consiglio Nazionale delle Ricerche (Italian National Council of Research)
- Demorazza (See DGDR)
- DGDR = Direzione Generale per la Demografia e la Razza (General Administration for Demography and Race)
- MI = Ministero dell'Interno (Ministry of the Interior)
- PCM = Presidenza del Consiglio dei Ministri (Presidency of the Council of Ministers)

AFG = Archivio Fondazione Giovanni Gentile, Rome
ANL = Accademia Nazionale dei Lincei, Rome
ASBo = Archivio di Stato (State Archive), Bologna
AScP = Académie des Sciences, Paris
ASUBo = Archivio Storico dell'Università degli Studi di Bologna (Historical Archives, University of Bologna)
UNESCO = United Nations Educational, Scientific and Cultural Organization, Paris

References

Alexanderson GL ed (1987) The Pólya picture album. Encounters of a mathematician. Birkhäuser, Basel.

Amaldi U (1954) Salvatore Pincherle. Commemorazione letta davanti alla Classe di Scienze fisiche, matematiche e naturali della Reale Accademia Nazionale dei Lincei nell'adunanza del 5 dicembre 1937. In Pincherle S. Opere scelte (Cremonese, Rome). Vol. 1: 3–16.

Bortolotti E (1937) Salvatore Pincherle. Bollettino dell'Unione Matematica Italiana 16: 37–60.

Bortolotti E (1938) Salvatore Pincherle. Annuario dell'anno accademico 1936–1937, R. Università di Bologna: 151–156.

Broche L (2005) Julien Luchaire, itinéraire d'un Français faussement « moyen » pendant la tourmente. In Perez, C and Mercier, P Déplacements, dérangements, bouleversement: Artistes et intellectuels déplacés en zone sud (1940–1944), Bibliothèque de l'Alcazar, Marseille, 3–4 juin 2005 < HYPERLINK "http://publications.univ-provence.fr/ddb/index.php">.

[39]"Personellement, bien qu'invité en ma qualité de représentant de l'Institut, je n'ai pas été reçu sur le même pied que les délégués des gouvernements, et les organisateurs du congès n'étaient nullement renseignés sur ma présence. Je l'ai fait remarquer discrètement en une ou deux occasions, mais je n'ai pas cru utile de m'en plaindre de façon officielle" (*ibid.*).

Broglie L de (1942) La vie et l'oeuvre de Émile Picard. Lecture faite dans la séance publique du 21 décembre 1942, Institut de France, Academie des Sciences. Gauthier-Villars, Paris.

Broglie L de (1957) Notice sur la vie et l'oeuvre de Émile Borel. Lecture faite en la séance annuelle des prix du 9 décembre 1957, Institut de France, Académie des Sciences. Palais de l'Institut, Paris.

Bru B (2003) Souvenirs de Bologne. Journal de la Société Française de Statistique 144: 135–226.

Bubkill JC (1964) Charles-Joseph de la Vallée Poussin. Journal of the London Mathematical Society 39: 165–175.

Buhl A (1928) Souvenirs de Bologne. L'Enseignement mathématique 27: 193–202.

Butzer P, Mawhin J (2000) A biography of Charles-Jean de La Vallée Poussin. In La Vallée Poussin, CJ de (2000) Collected works, edited by P Butzer, J Mawhin, P Vetro (Circolo matematico, Palermo; Académie royale de Belgique, Bruxelles): 3–9.

Capristo A (2015) Volterra, Fascism, and France. Science in Context 28: 637–674.

Cioli M (2016) Arte e scienza internazionale. Il "modernismo" fascista negli anni Venti. Dipartimento di Scienze politiche e sociali, Università di Bologna, Bologna.

Curbera GP (2009) Mathematicians of the world, unite! The International Congress of Mathematicians. A human endeavor. A.K. Perters, Wellesley.

Dauben JW (1980) Mathematicians and the World War I. The international diplomacy of G.H. Hardy and Gösta Mittag-Leffler as reflected in their correspondence. Historia Mathematica 7: 261–288.

Decleva E and Milza P (1996) La Francia e l'Italia negli anni Venti. Istituto per gli studi di politica internazionale, Milan.

Fehr H (1929) Le Congrés de Bologne 3–10 septembre 1928. L'Enseignement mathématique 28: 28–53.

Figà Talamanca A and Pepe L eds. (2014) Carlo Pucci tra scienza e impegno civile. Unione matematica italiana, Bologna.

Gennaro R (2013) Il Manifesto degli intellettuali fascisti e l'espansione culturale all'estero. La versione francese e due nuove liste di firmatari. Nuova Storia Contemporanea 17: 79–95.

Guerraggio A and Nastasi P (2005) Italian mathematics between the two world wars. Birkhäuser, Basel.

Guerraggio A and Nastasi P (2008) Roma 1908. Il congresso internazionale dei matematici. Bollati Boringhieri, Turin.

Guieu J-M (1998) L'engagement européen d'un grand mathématicien français. Émile Borel et la 'Coopération européenne' des années vingt aux années quarante. Bulletin de l'Institut Pierre Renouvin 5: 15–32.

Guieu J-M (2008) Le rameau et le glaive. Les militants français pour la Société des Nations. Presses de Sciences Po, Paris.

Guillen P (1991) L'échec d'un rapprochement franco-italien dans les années 1926–1929. In Migliazza A, Decleva E eds. Diplomazia e storia delle relazioni internazionali. Studi in onore di Enrico Serra (Giuffrè, Milan): 321–337.

Guillen P (1998) Franco-Italian relations in flux, 1918–1940. In Boyce R ed. French foreign and defence policy, 1918–1940. The decline and fall of a great power (Routledge, London and New York): 149–163.

Guiraldenq P (1999) Emile Borel, 1871–1956. L'espace et le temps d'une vie sur deux siècles. Albert Blanchard, Paris.

Hadamard J (1942) Emile Picard, 1856–1941. Obituary Notices of Fellows of the Royal Society 4(11): 129–150.

International Congress of Mathematicians Proceedings (1929–1932) Atti del Congresso Internazionale dei Matematici. Bologna, 3–10 settembre 1928. N. Zanichelli, Bologna.

Jeannelle JL (2011) Julien Luchaire. Coopération internationale et nationalisme littéraire. In Compagnon A ed. La République des Lettres dans la tourmente (1919–1939) (CNRS, Paris): 151–165.

Laqua D (2011) Internationalisme ou affirmation de la nation? La coopération intellectuelle transnationale dans l'entre-deux-guerres. Critique internationale 52: 51–67.
Lehto O (1998) Mathematics without borders. A history of the International Mathematical Union. Springer, New York.
McKinnon Riehm E and Hoffman F (2011) Turbulent times in mathematics. The life of J.C. Fields and the history of the Fields medal. American Mathematical Society, Providence, Rhode Island and The Fields Institute, Toronto, Ontario.
Marbo C (1968) A travers deux siècles. Souvenirs et rencontres (1883–1967). B. Grasset, Paris.
Markwell, DJ (1986) Sir Alfred Zimmern revisited. Fifty years on. Review of international studies 12: 279–292.
Mazliak L (2010) From Markov to Doeblin. Events in chain http://www.proba.jussieu.fr/~mazliak/Markov_Rouen.pdf.
Mazliak, L (2015). Volterra and the journeys of French students to Italy in the 1910s. Revista Brasileira de História da Matemática, 14(29): 1–30.
Mazliak L and Catellier R (2012) The emergence of French statistics. How mathematics entered the world of statistics in France during the 1920s. Revue d'histoire des mathématiques 18: 271–335.
Mazliak L and Tazzioli R (2009) Mathematicians at war. Volterra and his French colleagues in World War I. Springer, Dordrecht-New York.
Mazliak L and Tazzioli R (2015) To each according to his trade. Italian mathematicians and the Great War. Lettera matematica International 3, 1–2, 3–15.
Melis G (2000) Giannini, Amedeo. In Dizionario Biografico degli Italiani (Istituto della Enciclopedia italiana, Rome), Vol. 54: 485–489.
Milza P (1987) Le fascisme italien et la presse française, 1920–1940. Ed. Complexe, Bruxelles.
Minesso M (2012) Giuseppe Belluzzo. Tecnico e politico nella storia d'Italia, 1876–1952. Franco Angeli, Milan.
Montel P (1962) Notice nécrologique sur Ch.-J de La Vallée Poussin. Comptes rendus hebdomadaires des séances de l'Académie des Sciences 254: 2473–2476.
Munster JF (1999) Inventaire des papiers d'Etienne de la Vallée Poussin (1921–1955). Université catholique de Louvain, Louvain-la-Neuve.
Nastasi P (1998) Il contesto istituzionale. In Di Sieno S, Guerraggio A, Nastasi P eds. La matematica italiana dopo l'Unità. Gli anni tra le due guerre mondiali (Marcos y Marcos, Milan): 817–943.
Nastasi P (2001) La matematica. In Simili R, Paoloni G eds. (2001). Vol. 1: 243–280.
Nastasi P and Tazzioli R (2013) I matematici italiani e l'internazionalismo scientifico (1914–1924) La Matematica nella Società e nella Cultura 6: 355–405.
Paoloni G ed. (1990) Vito Volterra e il suo tempo (1860–1940). Mostra storico-documentaria. Accademia nazionale dei Lincei, Consiglio nazionale delle ricerche, Archivio centrale dello Stato, Rome.
Papa ER (1958) Storia di due manifesti. Il fascismo e la cultura italiana. Feltrinelli, Milan.
Pertici R (2001) Giuliano, Balbino. In Dizionario Biografico degli Italiani (Istituto della Enciclopedia italiana, Rome) Vol. 56: 770–776.
Poupault C (2009) Les voyages d'hommes de letters en Italie fasciste. Espoir du rapprochement franco-italien et culture de la latinité. Vingtième siècle 104: 67–79.
Poupault C (2014) À l'ombre des faisceaux. Les voyages français dans l'Italie des chemises noires (1922–1943). École Française de Rome, Rome.
Pucci C (1986) L'Unione Matematica Italiana dal 1922 al 1944. Documenti e riflessioni. Symposia Mathematica 27: 187–212.
Rasmussen A (2007) Réparer, réconcilier, oublier. Enjeux et mythes de la démobilisation scientifique, 1918–1925. Histoire@politique 3: 8–8.
Renoliet J-J (1999) L'Unesco oubliée. La Société des nations et la coopération intellectuelle (1919–1946). Publications de la Sorbonne, Paris.

Salustri S (2010) Un ateneo in camicia nera. L'Università di Bologna negli anni del fascismo. Carocci, Rome.
Santoro S (1999) La diplomazia italiana di fronte all'epurazione. Il caso Amedeo Giannini. Italia contemporanea 216: 529–540.
Schroeder-Gudehus B (1973) Challenge to transnational loyalties. International scientific organization after the first World War. Science studies 3: 93–118.
Segal SL (2002) War, refugees, and the creation of an international mathematical community. In Parshall KH and Rice AC eds. (2002): 359–380.
Segal SL (2003) Mathematicians under the Nazis. Princeton University Press, Princeton and Oxford.
Shorrock W (1988) From ally to enemy. The enigma of fascist Italy in French diplomacy, 1920–1940. The Kent State University Press, Kent, Ohio and London.
Siegmund-Schultze R (2011) Opposition to the boycott of German mathematics in the early 1920's. Letters by Edmund Landau (1877–1938) and Edwin Bidwell Wilson (1879–1964). Revue d'Histoire des Mathématiques 17: 139–165.
Siegmund-Schultze R (2014) One hundred years after the Great War (1914–2014). A century of breakdowns, resumptions and fundamental changes in international mathematical communication. In Proceedings of the International Congress of Mathematicians Seoul 2014. Vol. 4: 1231–1253.
Siegmund-Schultze R (2016) Mathematics knows no races. A political speech which David Hilbert planned for the ICM in Bologna 1928. Mathematical Intelligencer 38: 56–66.
Simili R (2001) La presidenza Volterra. In Simili R, Paoloni G eds. (2001). Vol. 1: 72–127.
Simili R and Paoloni G eds. (2001) Per una storia del Consiglio nazionale delle ricerche. Laterza, Rome and Bari.
Supino G (1946) Necrologio di Umberto Puppini. Bollettino dell'Unione Matematica Italiana ser. 3, vol. 1: 69–70.
Talenti G (2005) Necrologio di Carlo Pucci. Bollettino dell'Unione Matematica Italiana ser. 8, vol. 8-A: 357–377.
Tonelli L (1937) Salvatore Pincherle. Annali della Scuola Normale Superiore di Pisa. Classe di Scienze, 2. serie, 6: 1–10.
Union interparlementaire (1928) Compte rendu de la XXVme conférence tenue à Berlin du 23 au 28 août 1928. Payot & cie, Lausanne [etc.].
Urso S (2003) Guadagnini, Giuseppe. In Dizionario Biografico degli Italiani (Istituto della Enciclopedia italiana, Rome) Vol. 60: 82–84.
Valderrama F (1995) A history of UNESCO. UNESCO, Paris.
Van Dalen D (2005) Mystic, geometer, and intuitionist. The life of L.E.J. Brouwer. Vol. 2: Hope and disillusion. Clarendon Press, Oxford.
Van Dalen D ed. (2011) The selected correspondence of L.E.J.Brouwer. Springer, London.
Vivanti G (1936) Salvatore Pincherle. Rendiconti dell'Istituto lombardo di scienze e lettere, 69: 956–957.
Whitehead JHC (1952) Elie Joseph Cartan, 1869–1951. Obituary Notices of Fellows of the Royal Society 8(21): 71–95.
Zoller I (1930) Pincherle Salvatore. In Jüdisches Lexikon (Jüdischer Verlag, Berlin). Vol. 4: 944–945.

Author Index

F. Brechenmacher et al. (eds.), *Images of Italian Mathematics in France*,
Trends in the History of Science, DOI 10.1007/978-3-319-40082-2

CPSIA information can be obtained
at www.ICGtesting.com
Printed in the USA
BVOW10*0656181016
465323BV00008B/24/P

9 783319 400808